MODERN ASPECTS OF ELECTROCHEMISTRY

No. 41

Modern Aspects of Electrochemistry

Topics in Number 40 include:

- Polymer Electrolyte Membrane (PEM) fuel cell bipolar plates, discussion of the difficulties associated with confronting bipolar plate development
- The use of graphs in electrochemical reaction networks with focus on analysis of variance (ANOVA) observation methods
- Nano-materials in lithium ion battery electrode design, presentation of a plasma-assisted method to create a carbon replica of an alumina template membrane
- Direct methanol fuel cells, extensive discussion and review of various types of fuel cells and advances made in the performance of DMFC's since their inception
- Direct simulation of polymer electrolyte fuel cell catalyst layers, presentation of a systematic development of the direct numerical simulation

Topics in Number 39 include:

- Approaches to Solute-Solvent Interactions including two approaches to computational determination of solution properties, and several other procedures that establish correlations between properties of interest and certain features of the solute and/or solvent molecules
- Porous Silicon, including its morphology and formation mechanisms, as well as anodic reaction kinetics
- Modeling Electrochemical Phenomena via Markov Chains and Processes gives an introduction to Markov Theory, then discusses applications to electrochemistry, including modeling electrode surface processes, electrolyzers, the repair of failed cells, analysis of switching-circuit operations, and other electrochemical systems
- Fractal Approach to Rough Surfaces and Interfaces in Electrochemistry, from a review of Fractal Geometry to the application of Fractal Geometry to the classification of surfaces and Electrochemistry
- Phenomenology and Mechanisms of Electrochemical Treatment (ECT) of Tumors, starting from fundamentals and proceeding to electrochemical treatment of tumors in animals and then in humans

MODERN ASPECTS OF ELECTROCHEMISTRY

No. 41

Edited by

CONSTANTINOS VAYENAS

University of Patras
Patras, Greece

RALPH E. WHITE

University of South Carolina
Columbia, South Carolina, USA

and

MARIA E. GAMBOA-ALDECO

Managing Editor
Superior, Colorado, USA

Springer

Constantinos Vayenas
Department of Chemical Engineering
University of Patras
Patras 265 00
Greece
cat@chemeng.upatras.gr

Maria E. Gamboa-Aldeco
1107 Raymer Lane
Superior, CO 80027
USA
mariagamboa06@msn.com

Ralph E. White
Department of Chemical Engineering
University of South Carolina
Columbia, SC 29208
USA
white@engr.sc.edu

Library of Congress Control Number: 2006938037

ISBN-10: 0-387-46107-8 ISBN 0-387-46108-6 (eBook)
ISBN-13: 978-0-387-46107-6 ISBN 978-0-387-46108-3 (eBook)

Printed on acid-free paper.

9 8 7 6 5 4 3 2 1

springer.com

Preface

This volume contains four chapters. The topics covered are solid state electrochemistry devices and techniques; nanoporous carbon and its electrochemical application to electrode materials for super-capacitors; the analysis of variance and covariance in electro-chemical science and engineering; and the last chapter presents the use of graphs in electrochemical reaction networks.

Chapter 1 by Joachim Maier continues the solid state electrochemistry discussion that he began in Volume 39 of the Modern Aspects of Electrochemistry. He begins by introducing the reader to the major electrochemical parameters needed for the treatment of electrochemical cells. In section 2 he discusses various sensors: electrochemical (composition), bulk conductivity, surface conductivity, galvanic. He also discusses electrochemical energy storage and conversion devices such as fuel cells.

Gyoung-Ja Lee and Su-Il Pyun in chapter two review the synthesis and characterization of nanoporous carbons and their electrochemical application to electrode material for super-capacitors in relationship to the key role nanoporous carbons have played, for example, in the purification of liquids and carbons, catalytic reaction, and the storage of energy. Activation and templating methods are discussed in section II. They survey the structural characteristics of the porous carbons by the use the gas adsorption method in Section III. Section IV discusses the molecular probe method and the image analysis method for quantitative characterization of the pore surface irregularity and the size distribution irregularity based on the fractal theory. The investigation of the electrochemical performance of the porous electrodes for electric double layer capacitors considering the effects of geometric heterogeneity and surface inhomogeneity on kinetics of double-layer charging/discharging occurs in section V.

Joseph D. Fehribach reviews and discusses in Chapter 3 the uses of graphs in the study of chemical reaction network, particularly electrochemical reaction networks for electrochemical systems. He defines any graph used to study a reaction network as a reaction graph. He mentions three categories that cover the uses of graphs: (1) reaction species graphs, (2) reaction mechanism graphs, and (3) reaction route graphs.

Keith Scott and Yan-Ping Sun review and discuss three dimensional electrode structures and mathematical models of three dimensional electrode structures in chapter four. Conductivity limitations of these three-dimensional electrodes can cause the current overpotential to be non-uniform in structure. Adomian's Decomposition Method is used to solve model equations and approximate analytical models are obtained. The first three to seven terms of the series in terms of the nonlinearities of the model are generally sufficient to meet the accuracy required in engineering applications.

C. Vayenas

University of Patras
Patras, Greece

R. E. White

University of South Carolina
Columbia, South Carolina, USA

Contents

Chapter 1

SOLID STATE ELECTROCHEMISTRY II: DEVICES AND TECHNIQUES

Joachim Maier

Chapter 2

SYNTHESIS AND CHARACTERIZATION OF NANOPOROUS CARBON AND ITS ELECTROCHEMICAL APPLICATION TO ELECTRODE MATERIAL FOR SUPERCAPACITORS

Gyoung-Ja Lee and Su-Il Pyun

Chapter 3

THE USE OF GRAPHS IN THE STUDY OF ELECTROCHEMICAL REACTION NETWORKS

Joseph D. Fehribach

Chapter 4

APPROXIMATE ANALYTICAL SOLUTIONS FOR MODELS OF THREE-DIMENSIONAL ELECTRODES BY ADOMIAN'S DECOMPOSITION METHOD

Keith Scott and Yan-Ping Sun

LIST OF CONTRIBUTORS

JOSEPH D. FEHRIBACH
Fuel Cell Center
Departments of Chemical Engineering and Mathematical Sciences
Worcester Polytechnic Institute
100 Institute Road
Worcester, MA 01609-2247

GYOUNG-JA LEE
Department of Materials Science and Engineering
Korea Advanced Institute of Science and Technology
373-1 Guseong-dong, Yuseong-gu
Daejeon, 305-701
Republic of Korea

JOACHIM MAIER
Max-Planck-Institut für Festkörperforschung
Heisenbergstraße, 70569 Stuttgart,
Germany

SU-IL PYUN
Department of Materials Science and Engineering
Korea Advanced Institute of Science and Technology
373-1 Guseong-dong, Yuseong-gu
Daejeon, 305-701
Republic of Korea
Tel: +82-42-869-3319
Fax: +82-42-869-3310
E-mail: sipyun@webmail.kaist.ac.kr

KEITH SCOTT
School of Chemical Engineering and Advanced Materials
University of Newcastle upon Tyne, Newcastle, NE1 7RU,
Great Britain

YAN-PING SUN
Chemical Engineering Department, Taiyuan University of Technology, Taiyuan, Shanxi 030024, China

1

Solid State Electrochemistry II: Devices and Techniques

Joachim Maier

Max-Planck-Institut für Festkörperforschung
Heisenbergstraße 1, 70569 Stuttgart, Germany

I. INTRODUCTION

This second part of systematic treatment of solid state electrochemistry deals with solid state electrochemical cells which serve as devices in the context of a specific application, or are designed for measuring thermodynamic or kinetic properties. Generally, electrochemical cells allow one to transform electrical signals into chemical signals and vice-versa. The relation between chemistry and electricity is explicitly anchored in the definition of the electrochemical potential[1] as $\tilde{\mu} = \mu + zF\phi$ with μ being the chemical potential and $zF\phi$ the electrical potential times the molar charge. In order to work out electrochemical principles on one hand and expose characteristic features of the solid state on the other hand, the consideration of characteristic issues is emphasized rather than an exhaustive survey being attempted. In Part I[2] on which this contribution is based, the nature of the ionic and electronic charge carriers in solids has been highlighted and their thermodynamic and kinetic behaviors in the bulk or at boundaries have been treated.

Before we start with the consideration of measurement techniques and applications, let us briefly introduce the major electrochemical parameters needed for the treatment of electrochemical cells. A typical electrochemical cell that we want to

Modern Aspects of Electrochemistry, Number 41, edited by C. Vayenas et al., Springer, New York, 2007.

consider, consists of a series of phases with Cu terminals (the leads) on both sides of the arrangement which electrochemically connect the cell with the outer electrical circuit components. One phase (MX in Figure 1) exhibits a significant portion of ionic conduction (called central phase) and may be contacted on both sides by other electronically and/or ionically conducting phases. In addition to imposing and measuring electrical fluxes and forces, chemical fluxes or forces can also be applied and measured. If we, e.g., consider oxides, the latter may be achieved by imposing or measuring the outer oxygen partial pressure (see Figure 1).

The total current (I) flowing through such a cell is composed of displacement (I_C) and conduction contributions (I_R), i.e., we neglect inductive effects and can concentrate on capacitive and resistive elements. For mechanistic considerations we consider current densities, which in the quasi one-dimensional (laterally homogeneous) case are connected with the current via $i = I/a$ (a = area), while in the general case this connection reads $\mathbf{i} = \partial I/\partial \mathbf{a}$ ($\mathbf{a}$: area vector, $\partial/\partial \mathbf{a}$: gradient operator with respect to the $\mathbf{a}$-coordinates).[4] The continuity equation of the charge density (ρ)

$$\dot{\rho} = -\text{div}\,\mathbf{i}_R \tag{1}$$

expresses the fact that the negative divergence of the current density is equivalent to a local charge accumulation. This charge accumulation can also be expressed in terms of a divergence of a capacitive current density ($\dot{\rho} = \text{div}\,\mathbf{i}_C$) which in a given medium is represented by the displacement current density $\dot{\mathbf{D}} \equiv \partial/\partial t(\varepsilon\mathbf{E})$ (ε: dielectric permittivity, $\mathbf{E}$: electrical field vector) (see e.g., Ref.[5]). Hence

$$\text{div}\left(\mathbf{i}_R + \mathbf{i}_C\right) = 0 \tag{2}$$

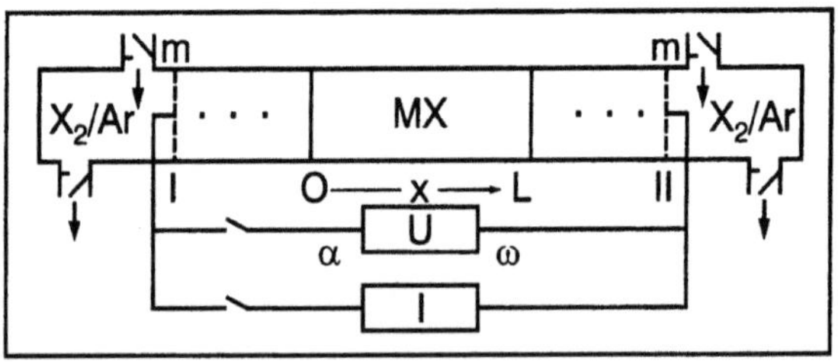

Figure 1. Sketch of a general electrochemical cell, as referred to in the text, with the mixed conductor MX as the central phase. The "current-collecting" metal is denoted by "m".[3] Reprinted from J. Maier, *Z. Phys. Chem. NF* (1984) 191–215. Copyright © 1984 with permission from Oldenbourg Verlagsgruppe.

indicating that the total current is positionally constant. The conduction currents can be decomposed into electronic and ionic contributions ($\mathbf{i}_{eon}, \mathbf{i}_{ion}$) each of them being composed of contributions from individual charge carriers k (such as conduction electrons and holes or vacancies and interstitial defects). The individual fluxes are proportional to gradients in the electrochemical potentials. Neglecting crosseffects the particle current densities read

$$\mathbf{i}_k = -\frac{\sigma_k}{z_k \mathrm{F}} \nabla \tilde{\mu}_k. \tag{3}$$

If only gradients in the electric potential occur (i.e., $\nabla \mu_k = \mathbf{0}$ and $\mathbf{i}_k = -\sigma_k \nabla \phi = -\sigma_k \mathbf{E}$), Eq. (2) can be rewritten as

$$\mathrm{div}\left[\left(\varepsilon \frac{\partial}{\partial t} + \sigma\right)\mathbf{E}\right] = 0 \tag{4}$$

or, using the Maxwell equation $\mathrm{div}\mathbf{D} = \rho$ and assuming ε and σ to be constant, as

$$\dot{\rho} + \rho/\tau = 0. \tag{5}$$

In Eq. (5) τ defines the dielectric relaxation time ($\tau = \varepsilon/\sigma$) according to which obviously a charge perturbation decays exponentially in a conductor. This defines a parallel R–C circuit as a good approximation of a homogeneous conductor (see Section III). In the following part of this section we consider the steady state, in which the conduction current represents the total current and capacitive contributions have vanished.

Let us now separately consider the continuity equations for ions and electrons

$$\frac{\partial c_{ion}}{\partial t} = -\frac{\mathrm{div}\,\mathbf{i}_{ion}}{z_{ion}\mathrm{F}} + \nu_{ion}\mathcal{R}, \tag{6}$$

$$\frac{\partial c_{eon}}{\partial t} = +\frac{\mathrm{div}\,\mathbf{i}_{eon}}{\mathrm{F}} + \nu_{eon}\mathcal{R}, \tag{7}$$

(where $\nu\mathcal{R}$ takes account of carrier generation by a local reaction) and apply them to a quasi-one-dimensional electrochemical cell with electrodes attached to pure electrolytes connected by an outer circuit. Generalizations are straightforward. In Eqs. (6) and (7), $\mathcal{R}$ is the rate of the electrode reaction which connects ions and electrons according to

$$\nu_{ion}\mathrm{ion} + \nu_{eon}\mathrm{eon} + \cdots \rightleftharpoons \mathrm{Null} \tag{8}$$

(where the stoichiometric numbers ν are coupled via $\nu_{eon}/\nu_{ion} = z_{ion}$; z_{ion} being the charge number of the ion). Within the electrolyte (where $(\partial/\partial x)\tilde{\mu}_{\text{ion}} \propto i_{\text{ion}}/\sigma_{\text{ion}} \simeq 0$) the ionic transport proceeding via

$$\text{ion}(x) \rightleftharpoons \text{ion}(x') \tag{9}$$

is decisive, while in the electronic connections (where $(\partial/\partial \text{x})\tilde{\mu}_{\text{eon}} \propto i_{\text{ion}}/\sigma_{\text{eon}} \simeq 0$) the transport "reaction"

$$\text{eon}(x) \rightleftharpoons \text{eon}(x') \tag{10}$$

determines the flux. In the homogeneous phases it holds that $\mathcal{R} = 0$, and the steady-state condition simplifies to a constancy of the respective current (determined by Eqs. (9) and (10)). At the electrode/electrolyte interface $(\partial/\partial x)i_{\text{eon}} = -i_{\text{eon}}(\text{outer circuit})/\Delta x$; $(\partial/\partial x)i_{\text{ion}} = +i_{\text{ion}}(\text{electrolyte})/\Delta x$) the outer electronic current is converted into an internal ionic current via the electrode reaction Eq. (8) (with rate $\mathcal{R}$). In short,

$$i_{\text{ion}} = \begin{cases} 0 & \text{outer circuit,} \\ I/a = \nu_{\text{eon}}\text{F}\mathcal{R}\Delta\text{x} & \text{electrolyte,} \end{cases} \qquad i_{\text{eon}} = \begin{cases} I/a = \nu_{\text{eon}}\text{F}\mathcal{R}\Delta x & \text{outer circuit,} \\ 0 & \text{electrolyte.} \end{cases}$$

The fact that $|i_{\text{ion}}| = |i_{\text{eon}}| \propto I \propto \mathcal{R}$, also provides Faraday's law $(\dot{\text{n}} \propto \dot{\text{Q}})$. Generalizations for nonzero electronic contributions in the interior of the electrolyte or nonzero ionic contributions in the electrode are straightforward.

While partial conduction currents are driven by the gradients in the respective electrochemical potentials (according to Eq. (3)), the external voltage that we measure is determined by the difference of the electrons' electrochemical potentials ($\tilde{\mu}_{\text{eon}}$) or Fermi-levels ($\tilde{\mu}_{\text{eon}}/\text{N}_{\text{m}}$) at both terminals

$$U = -\Delta\tilde{\mu}_{\text{eon}}/\text{F}, \tag{11}$$

which can be decomposed into contributions from different phases. The voltage drop within the central phase can be directly obtained by integration of the electrochemical potential gradient. The voltage (U) can also be decomposed into the open circuit voltage which under equilibrium conditions corresponds to the emf (E) and the losses due to current flow (η). In the steady state these overvoltages (η) are caused by resistances (see textbooks of electrochemistry[7,8]), and for a series combination, we can write

$$U(I) = E + \Sigma_\alpha I_\alpha \bar{R}_\alpha(I) = E + \Sigma_\alpha \eta_\alpha. \tag{12}$$

$\bar{R}_\alpha(I)$ is the integral resistance of the process α, while the differential resistance $R_\alpha(I) \equiv \partial\eta_\alpha/\partial I$.

Similarly, we distinguish between an integral capacitance $\bar{C}$ and differential capacitances $C = \partial Q/\partial\Delta\phi$ (Q being the charge). Similarly as $\bar{R}$ refers to η and not to U (note that $U(I=0) = E \neq 0 = \eta(I=0)$), C is not inversely proportional to $\Delta\phi$ but to $\Delta\phi - \Delta\phi_{pzc}$ (note that $Q(\Delta\phi = 0) \neq 0 = Q(\Delta\phi = \Delta\phi_{pzc})$) where $\Delta\phi_{pzc}$ denotes the Galvani potential change for zero-charge conditions.

Figure 2 gives an overview on the definition of and relation between quantities used in surface science and in electrochemistry such as work function or surface potential. In the following we distinguish between (i) cells to which we apply a current ($I > 0$, the current direction being opposite to the short-circuit current direction, $U = E + \Sigma_\alpha I\bar{R}_\alpha > E$), named polarization cells (cells under load), (ii) cells from which we extract current ($I < 0$, in short-circuit direction, $U < E$), named current-generating cells, and finally (iii) open-circuit cells ($I \simeq 0, U \simeq E$). In all cases we

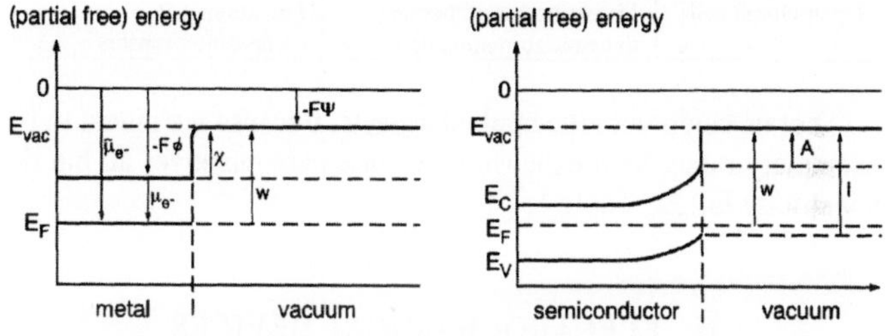

Figure 2. The interdependencies of the parameters work function (w), Galvani potential (ϕ), Volta potential (ψ), and surface potential (χ), work function, electron affinity (A), and ionization energy (I) for metal (*left*) and semiconductor (*right*). E_{vac} is the energy of the electron in vacuum immediately in front of the surface. The difference between I and A corresponds to the band gap according to Ref.[6] Note that A and I are referred to the band edges while w refers to E_F. (Please do not confuse the symbol for energy with the symbol E used in the text for emf.) Also pay attention to the fact that at finite temperatures local entropy effects are to be considered (entering the "energy levels") and hence the free energy must be addressed. In fact even more accurately one has to refer to partial free energies, i.e., to electrochemical potentials. Band edges are related to the nonconfigurational part of the electrochemical potential of conduction electrons and holes, respectively. The Fermi-level is the full electrochemical potential of the electrons. The vacuum level is the electrochemical potential of the electrons immediately outside the solid (it does naturally not contain configurational contributions). Reprinted with permission from J. Maier, *Physical Chemistry of Ionic Materials. Ions and Electrons in Solids* (2004). Copyright © 2004 with permission from John Wiley & Sons, Ltd.

can distinguish between electrochemical applications and electrochemical measurements for the purpose of measuring fundamental kinetic and thermodynamic parameters (as detailed in Table 1). Priority is given to those aspects in which solid state problems are to the fore. The reader is also advised to the conventional or general electrochemical literature as regards measurement techniques and application.[9]

Table 1.
Overview on Electrochemical Devices and Measurement Techniques Based on Various Cell Types.

Cell type	Measurement technique	Technological application
Polarization cell	Measurement of kinetic data by polarization	Electrochemical composition actors (electrolyzers, pumps, windows), electrochemical composition sensors (ampèrometric, conductometric)
Current-generating cell	Measurement of kinetic data by depolarization	Electrochemical energy storage and conversion devices (batteries, fuel cells, supercapacitors)
Open-circuit cell	Measurement of thermodynamic formation data	Potentiometric composition sensors

Let us begin with the applied aspects. Because we intend to be exemplary rather than exhaustive we orientate ourselves at, but do not strictly follow, Table 1.

II. ELECTROCHEMICAL DEVICES AND APPLICATIONS

There is an extended special literature[3,10–16] on applications of solid state electrochemistry and even more on electrochemical devices. According to our objective, in this section applications will be emphasized in which migration and diffusion in the solid state are decisive processes (as discussed in Part I[2]). We intend to subsume such applications under the headlines composition sensors, composition actors, and energy storage or conversion devices.

Electrochemical devices allow for the conversion of chemical energy or information into electrical energy or information, or vice versa. One characteristic feature of the solid state in this respect is the thermal and mechanical stability which allows performance at high temperatures as well as an easy integration and miniaturization, another the selective mobility of certain ions which can be favorable, e.g., in the context of sensors.

We will begin with a description of electrochemical sensors or more specifically composition sensors based on electrochemical principles (i.e., we refer to an electrochemical detection of composition). Another group of applications refers to devices in which the transference of mass and charge is used primarily to change composition or produce chemicals (electrochemical pumps and electrochemical reactors, or electrochemical filters); we will term such devices *composition actors*. At the end we will discuss energy conversion and storage devices (which we do not subsume under the term composition actors as here the energy aspect is to the fore).

1. Electrochemical (Composition) Sensors

To be exemplary, let us concentrate on gas sensors as the most important applications in this context. We will begin with sensors for redox active gases and, to be even more specific, mostly refer to oxygen detectors. Let the active phase have the gross composition MO. Sensor principles that rely on partial or even total equilibrium conditions with regard to the mixed conducting sensing phase MO, can—referring to Table 2—be classified as follows (cf. Ref.[17,18]):

1. At sufficiently high temperatures complete equilibrium (i.e., $\nabla\mu_O = \nabla\tilde{\mu}_{e^-} = \nabla\tilde{\mu}_{O^{2-}} = \mathbf{0}$) of MO is achieved with the gas phase, and the equilibrium composition can be usually

Table 2.
Three Major Classes of Electrochemical Solid State Sensors for Redox-Active Gases Relying on Full or Partial Equilibria[a].

Electrochemical sensor	Solid phase	$\nabla\tilde{\mu}_{ion}$	$\nabla\tilde{\mu}_{eon}$
Bulk conductivity sensor	Mixed conductor	**0**	**0**
Surface conductivity sensor[a]	Electronic conductor	$\neq \mathbf{0}$	**0**
Potentiometric sensor	Ionic conductor	**0**	$\neq \mathbf{0}$

[a] Experiments on SnO_2[19] show that the picture may be more subtle for the surface conductivity sensor: The subset of the interfacial reaction steps succeeding the sensor action may be sluggish, and not necessarily the diffusion step. (Then the statement $\nabla\tilde{\mu}_{e^-} = \mathbf{0}$ can be violated.) Cf. Section II.1.ii

sensitively (cf. MO be electronically conducting) detected by conductivity experiments, i.e., we make use of the typical power-law dependencies discussed in Part I (Section IV). We call such a detector a bulk conductivity sensor.

2. At low temperatures the ionic equilibrium will freeze and hence $\nabla\mu_{\mathrm{O}} \neq \mathbf{0} \neq \nabla\mu_{\mathrm{O}^{2-}}$ while the electronic contact equilibrium may still be fast enough, i.e., $\nabla\tilde{\mu}_{\mathrm{e}^-} = \mathbf{0}$. This is the principle of the surface conductivity sensor (or Taguchi sensor) in which oxygen is adsorbed on the surface, trapping electrons out of the space charge zone and causing depletion of conduction electrons, i.e., in the case of an n-conducting oxide such a SnO_2 increasing the surface resistance. (Yet, this simple picture has to be modified* as shown later in Section II.1.ii.) The resistance variation can be easily measured. This sensor is typically faster but less selective than the equilibrium sensor. While the bulk conductivity sensor only detects the oxygen partial pressure (P_{O_2}), the latter responds to virtually all redox-active gases.[2]
3. The third variant is realized with solid electrolytes (emf sensor) in which the lacking electronic conductivity prevents the establishing of the electronic equilibrium ($\nabla\tilde{\mu}_{e-} \neq \mathbf{0} \neq \nabla\mu_{\mathrm{O}}$ but $\nabla\tilde{\mu}_{\mathrm{O}^{2-}} = \mathbf{0}$) while owing to the high ionic conductivity now the "ionic Fermi-level" ($\tilde{\mu}_{\mathrm{O}^{2-}}$) is constant. This is the principle of a galvanic cell under open circuit conditions. The oxygen partial pressure (P_{O_2}) on the measuring side can be determined, if the reference P_{O_2} is known.

Before we begin with a more detailed discussion, let us consider important parameters characterizing sensor performance. It is rather the rule than the exception that one obtains a change in a given material's property such as electrical conductivity on variation of the composition of the ambient gas phase. In other words, obtaining a signal is only a necessary condition, in addition, a useful sensor signal must be unambiguous, sensitive, and selective. Preferably, it should also be drift-free, quickly established, long-time stable, and easy to record. It goes without saying that aspects such as cost, size, and environmental benignity of the device are of course also decisive for commercialization.

Let us consider a signal S which is a function of the concentration (c) of the species j to be detected. Since the sensor will also respond to other species, i.e.,

$$S = S(c_1, c_2, \ldots c_j, \ldots c_n), \tag{13}$$

the total change of the signal is given by

$$\mathrm{d}S = \Sigma_{i \neq j} \frac{\partial S}{\partial c_i} \mathrm{d}c_i + \frac{\partial S}{\partial c_j} \mathrm{d}c_j. \tag{14}$$

In case of an ideally selective sensor, the term $\left| \frac{\partial S}{\partial c_j} \right|$, i.e., the sensitivity, should be as high as possible, whereas the other derivatives should be as small as possible. For a set of signals 1, 2, …, n responding to the species 1, … j, …, n, the diagonal terms only should be nonzero for a selective sensing procedure. The procedure is said to be specific with respect to j if only S_{jj} is nonzero.[20] (The terms selectivity and specificity can also be defined as continuous quantities.[21] The sensitivity is then given by the determinant of the response matrix.)

It is instructive[18] to consider the time dependence of $S(c(t),t)$ and to extend Eq. (13) to

$$\frac{\mathrm{d}S}{\mathrm{d}t} = \left.\frac{\partial S}{\partial c}\right)_t \frac{\mathrm{d}c}{\mathrm{d}t} + \left.\frac{\partial S}{\partial t}\right)_c. \tag{15}$$

In the ideal case the explicit time dependence $\left.\frac{\partial S}{\partial t}\right)_c$ should disappear. This does not hold for the implicit time dependence, rather we wish to follow the time change of the local concentration $c(t)$ instantaneously $\left(\text{cf.} \frac{\partial S}{\partial c}\frac{\mathrm{d}c}{\mathrm{d}t}\right)$. If we identify c in Eq. (15) with the local concentration where the signal is taken, the explicit time dependence refers to the drift of the signal. If c refers to the concentration in the gas phase it also includes response time. If response and drift processes occur at different time scales, the concentration can be reliably deduced from the signal. Thus, we assume that there is a fast response of the order of τ_R, and then after having obtained a pseudo-steady-state, a possible signal variation on a long-time scale ($\gg\tau_R$). Then the response time τ_R is to be defined by the time needed to reach the pseudo-steady-state $\Big($e.g., $\left| \frac{S(\tau_R) - S(\text{“}\infty\text{”})}{S(\text{“}\infty\text{”})} \right| \simeq 1\%\Big)$, and the drift by the change for $t \gg \tau_R \left(\dot{S}(t \gg \tau_R)\right)$

Let us discuss the detection of redox active gases, and consider an elemental anion-forming gas E_2 ($E_2 = \mathrm{Cl}_2, \mathrm{O}_2$, etc.), i.e.,

$$\frac{1}{2}E_2 + \varepsilon \mathrm{e}^- \rightleftharpoons \mathrm{E}^{\varepsilon-} \tag{16}$$

for which in equilibrium

$$\frac{1}{2}\mu_{E_2} + \varepsilon\tilde{\mu}_{\mathrm{e}^-} = \tilde{\mu}_{E^{\varepsilon-}} \tag{17}$$

holds. Only in the case of the bulk conductivity sensor (mode 1) overall equilibrium is reached after a certain time, while in other cases Eq. (17) is violated or may only hold locally.

(*i*) *Bulk Conductivity Sensor (Mode 1)*

In a mode 1 sensor, Eq. (17) applies throughout the system. To be more specific, let us formulate the overall introduction of E atomistically as

$$\frac{1}{2}E_2(\mathrm{g}) + \mathrm{v}_\mathrm{E}^{\varepsilon\cdot} + \varepsilon \mathrm{e}' \rightleftharpoons E_\mathrm{E}, \tag{18}$$

whereby $\mathrm{v}_\mathrm{E}^{\varepsilon\cdot}$ denotes a vacancy with the effective charge given as upper index, and e' denotes a conduction electron. $E_\mathrm{E}^\times$ stands for a a regular anion in the crystal in (charge-)relative notation. For low carrier concentrations the mass action law states that $\left[\mathrm{v}_\mathrm{E}^{\varepsilon\cdot}\right][\mathrm{e}']^\varepsilon = K_\mathrm{E}^{-1}P_{\mathrm{E}_2}^{-1/2}$. Based on this relation the partial pressure dependence and the T-dependence of the signals can then be directly derived by considering the electroneutrality condition. We assume that the electronic conductivity ($\propto [\mathrm{e}']$) is measured ($u_\mathrm{eon} \gg u_\mathrm{ion}$, u: mobility). As long as we can assume that the defect chemistry is simple, i.e., only $\mathrm{v}_\mathrm{E}^{\varepsilon\cdot}$, e' or dopants have to be considered as defects, the signal change $\delta \ln\sigma$ is proportional to $\delta \ln P_{\mathrm{E}_2}$ (cf. Part I[2]). In the intrinsic case, (i. e., $\varepsilon\left[\mathrm{v}_\mathrm{E}^{\varepsilon\cdot}\right] = [\mathrm{e}']$) the proportionality constant which determines sensitivity, is given by $-\frac{1}{2(\varepsilon+1)}$, in the acceptor doped case $\left(\left[\mathrm{v}_\mathrm{E}^{\varepsilon\cdot}\right] = \mathrm{const}\right)$ it is $-\frac{1}{2\varepsilon}$ (the same would be valid for overwhelming ionic disorder, e.g., $\left[\mathrm{v}_\mathrm{E}^{\varepsilon\cdot}\right] = \left[\mathrm{E}_\mathrm{i}^{\varepsilon'}\right]$) and 0 in the donor-doped case ($[\mathrm{e}'] = \mathrm{const}$). Obviously sensitivity is highest in the acceptor doped case. Here, however, the temperature dependence is also highest, namely determined via $\Delta_E H^\circ/\varepsilon$ compared to $\Delta_E H^\circ/(\varepsilon+1)$ and 0, respectively ($\Delta_E H^\circ$: enthalpy change associated with reaction 18). In the diffusion-controlled regime, the response time ($\tau_\delta \propto L^2/D^\delta$) is determined (besides sample thickness L) by the chemical diffusion coefficient (D^δ) which also differs for the different defect chemical situations just discussed (see Section IV.3, Part I). In the regime of interfacial control, effective rate constants (k^δ) are the decisive kinetic parameters ($\tau_\delta \propto 1/k^\delta$) (see Section VI.5, Part I). It is particularly worthy to note that the material must be as free as possible from deep dopants as to avoid a significant depression of the rate constants by trapping effects.[22]

$SrTiO_3$ may serve as a well-investigated material for such a bulk conductivity sensor. Its defect thermodynamics and also the relevant kinetic parameters have been discussed in detail in Part I.[2] In particular at low temperatures and at small sample thicknesses L, the kinetics of oxygen incorporation becomes surface reaction controlled, and $\overline{k}^\delta$ the decisive kinetic parameter.

Important advantages of the bulk conductivity sensor are, besides selectivity, its simplicity (no reference needed) and its selectivity; an important drawback is the T-dependence which can be quite significant (see above). Improvement via doping or by using a T-reference is straightforward, but partly at the cost of sensitivity, simplicity, or range of application.[23]

(ii) Surface Conductivity Sensors (Mode 2)

The second mode to be described relies on a purely electronic interaction, as expected for a purely electronically conducting oxide or more accurately for a mixed conductor under conditions under which the ions are immobile (or the incorporation reaction is kinetically not possible, see below). The material of choice is SnO_2. Its defect chemistry is determined by oxygen deficiency[24] and has also been considered in detail in Part I. In the pure oxide oxygen vacancies are compensated by conduction electrons $\left(2\left[V_O^{\bullet\bullet}\right]=\left[e'\right]\right)$. At high temperatures SnO_2 acts as a mixed conductor in that it readily dissolves oxygen, and can thus serve as an example of a mode 1 sensor, too. Under these conditions both the ionic and the electronic carriers are mobile; consequently the electronic bulk conductivity (which is still dominating the overall conductivity) changes with an oxygen partial pressure exponent of

$-1/6$ in pure SnO_2; in doped SnO_2 this is only true for very low P_{O_2}, at moderate or high P_{O_2} the characteristic exponent is $-1/4$ (if acceptor doped, i.e., $[V_O^{\cdot\cdot}] = \text{const}$) or 0 (if donor doped, i.e., $[e'] = \text{const}$).

The SnO_2-based surface conductivity sensor (Taguchi sensor), however, refers to the behavior at comparatively low temperatures.[25] There we might expect that we can neglect ionic motion. (Note that the high temperature equilibrium and the cooling conditions are important for the low temperature defect chemistry, especially in the presence of redox-active impurities.[26]) Then, oxygen cannot enter the bulk but remains adsorbed at the surface and, according to a pure electronic equilibrium, traps conduction electrons out of the space charge region, resulting in an increased surface resistance via the formation of a depletion layer. The signal is determined by the space charge conductivity relations derived in Section V of Part I.[2]

The response time is chiefly determined by the rate constants of the rate limiting surface step, while a great deal of the drift should be determined by the subsequent rate constants including bulk diffusion coefficient. The detailed analysis of the kinetics of SnO_2, which is developed in Ref.[19,27] and indicated in Figure 3, also taking account of trapping effects, transport in and through the space charge regions and the surface steps of the interaction, modifies the simplified picture of a negligible diffusion coefficient (cf. Figure 3); it rather suggests that one of the surface steps (after the steps which are indispensable for the sensor action) is severely hindered and thus enables the Taguchi-sensor performance.[19] Hence, not a negligible oxygen mobility but a sluggish incorporation reaction step at the surface (succeeding the signal determining electron transfer step) prevents SnO_2 from acting as a mode 2 sensor at low temperature. The chemical diffusion coefficient of oxygen for the usually very tiny SnO_2 sensor particles itself can be high enough to allow for equilibration in many cases. This questions the simple distinction in terms of the chemical potential (given in Table 1) at least in the case of SnO_2 nanoparticles or very thin films, but does not affect the essence of the presentation.

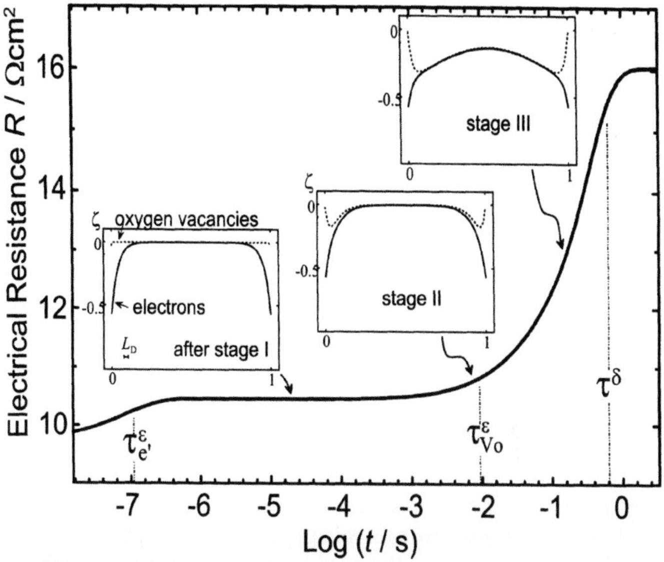

Figure 3. At least three processes with different time constants characterize resistance response to a sudden oxygen partial pressure change over (n-type conducting) SnO_2: electron transfer to adsorbed oxygen, oxygen vacancy distribution in the space charge zones, oxygen incorporation into space charge region and bulk.[28] Reprinted from J. Jamnik, B. Kamp, R. Merkle and J. Maier, *Solid State Ionics*, **150**, 157–166. Copyright © 2002 with permission from Elsevier.

Since many gases exhibit a redox-active behavior, and the selectivity of the lattice for dissolving oxygen does not matter, the selectivity of a mode 2 sensor is rather low. Not only are most elementary redox-active gases sensor-active, also complex gases such as H_2O, CH_4, CH_x, etc. NO_x interfere, the detailed mechanisms being different. H_2O induces a change in the surface chemistry leading to a crosseffect, e.g., in the case of O_2 or H_2 sensing. Direct electronic effects have also been claimed to occur.[29] NO_x gives an oxidizing signal while CH_x usually acts reducing with reaction products being alcohols, aldehydes, or carbon acids. In the literature several attempts have been made to increase the effective selectivity, e.g., by varying the operating temperature (since the sorption rates are differently T-dependent), or by using pattern recognition and applying calibrated sensor arrays. In the presence of metastable gas mixtures, such as CH_x and oxygen, the reaction may not be just a sorption reaction, SnO_2 may also serve as a heterogeneous catalyst.

Mode 2 devices which rely on a different detection principle are the Kelvin probe sensor and the CHEMFET. In the first case, a vibrating capacitor measures the change of the work function (see Figure 2), while in the second case the interaction is detected in the field-effect transistor mode.[29–31]

(iii) Galvanic Sensors (Mode 3)

In mode 3, the galvanic sensor mode, there is, as for mode 2, no global equilibrium with respect to Eq. (16) but now it is the ionic equilibrium that is established, i.e., $\nabla\tilde{\mu}_{E^{n-}} = 0 \neq \nabla\tilde{\mu}_{e^-} (\nabla\mu_E \neq 0)$: Owing to a negligible electronic conductivity ($\sigma_{eon} = 0$) in the electrolyte, the internal electronic current is zero and since $i = i_{ion} + i_{eon} = 0$, there is also no ionic current; consequently, $\nabla\tilde{\mu}_{E^{n-}}$ disappears because of the nonzero value of σ_{ion}. In this way nonzero differences of the electrochemical potential of the electrons, i.e., a nonzero cell voltage, can be established over the sample. In the case of the well-known λ-sensor,[32] which is extensively used in automobiles

$$\mathrm{Pt, O_2}(P_\mathrm{L})|\mathrm{ZrO_2(Y_2O_3)}|\mathrm{O_2}(P_\mathrm{R}),\mathrm{Pt}$$

we can express local equilibrium on both sides by

$$\frac{1}{2}\mu_{O_2} + 2\tilde{\mu}_{e^-}(\mathrm{Pt}) = \tilde{\mu}_{O^{2-}}(\mathrm{ZrO_2}). \quad (19)$$

We now use P instead of P_{O_2} to simplify notation. By calculating the difference of $\tilde{\mu}_{e^-}$ on both sides, Nernst's equation is obtained

$$E = \frac{RT}{4F}\ln\frac{P_\mathrm{R}}{P_\mathrm{L}} \quad (20)$$

($\Delta\tilde{\mu}_{O^{2-}} = 0$, $\Delta\tilde{\mu}_{e^-} = -FE$, E = open circuit cell voltage). If the P_{O_2} on one side is known (which is achieved by a mixture of oxygen with an inert gas or a reactive buffer mixture such as H_2/H_2O, CO/CO_2), this so-called λ-cell can measure the oxygen partial pressure on the other side (see Figure 4). The signal can also be used to control P_{O_2} directly.

A related cell is the formation cell

$$\mathrm{(Pt)Ag\,|\,AgCl\,|\,Cl_2,Pt}$$

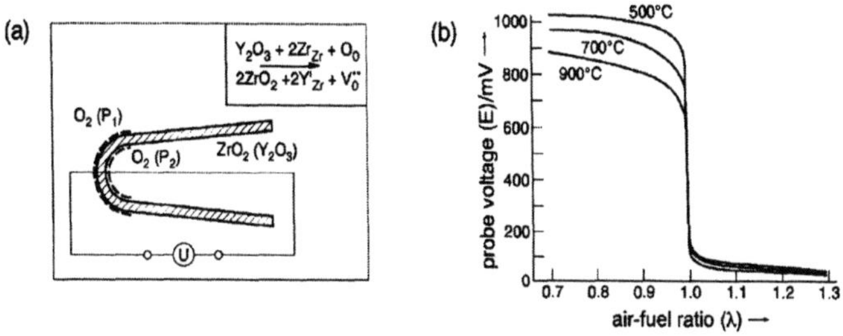

Figure 4. (a) Oxygen concentration cell with $ZrO_2(Y_2O_3)$ (termed YSZ) as the ceramic membrane, which finds application in O_2 sensors (λ probes), pumps, and fuel cells (in fuel cells the presence of the fuel (e.g., H_2) keeps the partial pressure on the anode side very low). Reprinted from J. Maier, *Physical Chemistry of Ionic Materials. Ions and Electrons in Solids*, Copyright © 2004 with permission from John Wiley and Sons, Ltd. (b) λ probe voltage as a function of air–fuel ratio. If the mixture is too fat $(\lambda < 1)$ or too lean $(\lambda > 1)$, the cell voltage deviates strongly from the value at the stoichiometric point $(\lambda = 1)$.[33] Reprinted from H. Dietz, W. Haecker, H. Jahnke, Electrochemical Sensors for the Analysis of Gases, in: *Advances in Electrochemistry and Electrochemical Engineering*, H. Gerischer, C.W. Tobias (eds.), J. Wiley and Sons. Copyright © 1977 with permission from John Wiley & Sons, Inc.

As on the chlorine side the contact $AgCl/Cl_2$ establishes a well-defined silver activity, it can also be regarded as a chlorine or silver activity cell; analogously the contact Ag/AgCl provides a well-defined partial pressure of chlorine.

Since silver chloride is an ionic conductor in itself, this chain represents a minimum phase scheme. Let us consider the thermodynamics in more detail; on the left-hand side we may write

$$\mu^{\circ}_{Ag} = \tilde{\mu}^{L}_{Ag^+}(AgCl) + \tilde{\mu}^{L}_{e^-}(Ag) = \tilde{\mu}^{L}_{Ag^+}(AgCl) + \tilde{\mu}^{L}_{e^-}(Pt) \quad (21)$$

and on the right-side hand

$$\frac{1}{2}\mu^{R}_{Cl_2} + \tilde{\mu}^{R}_{Ag^+}(AgCl) + \tilde{\mu}^{R}_{e^-}(Pt) = \mu^{\circ}_{AgCl}. \quad (22)$$

Forming the difference, $\tilde{\mu}_{Ag^+}$ cancels, whereas the difference in $\tilde{\mu}_{e^-}$ constitutes the cell voltage

$$EF = \mu^{\circ}_{Ag} + \frac{1}{2}\mu^{\circ}_{Cl_2} + RT\ln P^{R}_{Cl_2} - \mu^{\circ}_{AgCl} = -\Delta_f G_{AgCl}$$
$$= -\Delta_f G^{\circ}_{AgCl} + RT\ln P^{R}_{Cl_2}. \quad (23)$$

In other words, we added up

$$\begin{aligned} &\mathrm{Ag} \rightleftharpoons \mathrm{Ag}^+ + \mathrm{e}^-(\mathrm{L}) \\ &\mathrm{Ag}^+ + \frac{1}{2}\mathrm{Cl}_2 + \mathrm{e}^-(\mathrm{R}) \rightleftharpoons \mathrm{AgCl} \end{aligned} \tag{24}$$

to arrive at

$$\mathrm{Ag} + \frac{1}{2}\mathrm{Cl}_2 + \mathrm{e}^-(\mathrm{R}) \rightleftharpoons \mathrm{AgCl} + \mathrm{e}^-(\mathrm{L}) \tag{25}$$

yielding Eq. (23) directly by applying the electrochemical equilibrium condition $\Sigma_j \nu_j \tilde{\mu}_j = 0$ to Eq. (25); this again is equivalent to equating the electrical free enthalpy ($-zFE$) with the purely chemical free enthalpy ($\Delta_r G = \Sigma \nu_r \mu_r$), i.e., the ΔG-value of the purely chemical cell reaction

$$\mathrm{Ag} + \frac{1}{2}\mathrm{Cl}_2 \rightleftharpoons \mathrm{AgCl} \tag{26}$$

If the cell consists of more phases, a more local approach is advisable. Even though a well-established procedure in electrochemistry, this shall be briefly set out since in the field of solid state electrochemistry in general and in the field of sensors in particular, often differently conducting electrolytes are switched together without realizing the consequences ("double electrolyte cells"). The procedure is as follows: The difference in $\tilde{\mu}_{e^-}$(Pt) represents the electric potential difference across the cell (cell voltage) and can thus, as a state function, be obtained by summing up all ϕ-changes. The ϕ-changes of a given interface can be obtained from a local analysis. Let us consider the lhs (L) of the formation cell. Across the Ag|AgCl boundary, $\tilde{\mu}_{\mathrm{Ag}^+}$ is constant, thus $\Delta\phi = -\Delta\mu_{\mathrm{Ag}^+}/\mathrm{F}$. As the chemical potential of Ag^+ in Ag can be written as $\overset{\circ}{\mu}_{\mathrm{Ag}} - \mu_{e^-}(\mathrm{Ag})$, and as the latter is related to μ_{e^-}(Pt) via the ϕ-jump at the boundary Pt|Ag, the total electrode potential at the L-side follows as

$$\Delta\phi^{\mathrm{L}} \propto \mu^{\mathrm{L}}_{\mathrm{Ag}^+}(\mathrm{AgCl}) + \mu^{\mathrm{L}}_{e^-}(\mathrm{Pt}) - \mu^{\mathrm{L}}_{\mathrm{Ag}}. \tag{27}$$

In the same way, the electrode potential at the R-side is obtained as

$$\begin{aligned} \Delta\phi^{\mathrm{R}} &\propto -\mu^{\mathrm{R}}_{\mathrm{Cl}^-}(\mathrm{AgCl}) + \mu^{\mathrm{R}}_{e^-}(\mathrm{Pt}) + \frac{1}{2}\mu^{\mathrm{R}}_{\mathrm{Cl}_2} \\ &\propto -\overset{\circ}{\mu}_{\mathrm{AgCl}} + \mu^{\mathrm{R}}_{\mathrm{Ag}^+}(\mathrm{AgCl}) + \mu^{\mathrm{R}}_{e^-}(\mathrm{Pt}) + \frac{1}{2}\mu^{\mathrm{R}}_{\mathrm{Cl}_2}. \end{aligned} \tag{28}$$

On forming the difference between $\Delta\phi^{L}$ and $\Delta\phi^{R}$, the $\mu_{e^-}(\text{Pt})$-values cancel (the electronic concentration is essentially constant in Pt). Also μ_{Ag^+} cancels, provided the silver conductor has a sufficiently high defect concentration which is usually fulfilled. If this is not the case, the μ_{Ag^+}-change across the electrolyte leads to an additional $\Delta\phi$-term which has to be considered in the cell voltage. Equation (23) is the result in any case. Now, we consider a series switching of two silver halides

$$\text{Pt}|\text{Ag}|\text{AgX}'|\text{AgX}|\text{X}_2,\text{Pt}$$

The difference $\mu^{L}_{Ag^+} - \mu^{R}_{Ag^+}$ obviously only vanishes if we allow both the halides to equilibrate. For $X' = \text{Br}$ and $X = \text{Cl}$ a solid solution must form (Ag(Cl, Br)). For $X' = \text{Br}$ and $X = \text{I}$ the equilibrated contact is $\text{AgBr}_{sss} \mid \text{AgI}_{sss}$ corresponding to the formation of a miscibility gap (sss: saturated solid solution). In both cases the steady-state cell voltage does not represent the thermodynamics of the pure phases.

If we, however, switch between the two halides a solid electrolyte such as Ag^+-alumina which is only conductive for Ag^+ but blocks the anions, the difference $\mu^{L}_{Ag^+} - \mu^{R}_{Ag^+}$ vanishes ($\mu^{L}_{Ag^+}(\text{AgBr}) = \mu^{L}_{Ag^+}(\beta - \text{alumina}) = \mu^{R}_{Ag^+}(\beta - \text{alumina}) = \mu^{R}_{Ag^+}(\text{AgCl})$) albeit we refer to the nonequilibrated phases. This evidently allows the difference of the half cell potentials Ag/AgX and Ag/AgX′ to be measured.

If the two pure phases are contacted without such a membrane or equilibration, a (time dependent) diffusion potential has to be added to Eq. (23). In many cases diffusion potentials even if undefined, can be virtually constant and then taken account of by calibration. According to this, it is not surprising that e.g., changes in the O_2 potentials may be detected by using window glass (Na^+ conductor), or Cu potentials by applying silver conductors even if the cell is under-determined according to the phase rule.

Another complication arises if there are different, independent processes (1 and 2) that are both contributing to the electrode potential. Then, a mixed potential is generated (see, e.g., Ref.[8]). Let ε_1 be the potential for which $i_1 = 0$ $\left(\text{i.e., } |\vec{i}_1| = |\overleftarrow{i}_1|\right)$ and equally

$i_2(\varepsilon_2)=0$, then the potential ε_M at which the total current vanishes, lies in between ε_1 and ε_2 ($i(\varepsilon_M)=i_1(\varepsilon_M)+i_2(\varepsilon_M)=0$). (As at ε_M it holds that $|\vec{i}_1|\neq|\overleftarrow{i}_1|$ and $|\vec{i}_2|\neq|\overleftarrow{i}_2|$, a nonequilibrium situation is locally met.) It is obvious that the mixed potential is closer to the process that is characterized by the higher exchange current density (steeper $i(\varepsilon)$ curve). This implies that processes with lower exchange dynamics are not so important; this discrepancy may be augmented significantly if a perceptible current is drawn. Such mixed potential sensors can be explicitly used as sensing devices (see Ref.[31]). The discussion also highlights the fact that selectivity can be strongly varied by leaving the electrochemical equilibrium. Such amperometric devices based on zirconia electrolytes can be advantageously used for oxygen sensing. If the voltage is increased to a high enough value and hence significant currents are drawn, the cell acts as a gas pump which reaches a saturation current. This limiting current is proportional to the gas concentration.[34]

With regard to the potentiometric mode (mode 3) it may be concluded that potentiometric sensors are easily designed and often very sensitive. Their T-dependencies are typically moderate: If the activities involved do not change with temperature, the emf is simply proportional to T. If, however, in a λ-sensor a metal/metal oxide mixture (instead of air) is used as a simple possibility to establish a defined reference activity, the simplicity is bought at the expense of a strong T-dependence due to the chemical equilibrium constant. (Commercially miniaturized probes are available which contain Pd/PdO mixtures as reference electrodes.) Selectivities are usually not as good as for mode 1 but better than for mode 2. Drift processes are due to nonirreversibilities such as diffusion potentials, chemical reactions, or due to gas permeation because of non negligible electronic conductivities of the electrolyte. The latter is considered in greater detail in Section II.2.

(iv) Extension to Acid–Base Active Gases

The above treated gases were redox-active gases, i.e., gases whose interactions with electrons are the relevant one. Yet, the characteristic interaction of many complex gases such as NH_3, H_2O, CO_2 is the interaction with ions, i.e., they are acid–base active through

$$NH_3 + H^+ \rightleftharpoons NH_4^+$$

$$CO_2 + O^{2-} \rightleftharpoons CO_3^{2-}$$

$$H_2O + H^+ \rightleftharpoons H_3O^+$$

A detection of these gases via a redox change, may be possible, but would involve low selectivity. Let us explicitly follow Ref.[17] and systematically construct the analogues to the 3 modes discussed above, in terms of acid–base interactions.

(i) For a mode 1 sensor one has to use a material that is able to dissolve the complex gas under concern. For the detection of water such a material may be a perovskite that is conducting for both proton and oxygen ions (see Part I[2]).[35] By an ambipolar diffusion of these carriers, water transport is rendered possible. The proton conductivity (also the oxygen ion and the electronic conductors) changes with P_{H_2O}, what can be exploited as a sensor signal:

The chemical diffusion coefficient, $D^{\delta}_{H_2O} = f(c_{H_i^{\cdot}}, c_{V_O^{\cdot\cdot}}, \sigma_{H_i^{\cdot}}, \sigma_{V_O^{\cdot\cdot}})$ (which can be calculated just as done for $D^{\delta}_{O_2} = f(c_{h^{\cdot}}, c_{V_O^{\cdot\cdot}}, \sigma_{h^{\cdot}}, \sigma_{V_O^{\cdot\cdot}})$ in Part I), reads for negligible trapping effects as

$$D^{\delta}_{H_2O} = \frac{(2-x)D_{H_i} \cdot D_{V_O^{\cdot\cdot}}}{xD_{H_i} + 2(1-x)D_{V_O^{\cdot\cdot}}}, \tag{29}$$

where $x \equiv c_{H_2O}/c_{V_O^{\cdot\cdot}}$.[36] In relevant cases D^{δ} is rather low due to low $D_{V_O^{\cdot\cdot}}$, and thin films are required for a reasonable response time (see Figure 5). (As regards the k^{δ} value, i.e., the kinetics of the surface reaction, the situation strongly depends on the mechanism. The adsorption and dissociation steps may well be reasonably high owing to the polarity of the water molecule. In the presence of three carriers, peculiar kinetic situations may be met.[37]) Another obvious example are ammoniates for NH_3 sensing, for which the incorporation and transport kinetics are expected to be even less favorable.

(ii) This kinetic difficulty is not relevant in mode 2. Here, one has to search for an ionic conductor, whose ionic carrier concentration is influenced by the acid–base active gas. Basic gases may trap protons, or other cations, and accordingly influence the ionic surface conductivity, whereas acidic gases should show the opposite

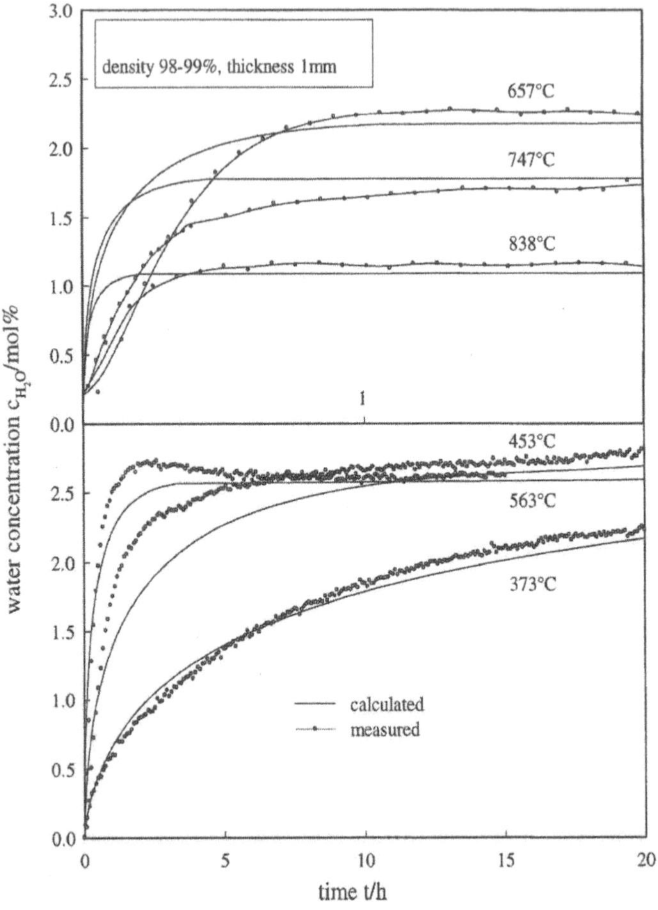

Figure 5. Water uptake of Gd-doped $BaCeO_3$. The time evolution is characterized by the chemical diffusion coefficient of H_2O.[36] Reprinted from K.-D. Kreuer, E. Schönherr, and J. Maier, *Solid State Ionics* **70/71** (1994) 278–284. Copyright © 1994 with permission from Elsevier.

effect. An ammonia sensor based on this novel sensor principle was introduced in Ref.[38] and uses AgCl. According to its basicity NH_3 is expected to trap Ag^+ and to increase the silver vacancy conduction, as was experimentally confirmed. The correspondence with the Taguchi (SnO_2) sensor becomes most obvious if we use the ionic and electronic (partial free) energy-level diagrams (see Figure 31 of Part I[2]). As shown in Ref.,[39] point electrodes can be used with advantage to reduce the response time and to increase (Figure 6) the sensitivity. Also Cu-conductors have been investigated as possible NH_3 sensors.[40]

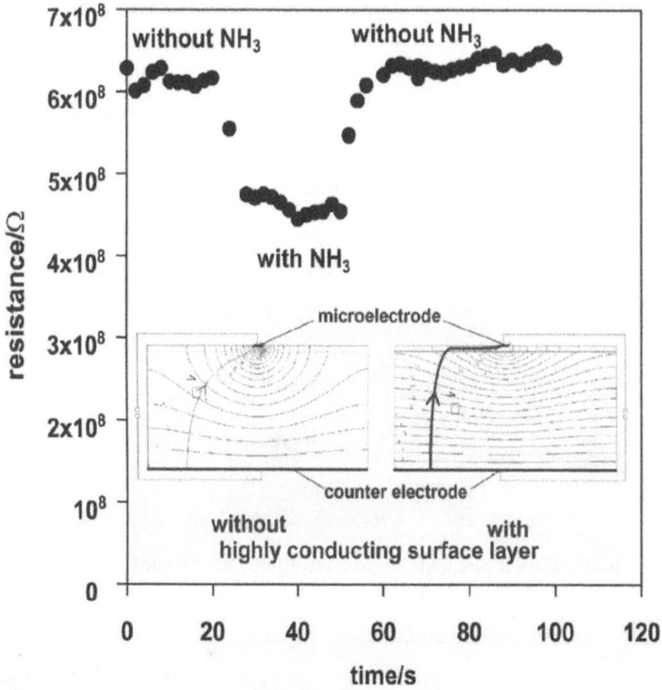

Figure 6. The conductivity response of AgCl on a contact to NH_3 (measured by microelectrodes).[39] Reprinted from M. Holzinger, J. Fleig, J. Maier, and W. Sitte, Ber. Bunsenges. Phys. Chem. **99** (1995) 1427–1432. Copyright © 1995 with permission from Deutsche Bunsen-Gesellschaft für Physikalische Chemie.

(iii) As regards mode 3, the correspondence cannot be complete since the electrode reactions are necessarily redox reactions. It is, however, possible to design cells, the overall reaction of which does not involve redox processes, but are of acid–base type.

This is achieved in the electrochemical CO_2 sensor to be described now. The gas sensitive electrode in the emf sensors for CO_2 detection is based on an alkaline or alkaline earth carbonate.[41–43] If Na_2CO_3 is to be used, Na- β'' -alumina or Nasicon (Part I[2]) are suitable electrolytes. A thermodynamically well-defined sensor is obtained by using a mixture of a binary and a ternary oxide such as[42]

$$Na_2CO_3, CO_2 \left| Na^+\text{-conductor} \right| Na_2MO_3, MO_2$$

with the whole cell being exposed to the same gaseous environment. It is immediately clear that the chemical potential of sodium oxide is fixed on both sides, e.g., $\mu^{L}_{Na_2O} = \mu_{Na_2CO_3} - \mu^{L}_{CO_2}$ (so on an oxidic level the cell is comparable to an oxygen concentration cell such as $O_2|O^{2-}$-conductor$|$Ni, NiO). As the Na^+ -conductor probes the sodium potential, a defined Na_2O activity is not sufficient, rather the sodium potential (cf. phase rule) has to be defined. Fixing the Na-potential by using, e.g., Na or Na-alloys, is not a good idea. Apart from the materials problems, it requires sealing. This is not only difficult from the aspect of compatibility, it is even fundamentally disadvantageous, as it results in an oxygen partial pressure dependence of the cell potential. In the approach to be discussed here, simply the oxygen of the atmosphere is used as the necessary third phase to fix μ_{Na} on the both sides. The cell reactions then read:

$$\text{lhs} \qquad Na_2CO_3 \rightleftharpoons 2Na^+ + 2e^- + \frac{1}{2}O_2 + CO_2$$

$$\text{rhs} \qquad 2Na^+ + 2e^- + \frac{1}{2}O_2 + MO_2 \rightleftharpoons Na_2MO_3.$$

In the overall chemical reaction

$$Na_2CO_3 + MO_2 \rightleftharpoons Na_2MO_3 + CO_2$$

or in the more precise electrochemical equilibrium formulation

$$Na_2CO_3 + MO_2 + 2e^-(L) \rightleftharpoons Na_2MO_3 + CO_2 + 2e^-(R)$$

P_{O_2} cancels, and the emf is solely dependent on CO_2, namely

$$FE = -\Delta_R G^\circ - RT \ln P_{CO_2}. \tag{32}$$

(An equivalent way to derive Eq. (32) is to split the chemical potential of the ternaries into $\tilde{\mu}_{Na^+}, \tilde{\mu}_{e^-}, \mu_{O_2}$ and μ_{CO_2} or μ_{MO_2}. Owing to the electrolytic nature of the electrolyte $\Delta\tilde{\mu}_{Na^+}$ is zero, while $\Delta\tilde{\mu}_{e^-}$ represents the cell voltage.) As CO_2 does not interact

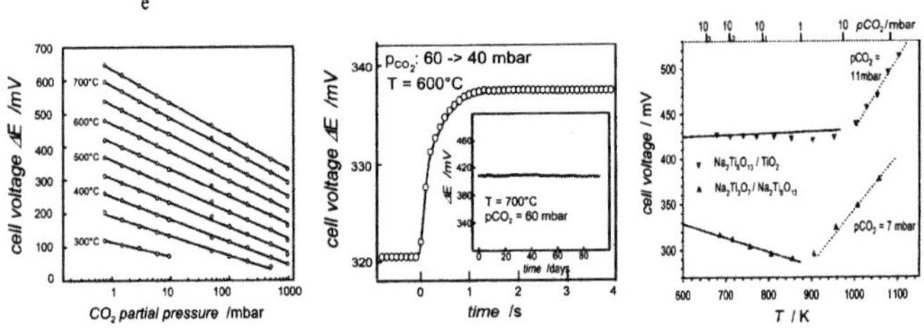

Figure 7. The performance of the open CO_2 sensor Au, O_2,CO_2,$|Na_2CO_3|$ Na-β-alumina$|O_2$,$Na_2Ti_6O_{13}$, TiO_2 is characterized by a signal[47] fulfilling the Nernst equation (*left*) (high sensitivity (*right*)), exhibiting negligible cross sensitivity with respect to O_2 (*left*), fast response time and long-time stability (*center*). The rhs figure shows that 1 ppm CO_2 is easily accessible; low P_{CO_2} are established by a $CaO/CaCO_3$ buffer; at high CO_2 the carbonate is used up and a constant P_{CO_2} is maintained.

with the reference side in the condition window of interest ($\Delta_R G < 0$), the sensor can be exposed to the same O_2 and CO_2 containing atmosphere at least for T>400°C.[44,45] The principle was first evidenced by using zirconates (i.e. M=Zr), which are, however, kinetically sluggish. Also stannates or silicates, and most advantageously titanates were used.[44,46] The sensor based on titanates:

$$\mathrm{Au}|\mathrm{CO_2,O_2,Na_2CO_3}|\mathrm{Na\text{-}\beta''\text{-}alumina}|\mathrm{(CO_2)O_2,Na_2Ti_6O_{13},TiO_2}|\mathrm{Au}$$

performs extremely well (Figure 7). It is very sensitive (ppm traces of CO_2 can be easily measured), its response time is very short (on the order of 1 s at 500–600°C), the signal is drift-free (we used one and the same sensor now for 5 years without perceptible change in the cell voltage), it exactly corresponds to the Nernst-equation at T>500°C and, in that temperature range, is independent of P_{O_2}. It is easy and inexpensive to construct, ecologically compatible and reflects all the advantages that mode 3 sensors can have. A disadvantage which is shared with the other competing emf solid state CO_2-sensors, is the elevated operating temperature and the reactivity of the carbonates with acidic gases. If the sensor is not always kept at high temperatures, also the hygroscopicity of Na_2CO_3 is a disadvantage. Li_2CO_3 or $SrCO_3$ are better candidates in this respect. Another challenge is the selective detection of complex gases which are both redox and acid–base active by electrochemical means, such as e.g., NO_x. The reader is referred to the specialists' literature for more details.

2. Electrochemical (Composition) Actors

In this section we will describe what may be called electrochemical composition actors, in which chemical action—and not mechanical action as usually described by the term actor (or actuator)—is caused by electrical excitation. Unlike electrochemical composition sensors which detect chemical composition via electrochemical methods, they cause changes in chemical composition. It shall not remain unmentioned that at present interesting work is pursued also to realize mechanical action on an electrochemical stimulus ("artifical muscles").[48]

Unlike in the case of basic storage devices (cf. the charging process of a battery) which we do not wish to subsume under this term and which we consider in the following section, here we will refer to situations in which the compositional changes caused by electricity, are meant to achieve more specific goals such as producing or supplying chemicals of interest or purifying mixtures, or causing optical effects. So we will refer to electrochemical pumps, filters, reactors as well as electrochemical windows.

Zirconia cells are powerful O_2 pumps in chemical engineering (see Figure 4), and serve not only to stabilize oxygen partial pressures but also to achieve substantial flow rates of pure oxygen. Furthermore, the tailored withdrawal or supply of oxygen or hydrogen via solid electrolytes is a promising tool to enable and direct chemical reactions such as hydrogenation or dehydrogenation (Figure 8) of hydrocarbons. The purification of metals from hydrogen and the syngas production using very pure O_2 are further examples. Related applications of future potential are the electrolysis of CO_2 to partly recover O_2 from exhaled air in space crafts and the high temperature water electrolysis by oxide or proton conductors. In this context, the high temperature proton and oxygen ion conductors are key materials and may become ingredients of a future hydrogen technology.

Electrochemical pumps have also been successfully used to promote reactions catalytically. If oxygen is pumped through a zirconia cell into a reaction chamber which is, e.g., filled with hydrocarbons, not only the oxygen that is transferred, reacts. The anode itself can act catalytically. It seems that this so-called NEMCA effect ("Nonfaradaic Electrochemical Modification of Catalytic Activity"[51]) relies on a hindered surface reaction as a consequence of which the applied potential is translated into a concentration polarization, as reflected by the enrichment of a not fully oxidized oxygen species (e.g., O^- or O^{2-}) at the interface.[51–53] For this species, redistribution equilibrium may be assumed that

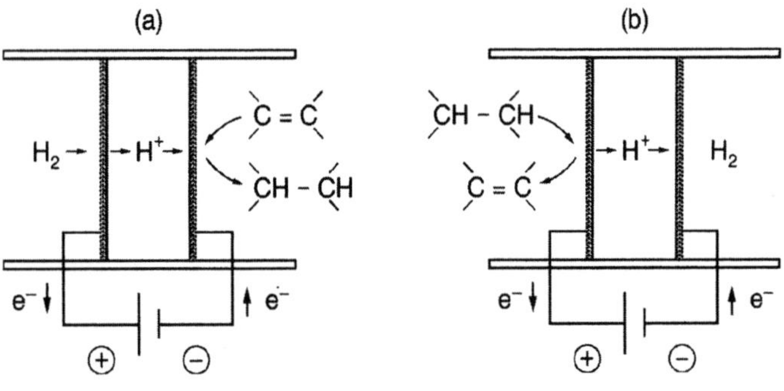

Figure 8. Proton conductors enable hydrogenating or dehydrogenating hydrocarbons without direct contact with H_2 gas.[49,50] Reprinted from H. Iwahara. *Solid State Ionics*, **77** (1995) 289–298. Copyright © 1995 with permission from Elsevier.

manifests itself in a spill-over to the electrode surface where it acts catalytically, in a direct or indirect way via a modified work function. The attraction of this effect lies in the vision to tune selectivity by the applied voltage.

Let us return to the pure mass transfer function and consider Figure 9. Instead of electrically driving a mass flow through a solid electrolyte, a pumping can also be achieved in a short-circuiting mode by applying an oxygen potential gradient (permeation) (Figure 9a). As above, this permeation can, owing to the selectivity of the solid, be used to filter out oxygen and to prepare highly purified gases. The short-circuiting electronic connection need not be an outer wire, it can also be internalized as in composites of negatively doped ZrO_2 and metals ("dual phases"; see Figure 9c). (Another possibility is to make use of the different conduction pathways within a given phase (Figure 9d); in slightly negatively doped CeO_2, e.g., grain boundaries can be n-conducting while the bulk is ionically conducting.[54,55]) More elegantly the filtering function is achieved by using a mixed conductor in which nonzero ionic and electronic conductivities are locally established (see Figure 9b).

While in the case of a solid electrolyte through which a current is electrically driven, the ionic conductivity alone is decisive, in the cases considered in Figures 9b–9c a neutral internal mass flux occurs from the side of the higher to the side of lower P_{O_2}, a

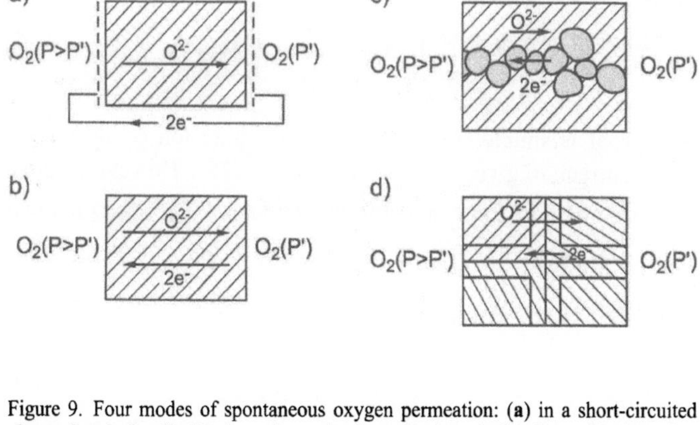

Figure 9. Four modes of spontaneous oxygen permeation: (**a**) in a short-circuited electrochemical cell; (**b**) through a mixed conducting single phase, (**c**) through a composite phase mixture comprising an ionic and an electronic conductor, and (**d**) through an ionic (electronic) conductor the grain boundaries of which are predominantly electronically (ionically) conducting.

process, for which both the electronic and ionic conductivities are important. Let us consider an O^{2-}-conductor with some degree of electronic conductivity (Figure 9d).

From the theory of ambipolar conduction (see, e.g., Ref.[56]) it follows immediately that given an oxygen potential difference $\Delta\mu_O$ between the lhs (μ_O) and the right-hand side, an oxygen flux which is of the form

$$j_O \propto -\sigma_O^\delta \frac{\partial \mu_O}{\partial x} \tag{33}$$

occurs where $\sigma_O^\delta = \sigma_{eon}\sigma_{ion}/\sigma$. In the steady-state j_O is positionally constant and

$$j_O \propto -\int_{\mu_O}^{\mu_O+\Delta\mu_O} \sigma_O^\delta \mathrm{d}\mu_O. \tag{34}$$

If we switch from a situation with uniform chemical potential ($\Delta\mu_O = 0$) to a situation in which on one side a different but constant P_{O_2} is established, a transient occurs during which the homogeneous stoichiometry profile changes to an approximately linear profile (see chemical polarization, see Appendix 3). As long as the electrode reactions are fast, the emf measured at such a sample is always determined by the invariant boundary values of the oxygen potential ($\mu_O, \mu_O + \Delta\mu_O$) but, owing to the internal virtually neutral short-circuit, lower than the Nernst-value. The result is, instead of Eq. (20),[56,57] now

$$E = \frac{RT}{4F}\int t_{ion}\mathrm{d}\mu_{O_2} = \frac{RT}{4F}\langle t_{ion}\rangle \ln\frac{P_R}{P_L}. \tag{35}$$

The proof is simple. As in the case of a pure ion conductor the ionic current is given by $i_{ion} = \sigma_{O^{2-}} \nabla\tilde{\mu}_{O^{2-}}/2F$. However, unlike the case of a pure ion conductor, the gradient $\nabla\tilde{\mu}_{O^{2-}}$ does not disappear, since $i_{ion} = -i_{eon} \neq 0$. By splitting it into $\frac{1}{2}\nabla\mu_{O_2} + 2\nabla\tilde{\mu}_{e^-}$, and replacing i_{ion} by $-\sigma_{e^-} \nabla\tilde{\mu}_{e^-}/F$,

$$\nabla\tilde{\mu}_{e} \propto \frac{\sigma_{O^{2-}}}{\sigma_{O^{2-}} + \sigma_{e^-}}\nabla\mu_{O_2} \tag{36}$$

is obtained, which, upon integration, yields Eq. (35).

Sr-doped La-manganates, cobaltates, nickelates, and ferrates are typical examples of such mixed conducting oxides. In fact, despite the high ionic conductivities of the latter ones, σ_{eon} prevails in all cases—i.e., $E = 0$ according to Eq. (35). As a consequence, σ_{ion} is the determining part for the steady-state permeation flux (note that $\sigma^{\delta} = \sigma_{\text{eon}}\sigma_{\text{ion}}/\sigma \simeq \sigma_{\text{ion}}$) (see Section II.3.iii.D). In the case of these mixed conducting membranes the presence of electrodes is not necessary. (In fact they are also typically good catalysts for the surface reaction.) While this is an advantage over electrolytic permeation cells, the latter have the benefit of allowing for an elegant flux control.

Both the flux equation (Eq. (34)) as well as the voltage equation (Eq. (35)) are more complicated if the ions can change their valence states. As shown in Ref.[3,58,59] the transport parameters then have to refer to "conservative ensembles"; in particular in Eq. (35), instead of t_{ion}, now $t_{\{\text{ion}\}}$ appears. If we take the example of a copper conductor with a Cu^{2+} and a Cu^{+} mobility, between which electronic equilibrium is established, the ratio between E and the Nernst-value is then

$$t_{\{\text{ion}\}} = \frac{\sigma_{Cu^{2+}} + 2\sigma_{Cu^{+}}}{\sigma_{Cu^{2+}} + \sigma_{Cu^{+}} + \sigma_{e^{-}}}. \tag{37}$$

The factor of two in the numerator, leads to a value different from one, even if $\sigma_{e^{-}} = 0$, which reflects the fact that a counter diffusion between Cu^{2+} and $2Cu^{+}$ can occur leading to an effective transport of electrons. The fact that $t_{\{\text{ion}\}}$ is greater than one, may appear puzzling but it is simply caused by the purposeful use of a charge number of 2 in Eq. (35). While these examples are academic in view of the prevailing electronic conductivity, valence changes have been shown to play a direct role for the emf in proton conductors.[60] In the general case—even if the ionic conduction predominates—a variety of species may contribute, allowing permeation of not only H, O but also H_2O (the simultaneous conductivity of O^{2-} and H^{+}), as already addressed in the context of sensors.

For the derivation of the emf of a solid that conducts H^{+}, O^{2-}, and e^{-}, one has to consider the coupling relations $\nabla\tilde{\mu}_{OH^{-}} = \nabla\tilde{\mu}_{H_2O} - \nabla\tilde{\mu}_{H^{+}}$; $\nabla\tilde{\mu}_{H^{+}} = \frac{1}{2}\nabla\mu_{H_2} - \nabla\tilde{\mu}_{e'}$; $\nabla\tilde{\mu}_{H_3O^{+}} = \nabla\mu_{H_2O} + \nabla\tilde{\mu}_{H^{+}}$, $\nabla\tilde{u}_{O^{2-}} = \frac{1}{2}\nabla\mu_{O_2} + 2\nabla\tilde{\mu}_{e^{-}}$ and the flux equations $i_k = -\frac{\sigma_k}{z_k F}\nabla\tilde{\mu}_k$. To demonstrate the procedure, let us consider a slightly simplified case for which $\sigma_{H_3O^{+}} = \sigma_{O^{2-}} = 0$.

Because of $\Sigma_k i_k = 0$, it is valid that $\frac{\sigma_{H^+}}{F}\nabla\tilde{\mu}_{H^+} - \frac{\sigma_{e^-}}{F}\nabla\tilde{\mu}_{e^-} - \frac{\sigma_{OH^-}}{F}\nabla\tilde{\mu}_{OH^-} = 0$. Using the coupling conditions between OH^-, H_2O, and e^-, and between H^+, H_2, and e^-, this can be rewritten as

$$\begin{aligned}&\frac{\sigma_{OH^-}+\sigma_{H^+}}{2F}\nabla\mu_{H_2} - \frac{\sigma_{OH^-}}{F}\nabla\mu_{H_2O} - \frac{\sigma_{e^-}+\sigma_{H^+}+\sigma_{OH^-}}{F}\nabla\tilde{\mu}_{e^-} = 0\\ &= \nabla\mu_{H_2} - \frac{\sigma_{OH^-}}{2F}\nabla\mu_{O_2} - \frac{\sigma_{e^-}+\sigma_{H^+}+\sigma_{OH^-}}{F}.\end{aligned} \tag{38}$$

A similar but slightly more complicated equation results if we also allow for nonzero values of $\sigma_{H_3O^{\cdot}}$ and $\sigma_{O^{2-}}$. The result is:

$$\begin{aligned}E &= \frac{RT}{4F}\left[t_{O^{2-}} + 2t_{OH^-} - 2t_{H_3O^+}\right]\Delta\ln P_{O_2} +\\ &-\frac{RT}{2F}\left[t_{H^+} - t_{OH^-} + t_{H_3O^+}\right]\Delta\ln P_{H_2}\\ &= \frac{RT}{4F}\left[t_{O^{2-}} + t_{H^+} + t_{OH^-} + t_{H_3O^+}\right]\Delta\ln P_{O_2} +\\ &-\frac{RT}{2F}\left[t_{H^+} - t_{OH^-} + t_{H_3O^+}\right]\Delta\ln P_{H_2O}.\end{aligned} \tag{39}$$

Equation (39), derived here in a straightforward way, has been discussed in detail by Norby and Kofstad (see, e.g., Ref.[60]) It is important to realize that the species OH^-, H_3O^+, H^+, and O^{2-} appear differently in the emf equation, expressing the fact that it matters how much oxygen and hydrogen is transferred by the species under consideration.

So far we dealt essentially with the permeation flux through the solid, but the storage itself is also important; in other terms, not only the chemical resistance is of significance ($R^\delta \propto 1/\sigma^\delta$) for applications, but also the chemical capacitance $\left(C^\delta \propto \frac{\partial n}{\partial \mu}\right)$ of solids is a relevant parameter.[15,61] In the context of permeation membranes, the chemical capacitance is of course a significant quantity as regards the transient processes ("time lag"). Generally this storage can be achieved chemically (change of μ_{O_2} in the ambient) or electrochemically (electrochemical titration). An example of the first kind was discussed in previous section, when we considered bulk conductivity sensors (response time). An example of the second kind are intercalation batteries (charge/discharge rates) which are to be considered in the upcoming section. Besides the energy storage aspect, such stoichiometry changes in mixed conductors can be used to alter other properties such as optical absorption. Reversible electrochemical variation of the oxygen content in e.g., tungsten bronzes by dissolving or ex-solving O, H, or Li leads to color changes that may be used to condition windows or mirrors in situ.[62]

3. Electrochemical Energy Storage and Conversion Devices

In the following we adopt the usual distinction between devices that convert chemical in electrical energy but cannot be electrically charged again (primary batteries), "metabolistic" cells in which active masses are continually supplied and removed by gas flow (fuel cells) and cells that can be recharged electrically (secondary batteries). In all these cases great technological advances have been made and there exists a rich literature on these topics.

There is a considerable interest in such systems,[63–66] especially since they represent a means of locally supplying electrical energy, necessary for, e.g., household appliances, telephones, clocks, laptops, or electrical vehicles.

The chemical storage of electrical energy is also essential for meeting peak power demands. Figure 10 gives an overview of mass specific energy and power densities as parameters for some relevant storage systems. We will only exemplarily discuss a few selected examples that reveal the solid state electrochemical background. In realistic battery systems at least one phase is a solid material. Even though solid–liquid interfaces provide lower contact resistances, there is still a strong driving force for using all-solid state cells (and hence for replacing liquid electrolytes by solid electrolytes). Solid electrolytes have the advantage of allowing for a better handling and miniaturization due to the higher mechanical and thermal strengths as well as for a far-reaching avoidance of side-reactions because of a high transport selectivity. Typically solid electrolytes require higher temperatures; this is not necessarily a trade-off, in many cases the high thermal stability (in contrast to liquid electrolytes) can be an inestimable advantage, for electrode reactions can be exploited that would not be reversible within the stability range of liquid electrolytes.

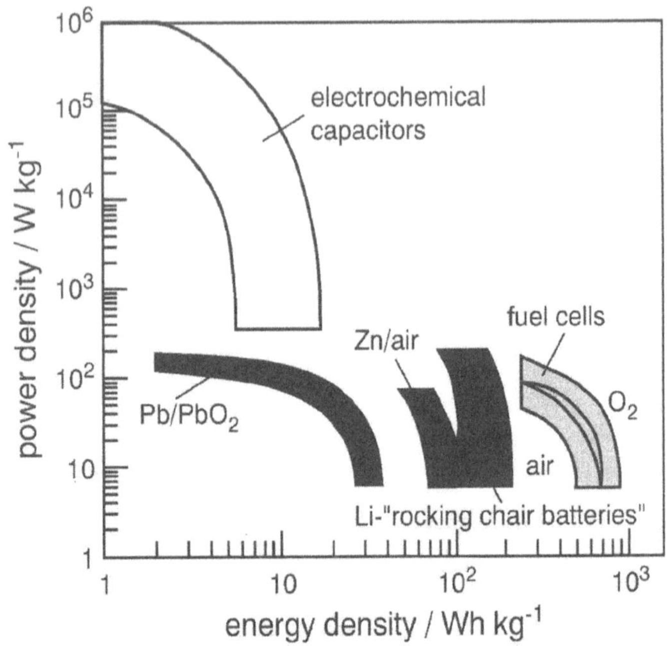

Figure 10. Power densities and energy densities for a variety of electrochemical energy conversion devices. The question of how much and how rapidly energy per unit mass can be "called up" is decisive for electrochemical applications.[65] Reprinted from M. Winter, J.O. Besenhard, M.E. Spahr, P. Novák, *Adv. Mater.*, **10** (1998), 725–763. Copyright © 1998 with permission from Wiley-VCH Verlag GmbH.

Let us begin with fuel cells.

(*i*) *Fuel Cells*

Apart from the simplicity and the elegancy of the concept which enables a local energy conversion, it is the high efficiency of a direct transformation of chemical to electrical energy, which constitutes a large part of the fascination of galvanic elements in general and of fuel cells in particular. The reason for the high theoretical efficiency is the avoidance of thermal processes.[67] It is useful to compare the following cells:

$$\text{Cell a} \quad = \quad H_2 \mid O_2 \tag{40a}$$

$$\text{Cell b} \quad = \quad H_2 \mid (O^{2-}, e^-) \mid O_2 \tag{40b}$$

$$\text{Cell c} \quad = \quad H_2|(O^{2-})|O_2 \tag{40c}$$

Cell a brings H_2 and O_2 into direct contact (cell a), then (under kinetically favorable conditions) water is formed and the stored chemical enthalpy released ($\Delta_r H$). This is similar in cell b in which H_2 and O_2 come into contact via permeation through a mixed conductor (O^{2-}, e^-) (see Figure 9). In both cases a heat engine may be used to convert the enthalpy of reaction into mechanical energy and a dynamo to convert the mechanical energy into electrical energy. Assuming, in order to estimate the limits, the latter process to be loss-free, not more than a fraction w_c ($\equiv$ Carnot efficiency) of the chemical energy can be transformed into electrical energy (reversible limit)[68]

$$w = \left|\frac{\text{electrical energy}}{\Delta_r H}\right| \le w_C = \frac{T_2 - T_1}{T_2}, \tag{41}$$

(T_2 and T_1 are the operation temperatures of the Carnot process). Typical values for a steam engine are $T_2 = 583\text{K}$ (boiling point of water at 100 bar) and $T_1 = 313$ K, then w_c = 46%; real values are markedly less. If, however, a pure oxygen ion conductor separates H_2 and O_2 as in cell c, the electrons flow through the external leads, what allows for a direct transformation of the chemical energy into electrical energy. The maximum possible efficiency of this electrochemical process is given by w_g for which

$$w \le w_g = \frac{\Delta_r G}{\Delta_r H} = 1 - \frac{T\Delta_r S}{\Delta_r H} = 1 + \frac{T\Delta_r S}{|\Delta_r H|} \tag{42}$$

$(\Delta_r G < 0, \Delta_r H < 0)$. In the case of reactions with positive reaction entropy, w_g can obviously be greater than 100%[70] as also seen from Table 3 which lists the w_g values of some fuel reactions. For water formation from H_2 and O_2 w_g is about 80% at elevated temperatures.

Besides the high theoretical efficiency there are also other aspects which make fuel cells attractive: they are environmentally benign as regards pollutant output and noise level, they possess a high flexibility with regard to size and siting and offer the possibility of locally cogenerating electrical power and heat. They hence have the potential to be key elements of a decentralized energy household.

Table 3.
Thermodynamic data and efficiencies for important cell reactions.[69] Reprinted from F. von Sturm, *Elektrochemische Stromerzeugung*, Wiley VCH, Weinheim. Copyright © 1969 with permission from Wiley-VCH Verlag GmbH.

Cell reaction	Temp. (°C)	$\Delta_r H^\circ$ (J mol^{-1})	$\Delta_r S^\circ$ (J mol^{-1}K^{-1})	$\Delta_r G^\circ$ (J mol^{-1})	$-U$ (V)	dU/dT (mV^{-1}K^{-1})	$w_g^\circ = \frac{\Delta_r G^\circ}{\Delta_r H^\circ}$
$H_2 + \frac{1}{2}O_2 \rightarrow H_2O_{(l)}$	25	−285800	−162.40	−237400	1.23	+0.840	0.83
	60	−285050	−159.00	−231830	1.20	+0.820	0.81
	100	−283300	−155.00	−220370	1.17	+0.800	0.78
$H_2 + \frac{1}{2}O_2 \rightarrow H_2O_{(g)}$	25	−241830	−44.40	−228580	1.18	+0.230	0.945
	60	−242180	−45.60	−226990	1.18	+0.230	0.94
	100	−242580	−46.60	−225160	1.17	+0.240	0.93
	500	−246180	−55.10	−203530	1.05	+0.280	0.83
$NH_{3(g)} + \frac{3}{4}O_2 \rightarrow \frac{1}{2}N_2 + \frac{3}{2}H_2O_{(l)}$	25	−382510	−145.50	−339120	1.17	+0.500	0.89

Cell reaction	Temp. (°C)	$\Delta_r H^\circ$ ($J\ mol^{-1}$)	$\Delta_r S^\circ$ ($J\ mol^{-1} K^{-1}$)	$\Delta_r G^\circ$ ($J\ mol^{-1}$)	$-U$ (V)	dU/dT ($mV^{-1} K^{-1}$)	$w_g^\circ = \frac{\Delta_r G^\circ}{\Delta_r H^\circ}$
$NH_{3(g)} + \frac{3}{4}O_2 \rightarrow \frac{1}{2}N_2 + \frac{3}{2}H_2O_{(g)}$	25				1.13		
$N_2H_{4(l)} + O_2 \rightarrow N_2 + 2H_2O_{(l)}$	25	–621100	+5.10	–622600	1.61	–0.010	1.00
$2Na + H_2O_{(l)} + \frac{1}{2}O_2 \rightarrow 2NaOH_{(aq)}$	25	–653210	–174.90	–601050	3.11	+0.900	0.92
$C_{Gr} + \frac{1}{2}O_2 \rightarrow CO$	25	–110500	+89.10	–137080	0.71	–0.460	1.24
	500	–110800	+89.90	–180300	0.93	–0.460	1.63
$C_{Gr} + O_2 \rightarrow CO_2$	25	–393500	+2.87	–394350	1.02	–0.007	1.00
$CO + \frac{1}{2}O_2 \rightarrow CO_2$	25	–283000	–86.20	–257300	1.33	+0.440	0.91
$CH_3OH_{(l)} + \frac{3}{2}O_2 \rightarrow CO_2 + 2H_2O_{(l)}$	25	–726260	–76.50	–703700	1.21	+0.130	0.97

(continued)

Cell reaction	Temp. (°C)	$\Delta_r H^\circ$ (J mol^{-1})	$\Delta_r S^\circ$ (J mol^{-1}K^{-1})	$\Delta_r G^\circ$ (J mol^{-1})	$-U$ (V)	dU/dT (mV^{-1}K^{-1})	$w_g^\circ = \frac{\Delta_r G^\circ}{\Delta_r H^\circ}$
$CH_4 + 2O_2 \rightarrow CO_2 + 2H_2O_{(g)}$	25	−802400	−6.00	−800600	1.04	+0.007	1.00
	60	−802060	−4.90	−800420	1.04	+0.006	1.00
	100	−801700	−3.90	−800200	1.04	+0.005	1.00
	500	−800300	−1.70	−798900	1.03	+0.002	1.00
$CH_4 + 2O_2 \rightarrow CO_2 + 2H_2O_{(l)}$	25	−890200	−242.60	−817900	1.06	+0.310	0.92
$C_2H_4 + 3O_2 \rightarrow 2CO_2 + 2H_2O_{(l)}$	25	−1306320	−62.10	−1287810	1.11	+0.050	0.99
$C_2H_4 + 3O_2 \rightarrow 2CO_2 + 2H_2O_{(g)}$	25				1.09		
$C_3H_8 + 5O_2 \rightarrow 3CO_2 + 4H_2O_{(l)}$	25	−2218900	−374.00	−2107440	1.09	+0.190	0.95
$C_3H_8 + 5O_2 \rightarrow 3CO_2 + 4H_2O_{(g)}$	25	−2044000	+108.00	−2076380	1.07	−0.050	1.02
$n\text{-}C_4H_{10} + 6\frac{1}{2}O_2 \rightarrow 4CO_2 + 5H_2O_{(l)}$	25	−2878270	−438.00	−2747930	1.09	+0.170	0.955

There are, however, a variety of reasons why fuel cells are not yet established and well-developed constituents of our environment; these are in particular, the losses caused by current flow, materials problems reflected by deficiencies in chemical, electrical or mechanical strength, as well as additional specific reasons that contribute to the cost of cells, generator and balance-of-plant. The most important losses that lead to w_g not being reached in practice (Figure 11) are voltage losses due to the electrolyte resistance (the related overvoltage can be considered to be ohmic), the electrode over-potential (in the case of a transfer over-potential, it typically increases under-proportionally with increasing current), and the diffusion over-potential essentially caused by inhibition of gas transport (this over-voltage typically increases asymptotically with the current and acts current-limiting). (For conceptual details see Ref.[72])

Here we are not discussing fuel cells that are based on liquid electrolytes (molten salts, phosphoric acid, aqueous potassium hydroxide) and only touch upon fuel cells that are based on ion exchange membranes with liquid-like transport (see Table 4 for a brief overview). Our focus is on high temperature ceramic fuel cells exploiting the benefit of thermally stable solids, viz. to operate at temperatures at which electrode reactions even with natural gas or CO are sufficiently fast. As a consequence of a lower conductivity of solid matter, the contribution of the electrolyte resistance to the overall losses is significant. If H_2 is used, the electrode reaction resistances essentially stem from the cathode where oxygen is reduced, whereas the anode gains increasing importance if natural gases are used. Before we discuss the main functional parts of a solid oxide fuel cell (SOFC) in more detail, let us have a brief look at polymer fuel cells.

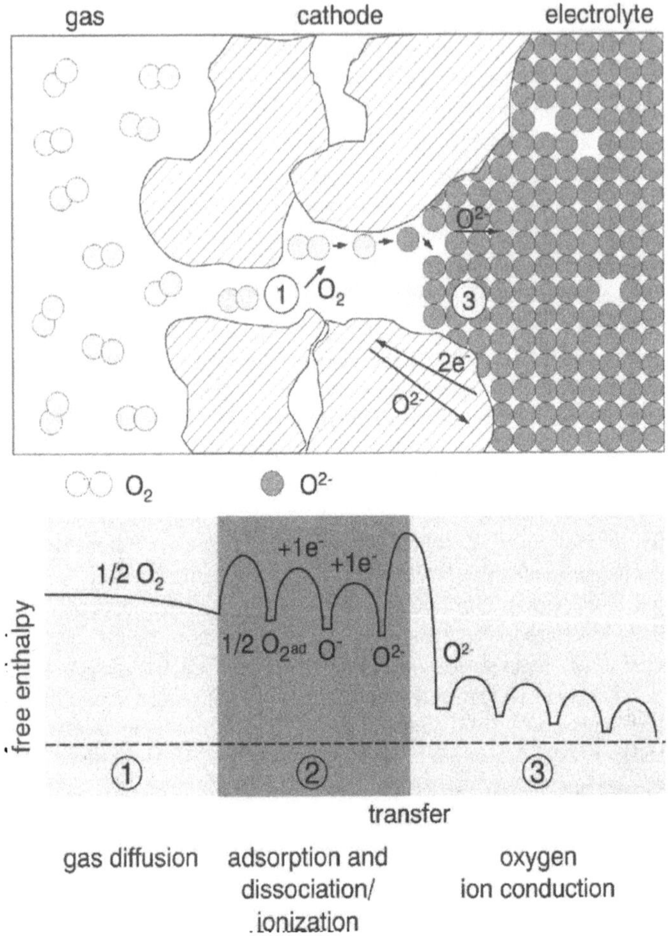

Figure 11. Cathode events in an SOFC fuel cell (schematic). (As regards the free enthalpy profile (1) the configuration effect is included in the free enthalpy in the case of gas diffusion (1), while otherwise the nonconfigurational value (2, 3) is shown.[71]) For the purpose of a simple presentation, it is assumed that the oxygen is completely ionized when it enters the electrolyte. Cf. text for a more specific discussion.[71]

Table 4.
Overview of important fuel cell types.[73] Reprinted from W. Gajewski, Spekt. Wissen., 7 (1995) 88–92. Copyright © 1995 with permission from Spektrum der Wissenschaft Verlagsgesellschaft mbH.

Name and abbreviation	Temperature in (°C)	Electrolyte	Fuel	Effiency (%)	Envisaged application
Alkali fuel cell (AFC)	80–90	Potash lye	Hydrogen	50–65	Transport, space flight, shipping
Polymer electrolyte membrane fuel cell (PEMFC)	80–90	Polymer membrane (Nafion)	Hydrogen, reformed methanol or methane	50–60	Transport, electro car, space flight, shipping
Phosphoric acid fuel cell (PAFC)	200	Phosphoric acid	Hydrogen, reformed methane	35–45	1 up to 100 MW power plants, 5–500 kW heating power station
Molten carbonate fuel cell (MCFC)	650	Calcium carbonate	Hydrogen, methane	45–60	1 up to 100 MW power plants, 5–500 kW heating power station
Solid oxide fuel cell (SOFC)	850–1000	Zirconia	Hydrogen, methane	50–60	1 up to 100 MW power plants, 5–500 kW heating power station

While high-temperature fuel cells are promising for localized power stations and for coupling with gas turbines, medium- or low-temperature fuel cells are the better candidates for nonstationary small-scale applications, as desired, e.g., for electrotraction. In these cases, polymer fuel cells are the candidates of choice (see Figure 12), and have already proved themselves in submarines and space vehicles. The standard electrolyte is Nafion.[74,75] Nafion consists of a perfluorinated hydrocarbon backbone with a high density of ether side chains terminated by sulfonic acid groups (see Figure 12). As a consequence of polar and nonpolar interaction, channels are formed which connect the acid groups and can be filled with water. On contact with the water in these channels, the protons of the acid groups are dissociated away. The proton transport in such an "acid sponge" is fluid-like with the important aspect that—since the counter ions are fixed at the interface—the protons move in the space charge zones (ion-exchange membrane electrolyte). The average diameter of these channels is typically smaller than the Debye-length, a fact that contributes towards a rather homogeneous transport.[76] Methanol–air fuel cells are promising energy providers of high capacity, suitable for local applications ranging from laptops to cars. A major driving force for electro-traction consists in the substitution of the liquid gasoline by another liquid phase (Figure 12). While chemically transforming the fuel into H_2 with the help of a reformer is state of the art, the direct electrochemical fuel conversion is much more demanding.

The disadvantages of present PEM cells lies not only in the electrolyte, in particular in its nontrivial preparation (cf. costs, environmental impact of fluorination), the necessary water management (water is not only necessary as a solvent, it is evaporating and dragged through the membrane), the solubility, and permeability of polar compounds such as H_2O or CH_3OH, but also in the fact that Pt-electrodes are required (which not only makes the cell more expensive but also leads to the problem of CO poisoning if reformed H_2 is used). World-wide there is intensive research on modifications and alternatives, such as polymers based on polyether-ketones (s. e.g., Ref.[77,78]). An interesting route consists in achieving proton transport via heterocycles or phosphonic acids and hence with proton donor–acceptor functions built in the organic matrix (see Figure 12).[79]

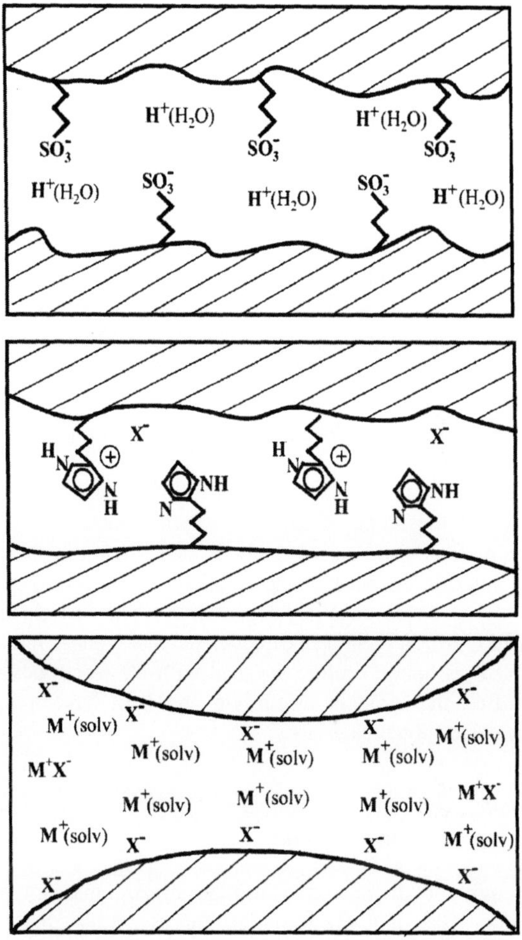

Figure 12. Three examples of complex, H^+ or M^+ (e.g., Li^+) conducting electrolytes. *Top*: Nafion consists of an organic backbone to which sulfonic acid groups are covalently bonded. The protons are dissociated away in contact with the water in the internal channels. *Center*: A covalent bonding of proton donor–acceptor molecules and a sufficiently dense stacking leads to a solvent free proton transport. *Bottom*: In the "soggy sand electrolytes" anions are absorbed at the surfaces of the insulating matrix (e.g., SiO_2). The respective cations (e.g., Li^+) are free while far away from the matrix essentially associated in form of ions pairs if the solvent is a weak dielectric.

As much as the low temperature performance is an engineering advantage, the fact that an operation temperature of ~80°C cannot be exceeded for stability reasons can—as already mentioned—also be a kinetic disadvantage, in particular as regards consumption of natural gas, methanol, or even CO. In such cases high temperature fuel cells appear to be the better choice.

The latter devices are fuel cells that consist of ceramic components which have to fulfill extremely demanding criteria with regard to thermal, mechanical, chemical, and electrical properties. Just consider the electrolyte: It does not only have to be thermally stable but also has to be mechanically and chemically compatible with the electrodes. It does not only have to be chemically stable over a very wide redox window but also has to maintain electrolyte properties within that window (redox stability). Owing to the high mobilities of the electronic carriers and the comparatively steep power law dependencies of their concentrations (see Part I), this requires an extremely high ratio of ionic versus electronic disorder at the reference point of p–n minimum (cf. Part I).[2]

In order not to be lost in engineering complexity, and to remain within the scope of the text we will consider only a few prototype materials and discuss them in the context of problems of principal interest. At present the standard SOFC works at 850–1000°C and is based on Y_2O_3-doped ZrO_2 (YSZ) as electrolyte, SrO-doped $LaMnO_3$ (LSM) as cathode, and a two phase mixture of Ni and YSZ as anode. Before we deal with the electrodes, let us consider first the zirconia electrolyte and then have a look at alternative electrolyte materials.

(*A*) *Stabilized ZrO_2: The Standard SOFC Electrolyte with a Large Electrolytic Domain*

Even though the dopant concentration in zirconia is typically greater than 10% and far too high for assuming ideal defect chemical conditions, the zirconia example is well suited to show the power of homogeneous doping for materials engineering. The excess charges generated by doping with several percent of a lower-valent metal oxide (Y_2O_3, Sc_2O_3, CaO, etc.) are essentially compensated by oxygen vacancies, the high vacancy concentration leading to an ionic conductivity of ≈0.1S cm^{-1} at 1000°C for a Y_2O_3 content of 8 mol%.[80,81] As shown in Part I, the introduction of lower-valent cations on Zr-sites leads to an increase of $[V_O^{\cdot\cdot}]$ and $[h^{\cdot}]$ while $[e']$ is depressed. Owing to the defect-chemical energetics the electronic concentrations and the resulting absolute changes in the electronic concentrations are small compared to $[V_O^{\cdot\cdot}]$.

Table 5.

Defect chemical parameters and transport coefficients of 9.5 mol% Y_2O_3-stabilized ZrO_2 [a] according to Ref.[84] The mass action constants are apparent values that include activity coefficients and vary with the Y-content.[84] Reprinted from K. Sasaki and J. Maier, *Solid State Ion.*, 134 (2000), 303–321. Copyright © 2000 with permission from Elsevier.

Charge carrier	Parameter/coefficient	Relation	Unit
Hole	Conductivity	$\sigma_{h^{\cdot}} = 1.8\times10^{1} \times P_{O_2}^{1/4} \times \exp(-1.5\ \mathrm{eV}/kT)$	$\mathrm{S\ cm^{-1}}$
	Concentration	$[h^{\cdot}] = 5.1\times10^{22} \times P_{O_2}^{1/4} \times \exp(-1.40\ \mathrm{eV}/kT)$	$\mathrm{cm^{-3}}$
	Mobility	$u_{h^{\cdot}} = (1.8/T)\times\exp(-0.1\ \mathrm{eV}/kT)$	$\mathrm{cm^2\ V^{-1}s^{-1}}$
	Diffusivity	$D_{h^{\cdot}} = 1.6\times10^{-4}\times\exp(-0.1\ \mathrm{eV}/kT)$	$\mathrm{cm^2\ s^{-1}}$
	Oxygen exchange equilibrium	$K^{*}_{O,P} \equiv [h^{\cdot}]^2\cdot\left[V_O^{\cdot\cdot}\right]^{-1}\cdot P_{O_2}^{-1/2}$	
		$= 1.0\times10^{24}\times\exp(-2.8\ \mathrm{eV}/kT)$	$\mathrm{cm^{-3}\ bar^{-1/2}}$
	Ionization equilibrium (Mn^{2+}/Mn^{3+})	$K_{Mn,P} \equiv \left[Mn_{Zr}''\right]\cdot\left[h^{\cdot}\right]\cdot\left[Mn_{Zr}'\right]^{-1}$	
		$= 1.6\times10^{24}\times\exp(-1.80\ \mathrm{eV}/kT)$	$\mathrm{cm^{-3}}$
	Ionization equilibrium (V^{3+}/V^{4+})	$K_{V,P} \equiv \left[V_{Zr}'\right]\cdot\left[h^{\cdot}\right]\cdot\left[V_{Zr}^{\times}\right]^{-1}$	
		$= 1.6\times10^{24}\times\exp(-1.94\ \mathrm{eV}/kT)$	$\mathrm{cm^{-3}}$

(continued)

Charge carrier	Parameter/coefficient	Relation	Unit
	Ionization equilibrium (Fe^{2+}/Fe^{3+})	$K_{Fe,P} \equiv \left[Fe''_{Zr}\right]\cdot\left[h^{\cdot}\right]\cdot\left[Fe'_{Zr}\right]^{-1}$	
		$= 1.6\times10^{24}\times\exp(-2.30\ eV/kT)$	cm^{-3}
Excess Electron	Conductivity	$\sigma_{e'} = 5.5\times10^{5}\times P_{O_2}^{-1/4}\times\exp(-3.72\ eV/kT)$	$S\ cm^{-1}$
	Concentration	$[e'] = 5.9\times10^{23}\times P_{O_2}^{-1/4}\times\exp(-3.17\ eV/kT)$	cm^{-3}
	Mobility	$u_{e'} = (6.3\times10^{3}/T)\times\exp(-0.55\ eV/kT)$	$cm^{2}\ V^{-1}\ s^{-1}$
	Diffusivity	$D_{e'} = 5.4\times10^{-1}\times\exp(-0.55\ eV/kT)$	$cm^{2}\ s^{-1}$
	Oxygen exchange equilibrium	$K^{*}_{O,n} \equiv \left[V_{O}^{\cdot\cdot}\right]\cdot[e']^{2}\cdot P_{O_2}^{1/2}$	
		$= 8.6\times10^{68}\times\exp(-6.34\ eV/kT)$	$cm^{-9}\ bar^{1/2}$
	Ionization equilibrium (Ce^{3+}/Ce^{4+})	$K_{Ce,n} \equiv \left[Ce^{\times}_{Zr}\right]\cdot[e']\cdot\left[Ce'_{Zr}\right]^{-1}$	
		$= 1.8\times10^{22}\times\exp(-2.13\ eV/kT)$	cm^{-3}
	Ionization equilibrium (Ti^{3+}/Ti^{4+})	$K_{Ti,n} \equiv \left[Ti^{\times}_{Zr}\right]\cdot[e']\cdot\left[Ti'_{Zr}\right]^{-1}$	
		$= 1.8\times10^{22}\times\exp(-1.47\ eV/kT)$	cm^{-3}

Hole + excess electron	Electronic disorder equilibrium	$K_B \equiv [h^{\cdot}]\cdot[e']$	
		$= 2.9\times10^{46} \times \exp(-4.57\ eV/kT)$	cm^{-6}
Oxygen vacancy	Conductivity	$\sigma_V \times T = 3.6\times10^{5} \times \exp(-0.83\ eV/kT)$	$SK\ cm^{-1}$
	Total concentration	2.5×10^{21}	cm^{-3}
	Mobility	$u_V = (4.5\times10^{2}/T)\times\exp(-0.83\ eV/kT)$	$cm^2\ V^{-1}\ s^{-1}$
	Diffusivity	$D_V = 1.9\times10^{-2} \times \exp(-0.83\ eV/kT)$	$cm^2\ s^{-1}$

[a] Note that $D_i = (kT|z|e)u_i$ and $\sigma_i = |z|e[i]u_i$ where $k = 8.618\times10^{-5}\ eV\ K^{-1} = 1.381\times10^{-23}\ J\ K^{-1}$ and $e = 1.602\times10^{-19}\ C$. Parameters may no longer be valid at much lower temperatures at which defect associations become very significant; they also vary with the impurity content as they are affected by interactions.

These high ionic defect concentrations also structurally stabilize the high temperature cubic phase, what explains the term "yttria stabilized zirconia" (YSZ) which is commonly used for these compositions. Lower doping levels (e.g., 2–3 mol% Y_2O_3) lead to a tetragonal phase being characterized by an improved fracture toughness and strength, but exhibiting lower ionic conductivities under fuel cell conditions.[82,83] The requirement of a sufficiently large electrolytic domain (see Figure 13), i.e., negligible electronic conductivity in the entire partial pressure range a solid oxide fuel cells[80,83] is exposed to, presupposes a high redox-stability, that is a comparatively high free energy necessary for introducing electronic defects (low mass action constants for oxygen incorporation reaction and electron–hole formation). The mass action constants and mobilities for 9.5 mol% ZrO_2 as re-analyzed by a recent treatment,[84] are given in Table 5 and the conductivity versus P_{O_2} isotherms for 800–1000°C in Figure 13.

In view of the large dopant concentration it may appear surprising that the P_{O_2} dependencies of $[V_O^{\bullet\bullet}]$, $[h^{\bullet}]$, $[e']$ (see Figure 13) behave ideally. The reason is that the activity coefficient mainly depends on the dopant and temperature and will not be affected by minor changes in the redox-chemistry. Hence the P_{O_2} dependencies are indeed expected to be ideal while severe

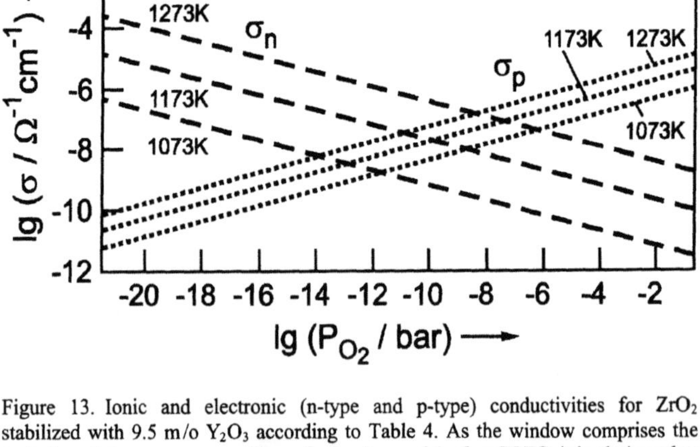

Figure 13. Ionic and electronic (n-type and p-type) conductivities for ZrO_2 stabilized with 9.5 m/o Y_2O_3 according to Table 4. As the window comprises the typical partial pressure range (lhs cathode, rhs anode) of an SOFC, it is obvious that the voltage is unaffected by electronic contributions (*v*: vacancy).

differences occur in the absolute values (when different doping contents are compared) as well as in the *T* and *C* dependencies (*C*: concentration of major dopant).

The latter is directly evident from the conductivity-level versus doping-level curves isotherms which exhibit maxima at dopant concentrations of around 10 mol% (Figure 14) and obviously depend on the nature of the dopant ion.[81,82,85,86]

This dependence on the dopants (at constant effective charge) can be understood in terms of interactions, ordering, and lattice relaxation phenomena. As a consequence, the local binding

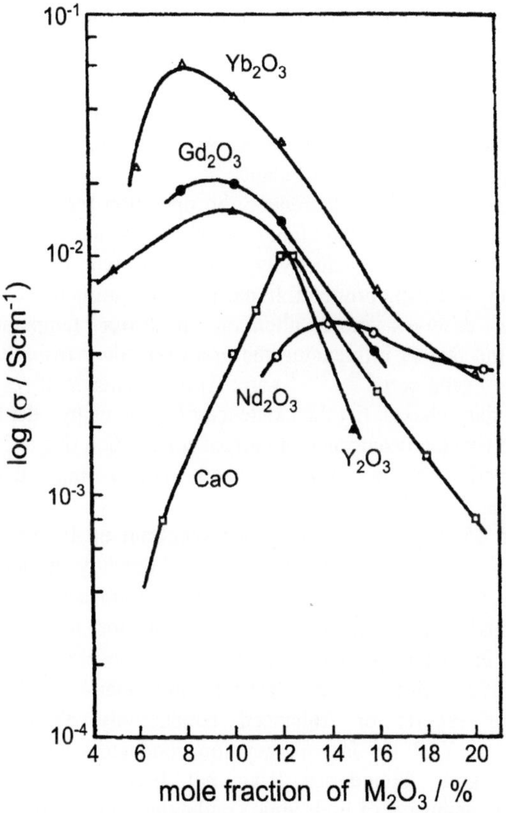

Figure 14. Conductivities of various ZrO_2–M_2O_3 systems at 800°C.[85] Reprinted from T. Takahashi, in: *Physics of Electrolytes. Thermodynamics and Electrode Processes in Solid State Electrolytes*, J. Hladik (ed.), Academic Press, London, Vol. 2, 980–1052, Copyright © 1972 with permission from Elsevier.

energies of the defect as well as the effective vacancy mobility depend on the polarizability, size, and concentration of the dopant.[87–89] The highest conductivity maximum (ca. 0.3 S cm^{-1} at 1000°C),[90,91] is achieved by doping with Sc^{3+} which is probably connected to the fact that its ionic radius is almost that of Zr^{4+}. Despite the higher conductivity of Sc-doped zirconia the cheaper Y_2O_3 dopant is usually used in SOFCs.

An often debated issue is the stationarity of the ionic conductivity which also addresses the question of the stationarity of the dopants' distribution.[91–93] A variety of different, partly contradicting reports in this respect have been given; recent studies indicate relaxation phenomena to be rather small for ca. 10% Sc doping.[92,93]

The cation conductivity is very small which on one hand leads to the high preparation temperatures but on the other to the favorable defect chemical stability as well as to negligible kinetic de-mixing under fuel cell performance.[94]

In view of the trend to reduce the operation temperature of fuel cells and in view of the highly thermally activated (ca. 0.8 eV at 800…1000°C[80,91]) conductivity, the electrolyte conductivity becomes an increasing problem. In particular the influence of grain boundaries deserves more attention at lower temperatures. Amongst the factors influencing the grain boundary resistance of zirconia[95–105] (and ceria[104, 106–109]) are: (i) the existence of resistive grain boundary phases, (ii) the existence of space charge depletion layers, and (iii) the occurrence of current constriction due to lateral inhomogeneities of the grain contacts. For a more detailed treatment see Ref.[110,111]

Figure 15 displays an impedance spectrum of 10 mol% Y-doped ZrO_2 revealing the significant grain boundary influence at reduced temperatures. The substantial grain boundary resistance has been analyzed in terms of the three reasons just mentioned.[111] As shown in Part I, grain boundaries are not necessarily highly resistive but can also constitute fast transport paths.[112–114] In spite of singular reports on enhanced conductivity along grain boundaries in YSZ for fast oxygen ion conductors such highly conductive grain boundaries have not been unambiguously identified as pathways of high ionic conduction in these materials. (As far as the carrier concentration is concerned, an enhanced conductivity is not expected in view of the high disorder.)

The simplest way to enhance the overall conductance is to decrease the electrolyte thickness, and there are many attempts to employ zirconia films of a few μm or even thinner ones. This

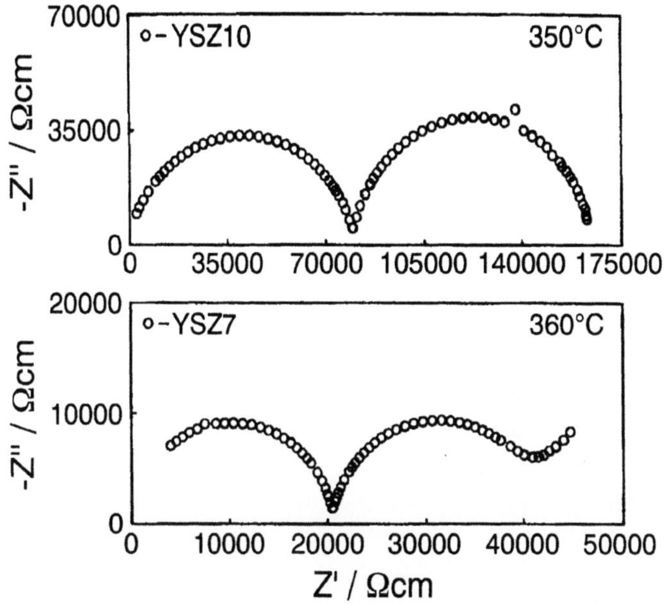

Figure 15. Impedance spectrum of 10 mol% Y_2O_3-doped zirconia according to Ref.[96] The rhs semicircle represents the grain boundary impedance. Reprinted from S.P.S. Badwal, Solid State Ionics **76** (1995), 67–80. Copyright © 1995 with permission from Elsevier.

approach is limited by the effective electrolyte resistance being dependent on the contact geometry at the electrode/electrolyte interfaces. Since current constriction occurs at the electrochemically active sites (e.g., three phase boundaries),[115–117] a thickness reduction does not significantly pay off in terms of resistance, if the constriction effects begin to dominate the electrolyte resistance. Hence it may be concluded that cells based on zirconia electrolytes require temperatures higher than ca. 700°C,[93] and intermediate temperature SOFCs (500–600°C) require other electrolyte materials such as CeO_2.

(B) CeO_2: A Mixed Conductor as SOFC Electrolyte

The ionic conductivity of CeO_2 doped with 10% Gd amounts to 0.25 S cm^{-1} at 1000°C and exceeds that of zirconia (of $\sigma \approx 0.1$ S cm^{-1} in YSZ); this difference is even greater at lower temperatures (2.5×10^{-2} S cm^{-1} [106] for CeO_2 versus ca. 3×10^{-3} S cm^{-1} for YSZ[81] at 600°C) owing to the lower activation energy (ca. 0.65 eV[106,118]). The discussion of the impact of the dopants and of defect–defect interaction is rather analogous to the previous chapter. Also here the influence of defect interactions results in a maximum in the conductivity-dopant concentration relation.[106] Also here, the defect interactions, ordering effects and cluster energies strongly vary with the nature of the dopant ion.[106,119,120]

The major disadvantage of ceria is the perceptible electronic conductivity occurring at low oxygen partial pressures[121] (Figure 16), corresponding to the comparatively easy reducibility of Ce^{4+} to Ce^{3+}. As displayed by Figure 16, CeO_2, unlike ZrO_2, becomes n-conducting at the cathode side. Within a fuel cell there still is an extended region distant from the cathode that truly behaves as a pure electrolyte. Accordingly the employment as a solid electrolyte is not impossible. However, owing to the mixed conduction region at the cathode side, there are losses to be envisaged during performance.

The effect of this electronic conduction on the electrochemical performance of a fuel cell has been studied in Ref.[122–125] Based on the current-equations given in Part I,[2] Riess[122] derived the following approximation for the power output close to the maximum value:

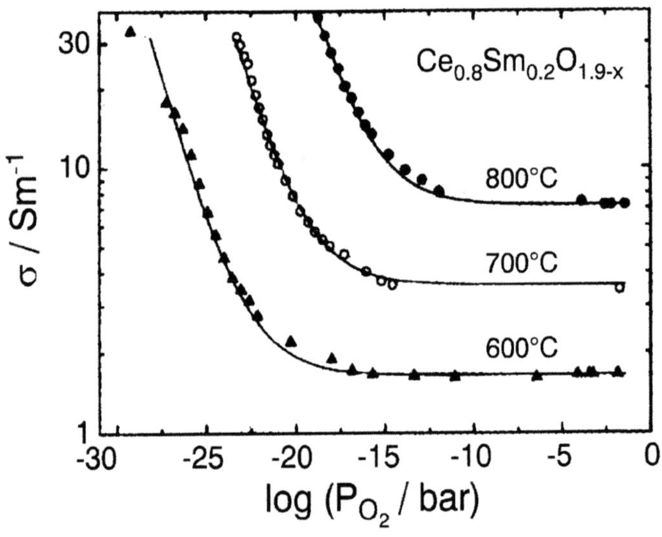

Figure 16. Oxygen partial pressure dependence of the electrical conductivity of doped CeO_2. The steep decrease is due to excess electrons, the flat behavior to oxygen vacancies. If we refer to typical oxygen partial pressures in an SOFC, viz. to 10^{-22} bar at the cathode and 0.2 bar at the anode, we see that the conductivity changes from ionic into n-type within a high temperature CeO_2 based fuel cell. Reprinted from M. Gödickemeier and L.J. Gauckler, J. Electrochem. Soc. **145** (1998) 414–421. Copyright © 1998 with permission from The Electrochemical Society, Inc.

$$\text{power} = \frac{U(E-U)}{4R_{\text{ion}}}\left[1-\left(\frac{1}{t_{\text{ion}}^{\text{red}}}-1\right)\exp\left[-\frac{z\text{F}}{\text{R}T}(E-U)\right]\right]. \quad (43)$$

(Eq. (43) assumes $E \gg \text{R}T/z\text{F}, E \equiv U(I=0)$, $t_{\text{ion}}^{\text{red}} \equiv \sigma_{\text{ion}}/\sigma$ at the reducing side.) If the fuel cell is operated close to maximum power conditions, losses are tolerable in the case of CeO_2 (typically 600°C). The calculation shows that at high temperatures requirements for efficiency and power output shrink the range of operation to a rather limited window. Attempts to reduce the electronic conductivity by adding small amounts of redox-active impurities that should trap off electronic carriers are doomed to fail in ideal defect chemical equilibrium. Since the electrons are minority carriers, their concentration is fully determined by P_{O_2}, T and major dopant concentration. Additions of redox-active impurities which are negligible in the electroneutrality condition cannot change the electron concentration. Rather the ratio of the redox-state of the dopant will adjust to the fixed electron concentration. (Only at high concentration the electron concentration may be changed, or even the overall electronic mobility may be varied by the formation of an impurity band.[126])

Owing to the activation energy, at intermediate temperatures the electrolytic domain increases and at e.g., 500°C the electronic conductivity plays only a minor role. In view of the ionic conductivity being still high enough at these temperatures, ceria appears to be an appropriate electrolyte for intermediate temperature fuel cells. Further information on the properties of ceria and its use in SOFCs can be found in Ref.[106,120]

(C) $LaGaO_3$ and the Search for Alternative Electrolytes

As already stated, the strategies to optimize ceramic fuel cells are essentially materials search strategies.

In view of the long-time operation we have to rely on thermodynamically stable structures and compounds, or on pronouncedly metastable situations. Under such conditions, given the nature of the constituents, the relevant control parameters are temperature T, component potentials or partial pressures (P), and doping content (C). For given operation conditions, T and P are fixed leaving the nature of the major chemical elements and the concentrations of dopants (C) as the only variable parameters. (In multinary oxides usually not all sublattices are mobile, with the consequence of having the additional freedom to varying the fine composition (typically the A/B ratio in A_xB_yO) by the processing or preparation temperature; this we subsume here under doping effects.).

We have shown in Part I that in addition to this homogeneous doping (tuning of the charge carrier concentrations by introducing aliovalent impurities), there is also the tool of heterogeneous doping.[127] This partly related technique affects transport properties by introducing metastable structure elements, but now of higher dimensionality. Conductivity variations are restricted to the neighborhood of the higher-dimensional defects and require percolation in order to become perceptible on a macroscopic level. Employing nanocrystalline solids or composites may be a sensible strategy for medium or room temperature applications, but probably not so much for high temperature fuel cells. One reason is the difficulty to maintain morphologic stability, another the lesser significance of the typically low activated boundary effects at high temperatures.

In summary: (i) the most important step is searching for sufficiently stable, mechanically strong, and electrochemically relevant SOFC compounds and structures, such as oxides in the fluorite, perovskite, pyrochlore, or brownmillerite structures as regards the electrolyte materials; (ii) the most important tool to optimize the electrical properties is then homogeneous doping (which includes all compositional variations within the stability range of the structure amongst which the introduction of aliovalent substantial effects is the most relevant), such as introducing lower-valent cations into MO_2 fluorites, which leads to a substantial oxygen conductivity via oxygen vacancies. The introduction of additional elements into the structure has its limits, not only because of internal interactions, but in particular also as chemical stabilities are expected to suffer because of the increased number of thermodynamic and kinetic reaction pathways. Thus, unsurprisingly, attempts to look for alternative electrolytes for SOFC have not been very successful.

One of the exceptions was the discovery of high ionic conductivity in appropriately doped $LaGaO_3$.[128,129] As in the other oxide ion conductors, its ionic conductivity depends on both the dopant level as well as on the nature of the dopant. A major difference to ceria and zirconia is the presence of two cations that can be substituted; the detailed defect chemistry of such solid solutions is far from being fully understood. Co-doping of Sr on A sites and Mg on B-sites leads to an ionic conductivity of ca. 0.12– –0.17 S cm^{-1} at 800°C,[130–133] which is similar to doped ceria but considerably exceeds the value of YSZ (ca. 0.03 S cm^{-1} at 800°C[80,81]). The activation energy also varies with composition and can be as low as ca. 0.6 eV.[130,131] At about 600–700°C, the conductivity of $LaGaO_3$ is still sufficiently high for being used in fuel cells. Below ca. 600°C, however, the observed increase in the activation energy makes Sr/Mg-doped $LaGaO_3$ inferior to doped ceria as far as σ_{ion} is concerned.[131,132] As regards the electronic conductivity, Sr/Mg-doped lanthanum gallate exhibits—like YSZ—a sufficiently large electrolytic domain to guarantee an overwhelming ion conductivity under all P_{O_2} values of interest.

The most critical point with respect to the use of $LaGaO_3$ in fuel cells has been reported to be its chemical stability. This concerns volatilization of gallium oxide during sintering[134,135]; phase separation after heating at 1000°C,[136] and chemical reaction during cofiring with typical electrode materials.[137] Nonetheless, recent reports on SOFCs based on $LaGaO_3$ are rather promising.[138]

(*D*) *Electrodes for Solid Oxide Fuel Cells*

The requirements that electrodes have to meet are different from the previous ones. Like the electrolyte, they have to be chemically and thermodynamically compatible with the neighboring phase, are however—unlike the electrolyte—not subject to substantial chemical potential gradients. Besides exhibiting high electronic conductivities, they must catalyze the electrode reaction, i.e., enable adsorption, dissociation, ionization, and charge transfer into the electrolyte to occur with sufficient reaction rates.

Mixed conductors are typically well-suited for such purposes, they provide the electronic carriers necessary for the connection with the outer circuit as well as for the redox-reaction; they also provide mobile ionic defects that represent acid–base active centers, and enable mass transport.

Materials of interest are negatively (typically Sr) doped perovskites based on lanthanum manganate, cobaltate, or ferrate. Recently $Ba_{0.5}Sr_{0.5}Co_{0.8}Fe_{0.2}O_{3-\delta}$ has been proposed as a very promising composition.[139] Here we will largely focus on the standard material LSM (Sr-doped $LaMnO_3$).

Figure 17 shows different mechanistic pathways for the oxygen reduction at the LSM cathode on YSZ electrolyte. The adsorbed, partially fully ionized oxygen may move along the surface to the three phase boundary where it is transformed into the electrolyte. (In principle it may also reach this place directly via the gas phase.) The oxygen may also reach the electrolyte by diffusion through the LSM bulk via a counter motion of O^{2-} and $2e^-$. Note that LSM sandwiched between Pt (serving as a reversible electrode) and YSZ

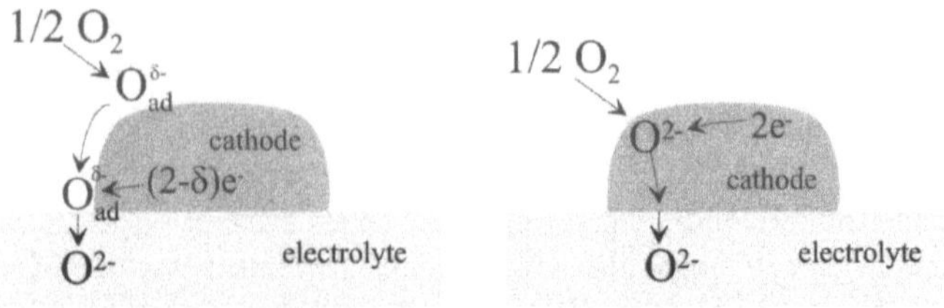

Figure 17. Transport pathways for oxygen reduction at a fuel cell cathode.[110] Reprinted from J. Fleig, K.D. Kreuer and J. Maier, in: *Handbook of Advanced Ceramics. Volume II: Processing and Their Applications*. S. Sōmiya, F. Aldinger, N. Claussen, R. M. Spriggs, K. Uchino, K. Koumoto and M. Kaneno (eds.), Elsevier Academic Press (2003) p. 59. Copyright © 2003 with permission from Elsevier.

(blocking the electron transport) constitutes a Wagner–Hebb polarization cell (see Section III.3), in the steady state of which a stationary electrochemical polarization is established with σ_{eon} being suppressed and σ_{ion} being the relevant transport coefficient. Only in the transients the ambipolar conductivity σ^{δ} is of significance as it enters the chemical diffusion coefficient. Nonetheless, because $\sigma_{eon} \gg \sigma_{ion}$ the ionic conductivity is always the decisive conductivity parameter (as long as σ_{eon} is dominant) as

$$\sigma^{\delta} \equiv \frac{\sigma_{e^-}\sigma_{O^{2-}}}{\sigma} \simeq \sigma_{O^{2-}}. \tag{44}$$

Figure 18 shows a simplified Kröger–Vink diagram for typical oxidic cathode materials (LSM, LSF, LSC, i.e., Sr-doped La-manganates, ferrates, cobaltates) the high P_{O_2} regions being

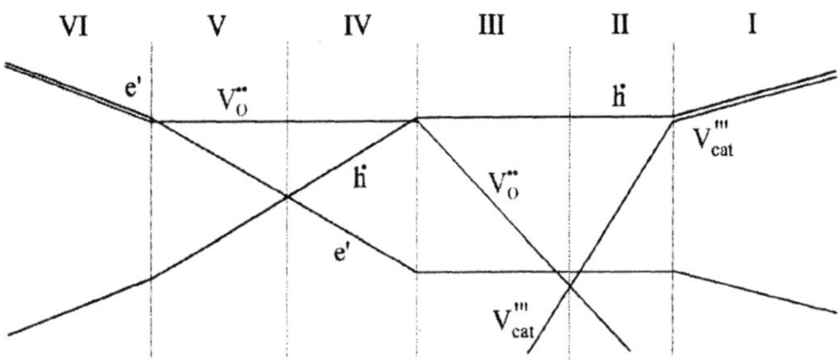

Figure 18. Kröger–Vink diagrams of a Schottky disordered negatively doped oxide (trivalent cations assumed). The ordinate refers to a logarithmic concentration axis, the abscissa to a logarithmic partial pressure axis.[110] Reprinted from J. Fleig, K.D. Kreuer and J. Maier, in: *Handbook of Advanced Ceramics. Volume II: Processing and Their Applications*, S. Sōmiya, F. Aldinger, N. Claussen, R. M. Spriggs, K. Uchino, K. Koumoto and M. Kaneno (eds.), Elsevier Academic Press (2003) p. 59. Copyright © 2003 with permission from Elsevier.

relevant and corresponding to high hole, anion, and cation vacancy concentrations of these materials (to be even more precise the different valencies of metal vacancies have to be taken into account). Figure 19 displays the ionic conductivity of these perovskites, suggesting that the bulk path of the oxygen incorporation scheme in Figure 17 becomes more favorable as we go from LSM to LSF and to LSC. SIMS studies showed the relevance of the three phase boundary if LSM is used under fuel cell conditions.[145] These measurements do not rule out that incorporation takes place according to the bulk path, provided it occurs close to the three phase contacts; the measurement also does not allow one to draw detailed conclusions about the mechanism under modified conditions (e.g., modified geometries). In Ref.[146] a detailed mechanistic study that enabled a far-reaching exploration of the relative rates of the three possibilities and the nature of the rate determining steps was conducted. There, simply the circumference and height of circular grain/LSM electrodes have been varied and the variation of the polarization resistance has been studied. The results confirm that indeed under normal fuel cell conditions and

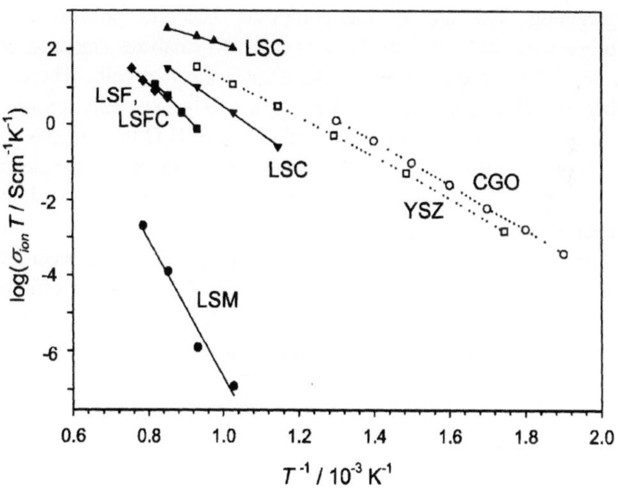

Figure 19. Oxygen ion conductivity of mixed conducting perovskites (LSC: (▼) $La_{0.5}Sr_{0.5}CoO_{3-\delta}$ (polycrystal, $pO_2 = 1$ bar, from D^*);[140] (▲) $La_{0.3}Sr_{0.7}CoO_{3-\delta}$ (polycrystal, $pO_2 = 0.21$ bar, from D^*),[141] LSF: $La_{0.75}Sr_{0.25}FeO_{3-\delta}$ (single crystal, $pO_2 = 0.065$ bar, from D^*),[142] LSFC: $La_{0.8}Sr_{0.2}Fe_{0.8}Co_{0.2}O_{3-\delta}$ (polycrystal, from permeation),[142] LSM: $La_{0.5}Sr_{0.5}MnO_{3-\delta}$ (polycrystal, $pO_2 = 1$ bar, from D^{*140}) in comparison with doped zirconia (YSZ: $Zr_{0.81}Y_{0.19}O_{2-\delta}$ (single crystal, AC conductivity)[143]) and ceria (CGO: $Ce_{0.69}Gd_{0.31}O_{2-\delta}$ (single crystal, AC conductivity)[144]).

morphologies, the three phase boundary mechanism prevails (resistance proportional to inverse circumference), but that already in the case of moderately thin films the bulk path may dominate (resistance proportional to inverse area and to height). Generally a positive bias favors the surface diffusion path (less $V_O^{\bullet\bullet}$), while the negative biases favor the bulk diffusion path (more $V_O^{\bullet\bullet}$).[146] An interesting effect that may be attributed to a varied cation distribution is the drastic performance enhancement caused by short bias pulses.[147]

Also for the anode side, mixed conducting perovskites such as donor-doped $SrTiO_3$ or acceptor-doped $LaCrO_3$ have been proposed.[148–150] Figure 20 shows the conductivity behavior of Ca-doped $LaCrO_3$ corresponding to the regions IV ($2[V_O^{\bullet\bullet}] \simeq [Ca'_{La}]$) and III ($[h^{\bullet}] \simeq [Ca'_{La}]$) in Figure 18. A reasonable anode performance also with respect to hydrocarbons as fuels has been reported in Ref.[150] for $(La_{0.75}Sr_{0.15})_{0.9}Cr_{0.5}Mn_{0.5}O_3$ (Figure 21) (see e.g., Ref.[151]).

Well-established anode materials are Ni cermets such as Ni/YSZ composites. The presence of the second phase increases the contact area and prevents the catalytically active Ni particles from aggregating. The use of the composite becomes problematic if hydrocarbons are to be directly converted: Ni catalyzes cracking, and the resulting carbon deposition deactivates the fuel cells. Therefore either pure H_2 has to be used or the fuel has to be externally reformed. A third way is internal conversion of CH_x with H_2O to synthesis gas. The necessary steam addition, however, reduces the overall efficiency. Another problem of Ni cermets, if they are to be used at lower temperatures, is a potential oxidation of the Ni. Alternatives are Cu/CeO_2 cermets in which Cu essentially provides the electronic conductivity and CeO_2 the catalytic activity. Note that an efficient current collecting property of the electrode presupposes a metal concentration above the percolation threshold.

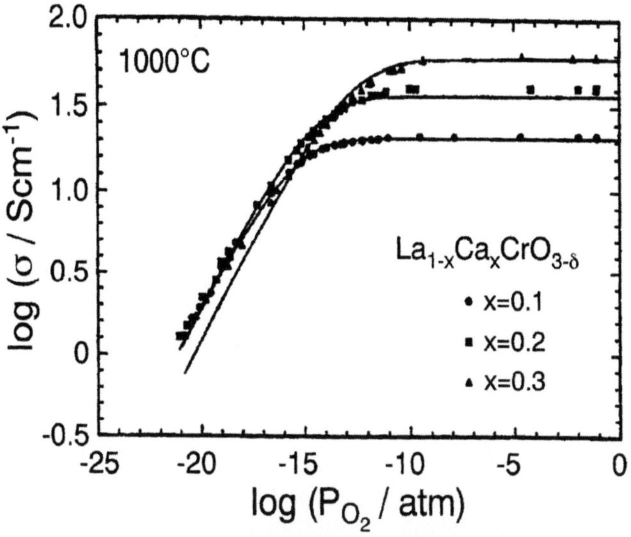

Figure 20. Conductivity of Ca-doped $LaCrO_3$ versus oxygen partial pressure at 1000°C.[149] Reprinted from I. Yasuda and T. Hikita, *J. Electrochem. Soc.* **140** (1993) 1699–1704. Copyright © 1993 with permission from The Electrochemical Society, Inc.

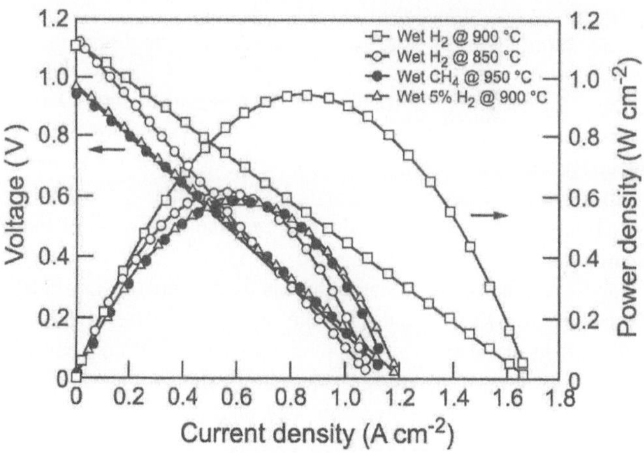

Figure 21. Performance of $(La_{0.75}Sr_{0.25})_{0.9}Cr_{0.5}Mn_{0.5}O_3$ as a fuel cell anode in different wet (3% H_2O) atmospheres. See Ref.[150] for corrections concerning losses caused by electrolyte and cathode.[150] Reprinted from S. Tao and J.T.S. Irvine, *Nature Mater.* **2** (2003) 320–323. Copyright © 2003 with permission from Macmillan Magazines Limited.

A decisive point as regards high power application is the appropriate combination of elementary cells to high power stacks (extreme cases are the tubular design in which the gas chambers are separated by the tubular form of the ceramic cell, and the planar design in which additional high temperature seals are necessary). As electrochemically conducting interconnect the above-mentioned $LaCrO_3$ may be used for high temperature fuel cells. Corrosion resistive alloys demand lower temperatures, which is, in view of the economic and technical advantages of such alloys, an important incentive to reduce operation temperatures.

(*a*) *Alternative SOFC Concepts*

In the following, alternative SOFC solutions shall be touched upon. The most attractive solution would be the use of one and the same compound that exhibits p-type conductivity at the anode, n-type conductivity at the cathode, and ionic conductivity in-between. Even though exactly this feature is expected from the Kröger–Vink diagram for many mixed conducting materials subject to a large P_{O_2} gradient (cf. Figure 18), the short-coming is that neither the absolute values of the conductivities are expected to be sufficient, nor the catalytic properties to be good enough (irrespective of power losses discussed in Section II.3.i.A).

A more realistic approach would be a monolithic fuel cell which relies on appropriate doping profiles of the same ground phase. In particular pyrochlores,[152] perovskites ($LaGaO_3$)[153] or brownmillerites[154,155] have been envisaged in this regard. However, meeting the many harsh conditions to be fulfilled for a fuel cell with one and the same ground material is difficult.

It has even been proposed to give up the separation of the gas chambers.[156–158] Then one has to completely rely on selective catalytic activity of electrocatalysts to activate reduction or oxidation processes. In Ref.[158] considerable power densities have been achieved in this way. This single chamber fuel cell concept might be indeed relevant for reduced temperature application.

Contrarily, also more complicated phase schemes than used in the conventional fuel cell arrangement have been proposed. Using two or more electrolyte materials indeed can have advantages, one is to avoid a contact between reacting phases (LSC | CeO_2 | YSZ instead of LSC | YSZ), another to avoid exposure to a redox-window in which the phase would be electronically or even chemically unstable (cathode | YSZ | CeO_2 | anode instead of cathode | CeO_2 | anode).[159–162]

In a recent approach a triode cell arrangement has been used in which the fuel cell kinetics was improved by an assisting voltage.[52] A promising SOFC concept gives up oxidation of fuels by oxygenation, rather it uses dehydrogenation; in other words hydrogen is removed with the help of ceramic proton conductors (see Part I,[2] Section IV.3). Such proton conductors can be produced by dissolving H_2O into acceptor-doped oxides according to

$$H_2O + V_{\ddot{O}} + O_O \rightleftharpoons 2OH \tag{45}$$

The proton conductivities achieved can be markedly higher than e.g., the oxygen conductivities in YSZ, in particular at reduced temperatures. An obvious advantage is the fact that H_2O is produced on the air side and is not diluting the fuel. Note, however, that for an oxidation of hydrocarbons to CO_2 via dehydrogenation the presence of water is needed at the fuel side. Here the simultaneous H^+ and O^{2-} conductivity of the proton conducting oxides offers a great conceptional advantage. As more H_2O is produced at the air side than needed at the fuel side, water can diffuse back via chemical diffusion enabled by nonzero $\sigma_{H^+}, \sigma_{O^{2-}}$. This operation mode was proposed by Coors[163] (see Figure 22). A material that is stable under fuel cell conditions and in particular against CO_2 is (Y-doped) $BaZrO_3$.[165] Its high grain boundary resistances are at present a serious problem that has to be solved. See Figure 23.

Mixtures of $BaZrO_3$ with small amounts of $BaCeO_3$ improve the overall conductivity without making this material chemically unstable against CO_2.[164] Hydrogen sulfates, hydrogen phosphates, etc. have also been proposed[167] but are typically prone to become reduced in the H_2 atmosphere.

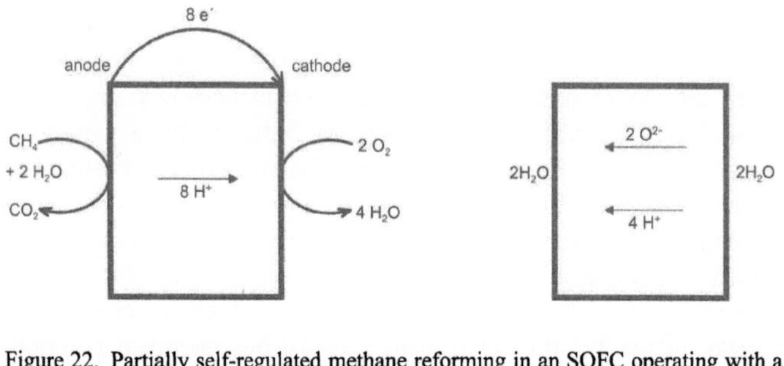

Figure 22. Partially self-regulated methane reforming in an SOFC operating with a proton conducting oxidic electrolyte as suggested by Coors[163] (*see text*): (*top*) the overall cell reaction and (*bottom*) the back diffusion step.[164] Reprinted from K.D. Kreuer, *Annu. Rev. Mater. Res.*, **33** (2003) 333–359. Copyright © 2003 with permission from Annual Reviews.

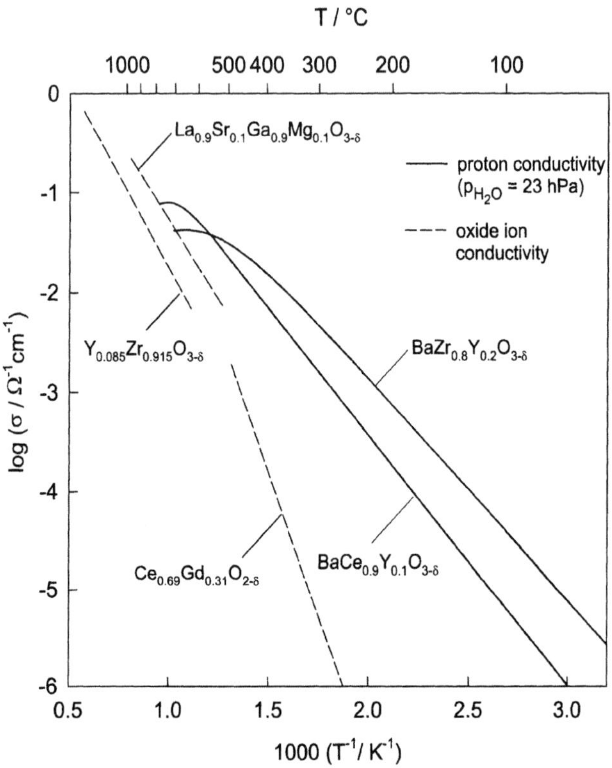

Figure 23. Proton conductivities of Y-doped $BaZrO_3$[165] and $BaCeO_3$[166] in comparison with the ion conductivity of the relevant solid oxygen ion electrolytes. Reprinted from K.D. Kreuer, St. Adams, W. Münch, A. Fuchs, U. Klock, and J. Maier, *Solid State Ionics*, **155** (2001) 295–306. Copyright © 2001 with permission from Elsevier.

(ii) *Batteries*

Even though secondary batteries become increasingly important, there is still a great interest in primary batteries for a variety of applications. Miniaturizable low power batteries of great reliability find important applications in measurement devices such as clocks or in medical devices such as heart pacemakers. In the Li–I_2 system which was initially used for the latter application, the LiI electrolyte is formed on contact which makes the battery self-healing. Cracks in the LiI do not lead to a disastrous chemical reaction, rather the electrolyte as reaction product is re-formed. In order to facilitate the handling and the electronic conductivity, I_2 is used in the form of an organic charge-transfer complex.[168] The accompanying decrease in the chemical potential and hence in cell voltage is tolerable. Heterogeneously doped LiI (LiI:Al_2O_3),[169] as a better conducting (heterogeneous) electrolyte (cf. Part I[2]), was also popular in this context for some time.

Owing to low electronegativity combined with the low mass, Li is predestined as negative electrode in high performance batteries. In the formerly very popular Li–$SOCl_2$ cell,[170] the liquid cathode material ($SOCl_2$) reacts with Li to give SO_2, sulfur and LiCl, hence again the Li^+-conducting solid electrolyte, is produced in situ. The deposition of LiCl at the porous "current collector" graphite at later stages limits the power production. The combination Li–MnO_2 is in use (see e.g., Ref.[65,63,64]) as a primary cell; cells with transition metal oxides are taken up again in the context of the secondary batteries.

As well-known, MnO_2 finds wide-spread applications in the classical Leclanché cell in which, put simply, Zn is oxidized to ZnO by MnO_2. Zn is a cheap anode material which is also employed in the Zn–air battery.† This intriguing battery concept represents a primary battery but exhibits similarities with a fuel cell.[63,64,69] Table 6 gives an overview on zinc-based primary elements.

Solid electrolyte batteries using silver systems have played a prominent role historically. The interest in silver systems was caused by the high conductivities of Ag electrolytes as well as by the typically high exchange current densities. The superionic conductor α-AgI in principle representing a suitable solid electrolyte for an Ag–I_2 battery, has the disadvantage of being stable only above 146°C (see Section VI.3.ii in Part I). In $RbAg_4I_5$ the superionic behavior is present even at room temperature, but this compound is unstable with respect to I_2 and reacts to RbI_3 and AgI. Stabilization is possible by using $(R_4N^+I^-)$–I_2 adducts what of course reduces the iodine activity, and thus the cell voltage. Again, the reaction product AgI increases the cell resistance,[168] which highlights the importance of thermodynamic and kinetic issues in particular the necessity to harbor the reaction product.*

Speaking about secondary batteries, the necessity of reversibly reversing the exothermic reaction imposes high demands on the

---

† In Zn–air cells the negative electrode is metallic zinc while the positive electrode is a porous air electrode, usually carbon-based. The product ZnO is dissolved in the circulating alkali electrolyte and washed away. During charging the ZnO-containing solution is washed back again which justifies the competition with a secondary cell.

* The same problem arises if AgI:Al_2O_3[172] is used, which exhibits very high Ag^+ conductivities.

Table 6.
Important zinc-based primary elements. The price per kW h increases monotonically within the table (from *top to bottom*) and is about an order of magnitude greater for the Ag–Zn cell than for the Leclanché cell (date 1996).[171]

Cell	Reaction	Working voltage (V)	Energy density	
			(Wh kg^{-1})	(Wh l^{-1})
Leclanché	$Zn + 2MnO_2 \xrightarrow{(NH_4Cl)} ZnO + Mn_2O_3$	1.25	50–60	100–120
Leclanché alkal.	$Zn + 2MnO_2 \xrightarrow{(KOH)} ZnO + Mn_2O_3$	1.1–1.2	80	210
mercury oxide cell	$Zn + HgO \xrightarrow{(KOH)} ZnO + Hg$	1.1–1.3	80–100	270–370
silver zinc cell	$Zn + AgO \xrightarrow{(KOH)} ZnO + Ag$	1.5	120–190	370

kinetics. One obvious solution is the use of liquid electrodes (Na–S-battery), another the use of solid mixed conductors that can dissolve or ex-solve components (intercalation electrodes in "Li-ion" batteries), and a most recent one to use extremely small particles.

A classical intercalation compound[173] is TiS_2 the Li content of which can be tuned over a wide range (see Figure 24 right). Figure 24, left, shows an example of an Li– TiS_2 microbattery[63] with glass electrolyte. Owing to the small dimensions, the electrolyte's conductivity is sufficiently high. Naturally the capacities are very small, but sufficient for low power applications; the right-hand side of Figure 24 proves that the discharge curves of the cell are adequately flat. In principle the cell

$$\mathrm{Li}\,|\,\mathrm{Li}^{+}\text{-electrolyte}\,|\,\mathrm{Li}\text{-intercalation compound}$$

can deliver electrical energy until the chemical potential at the rhs (positive electrode, i.e., cathode during discharge) becomes equal to that of the pure lithium (negative electrode, i.e., anode during discharge). In the present Li-batteries only a small window is considered, in which the homogeneity range of the intercalation compound is not exceeded and thus reversibility is at least partly guaranteed. A necessary criterion for a good reversibility is a high D^{δ}_{Li} value at the operating temperature.[174,175] Even though the Li-activity of the electrode will fall continuously with increasing Li content, still the discharge curves can be reasonably flat as shown in Figure 24. (A truly horizontal course is only expected if a nonvariant mixture is generated—in which further Li addition leads to a redistribution of the masses of the phases rather than to a compositional change—and if the Li addition is performed infinitely slowly (cf. Section III.5)).

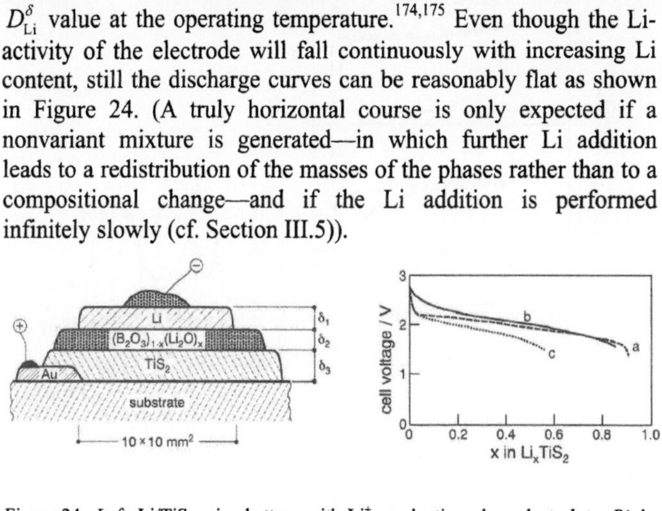

Figure 24. *Left*: Li/TiS_2 microbattery with Li^+ conducting glass electrolyte. *Right*: Discharge curves for Li/TiS_2 cells with amorphous TiS_2 film (*b*: Current density = 3μA cm^{-2}, *c*: 16μA cm^{-2}) and crystalline TiS_2 (*a*: 10 mA cm^{-2}) using liquid electrolyte.[63] Reprinted from Ch. Julien and G.-A. Nazri, *Solid State Batteries: Materials, Design and Optimization*, Kluwer Academic Publishers, New York (1994). Copyright © 1994 with permission from Springer.

At present, mixed conducting transition metal oxides such as the $LiMO_2$ phases of the α-$NaFeO_2$ type (M=Ni,Co,V),[176] or spinel phases such as $LiMn_2O_4$[177] are electrode materials of choice. The charge and discharge curves of such high-performance cells (the electrolyte is a nonaqueous liquid Li-salt solution) are shown in Figure 25a; information concerning the cell voltage, capacity, and energy density is provided by Figure 25b. We will meet such discharge curves, but then recorded in a strictly reversible manner, as coulometric titration curves in Section III.5.

Let us briefly consider the cobalt oxide-based cathodes. If the oxide is virtually Li free, the cell voltage with respect to Li is ca. 5 V. The Li-composition range (within the α-$NaFeO_2$ structure) in which typical batteries operate, is given in Figure 25a corresponding to an operable cell voltage of ca. 4 V.[178,179] This voltage has been shown to be increasing by substitution with Al[179] or Fe and Mn.[180–182] In the latter cases several individual redox levels can be distinguished (Co^{3+}/Co^{4+}, Mn^{3+}/Mn^{4+}, Fe^{3+}/Fe^{4+}) within the discharge/charge cycle. From the standpoint of environmental benignity $LiFePO_4$[183,184] is a promising candidate. The poor electronic conductivity requires small electronically connected particles or

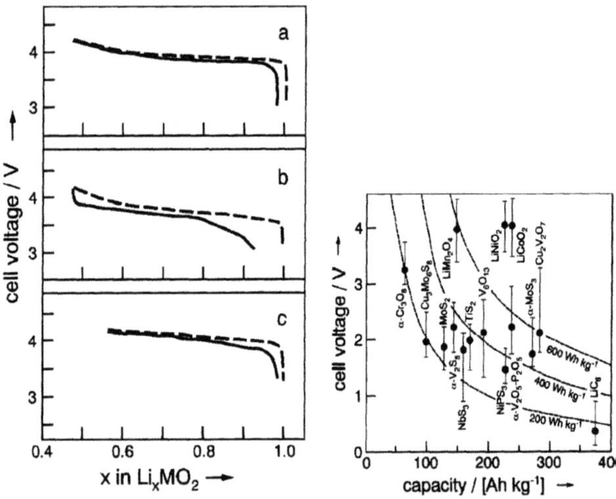

(a) (b)

Figure 25. (a) Charging and discharging curves of high performance cells with Li as anode, Li_xCoO_2 (*a*), Li_xNiO_2 (*b*), $Li_xMn_2O_4$ (*c*) as cathodes. The crystal structure of the latter is of the spinel type.[63] (b) Cell voltage and capacity of cathode materials for secondary Li intercalation cells.[63] Reprinted from Ch. Julien and G.-A. Nazri, *Solid State Batteries: Materials, Design and Optimization*, Kluwer Academic Publishers, New York, (1994). Copyright © 1994 with permission from Springer.

appropriate conditioning.[185–187] (It is still a matter of debate whether the dopants used increase the electronic conduction homogeneously or heterogeneously.)

A significant improvement of safety and corrosion problems is achieved also by using intercalation compounds on the anode side (or better of the negative electrode). If carbon is used, the Li activity and, hence, the cell voltage are only slightly reduced but the cycling properties are considerably improved. Such batteries with two intercalation electrodes are called "rocking chair" batteries (see Figure 10) as Li is rocked backwards and forwards between them,[188,65,63,64,180,181], e.g., in the case of Li_xC and Li_xMO_2 (M = Co, Ni, Mn; in the popular Sony cell M = Co)

$$Li_{x_1}C \,|\, \text{Li-electrolyte} \,|\, Li_{x_2}MO_2.$$

The ease of intercalation of Li in C is partly due to the layered structure.[189] When the Li content is changed from Li_0C to $Li_{1/6}C$ which is about the maximum range covered in a battery, the AB stacking sequence shown in Figure 26 (rhs) is changed into a situation where the Li separates layers that are arranged exactly on top of each other (AA in Figure 26 lhs); during this process the layer distance increases by about 10%.

If the Li content in the intercalation compound exceeds the solubility limit, a two- or multiphase mixture is produced, which in principle allows for a much greater capacity but at the expense of severe kinetic problems as regards charging and cycling.

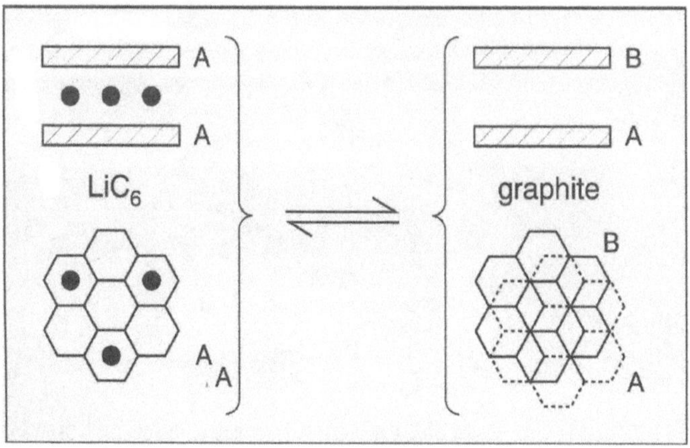

Figure 26. Displacement of the layer stacking sequence (*see text*) on intercalation of lithium in graphite.[11] Reprinted from J. Maier, *Physical Chemistry of Ionic Materials. Ions and Electrons in Solids*. Copyright © 2004 with permission from John Wiley & Sons, Ltd.

If the oxides are, however, nanocrystalline, the lithium can be almost reversibly stored via such a heterogeneous reaction as was first demonstrated using the example of CoO.[190] Upon discharge of macrocrystalline CoO, a two-phase mixture of nanocrystalline or even amorphous Li_2O and Co is formed in situ, which upon Li extraction is converted back into nanocrystalline CoO. The reversibility is at least partly due the tiny diffusion lengths (approximately 1 nm). An oxide whose reversibility is particularly pronounced,[191] is RuO_2. Here almost 100% of the initially consumed Li can be extracted in the first cycle (see Figure 27).[192,193] Also fluorides and nitrides have been shown to behave in this way.[180]

Owing to an excess term in the chemical potential which is approximately proportional to surface tension divided by mean curvature (Wulff crystal assumed), there are a variety of thermodynamic anomalies in nanocrystalline battery systems to be expected:[194] (i) The emf of nanocrystals is different from that of macrocrystals. (ii) Moreover, the discharge curves are not necessarily flat in virtually "nonvariant systems" owing to a variation in size or surface tension. (iii) Furthermore, nanocrystals of different size are expected to have different Li-stoichiometries at equal Li-potential. (iv) At interfaces of Li_2O with a metal, a novel storage mechanism is possible that can contribute markedly to the overall capacity.[194] While neither Li_2O nor a (nonalloying) metal can accept perceptible amounts of Li, a metal/ Li_2O interface can. This is by accommodating Li^+ in interstitial sites of Li_2O while the unwelcome electrons are accepted at the metal site. For more details on these novel aspects see Ref.[194] as well as the following section. Figure 28 displays the three modes of Li storage: homogeneous, intercalation, heterogeneous reaction, and interfacial storage.

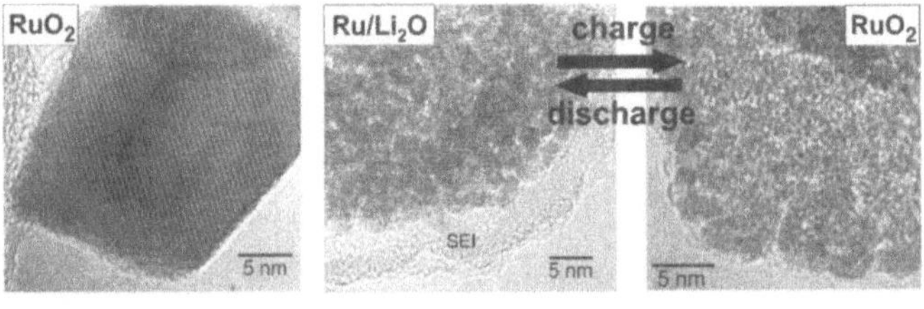

Figure 27. HRTEM images of RuO_2 electrodes: (a) initial, (b) fully discharged to 0.05 V—uptake of 5.6 Li and (c) charged to 4.3 V—extraction of 5.5.[191] Reprinted from P. Balaya, H. Li, L. Kienle and J. Maier, *Adv. Funct. Mater.*, **13** (2003) 621–625. Copyright © 2003 with permission from Wiley-VCH Verlag GmbH.

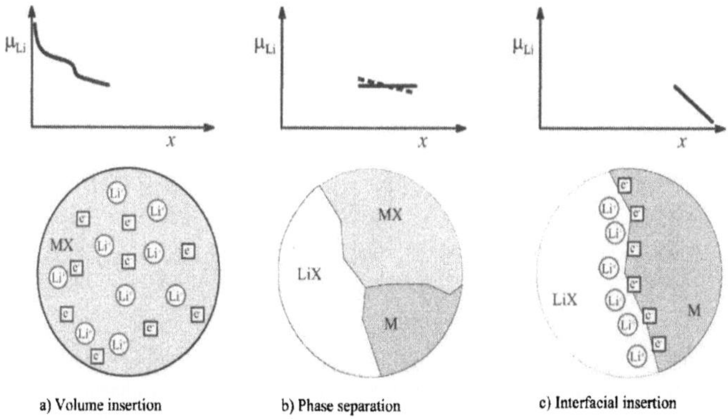

Figure 28. Three different stages of Li storage. (a) Insertion in the bulk of MX. (b) Compound reduction occurring if the solubility limit is exceeded. (c) Interfacial charging of the multiphase mixture. From Ref.[194] Reprinted from J. Jamnik and J. Maier, *Phys. Chem. Chem. Phys.*, **5** (2003) 5215–5220. Copyright © 2003 with permission from PCCP Owner Societies.

Because of the good contact and mechanical flexibility liquid electrolytes are generally used as ion conductors such as $LiPF_6$ in ethylene carbonate/dimethyl carbonate.[195] The (kinetic) stability of such cells is due to the formation of a passivating reaction layer at the electrode/electrolyte interface (SEI, "Solid electrolyte interface").[196] Immobilization of the liquid electrolytes by polymers such as PMMA (polymethyl methacrylate)[197] is able to improve the mechanical properties. The conductivity of pure polymeric electrolytes such as (Li^+-containing) PEO (see Figure 48, Part I[2]) is too low for many applications[198] and there are many attempts to increase it, e.g., by heterogeneous doping with oxide inclusions.[199] That this, at least partly, can be explained by the concept of heterogeneous doping outlined in Part I (Section V.2), has been shown by the recently discovered "soggy sand electrolytes" which are also good candidates for battery electrolytes themselves.[200,201] These are liquid nonaqueous solutions of Li salts but contain a high volume fraction of small SiO_2 particles. Owing to the surface interaction (adsorption of anions by SiO_2) almost all ion pairs are broken up leading to an increased conductivity. Beyond the increased conductivity these composites exhibit the favorable mechanical properties of soft matter. They are also expected to be safer than the pure liquid (Figure 29).

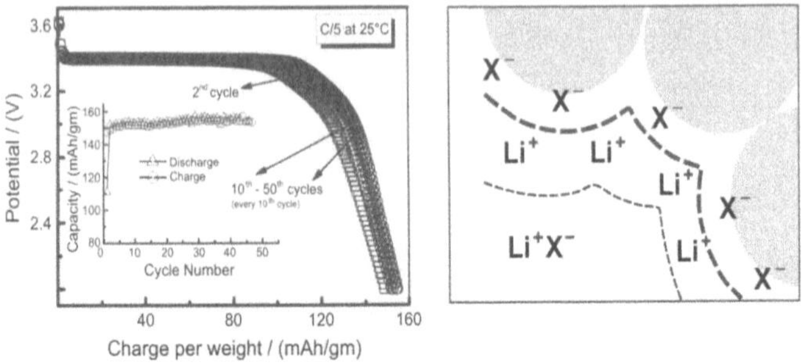

Figure 29. Lhs: Battery performance of a Li cell with $LiFePO_4$ as cathode and nano-SiO_2 (10 nm) in 1 M $LiCF_3SO_3$-EC:DMC (EC: ethylene carbonate, DMC: dimethyl carbonate).[202] Rhs: At the contact of anion adsorbing phases LiX is completely dissociated and Li^+ mobile in the space charge region.

If the lifetime of Li-based batteries (the term "lithium ion batteries" for batteries with polar Li-compounds as negative electrodes is very unfortunate) can be further enhanced, they will be also of importance for electrotraction. The classical battery type used in automobiles, viz. the lead–acid accumulator, is distinctly superior in terms of long-time stability but possesses too low an energy content per unit weight as to drive automobiles. Driving car of sensible size and performance with this alone requires a battery weight on the order of 1 t. (This problem is not removed by using Ni–Cd accumulators.) Much effort has been undertaken to develop a sodium–sulphur cell. In the Na–S cells[203]

$$\mathrm{Na} \mid \beta''\text{-}\mathrm{Al_2O_3}\left(\mathrm{Na_2O}\right) \mid \mathrm{S}$$

working at about 300°C the electrolyte is a β''-Al_2O_3–Na_2O ceramic ("β-alumina", cf. Part I[2]) which makes it possible to use liquid electrodes. The use of liquid sodium and sulfur as electrode phases ensures good contact; moreover the liquid sulfur can accommodate the reaction product formed according to

$$2\mathrm{Na} + x\mathrm{S} \rightleftharpoons \mathrm{Na_2S}_x.$$

Figures 30 and 31 refer to the excellent performance of this battery type (see Figure 31, high cell voltage of 3 V, ten times better mass-related energy content than for the lead–acid accumulator) such cells have not been commercialized up to now. This is not because of the problems to maintain the temperature—this is achieved by the waste heat—it is because the danger of crack formation and resulting catastrophic local chemical reactions that led to the fact that investigations with respect to electrotraction have been essentially abandoned.

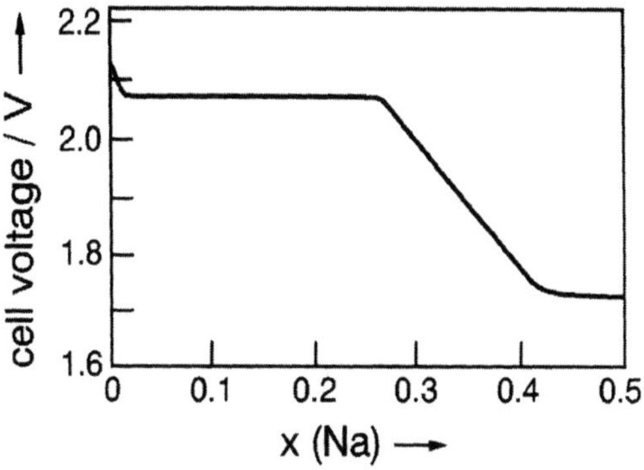

Figure 30. Open circuit voltage of the Na/S cell versus Na content (x) of the cathode.[63]

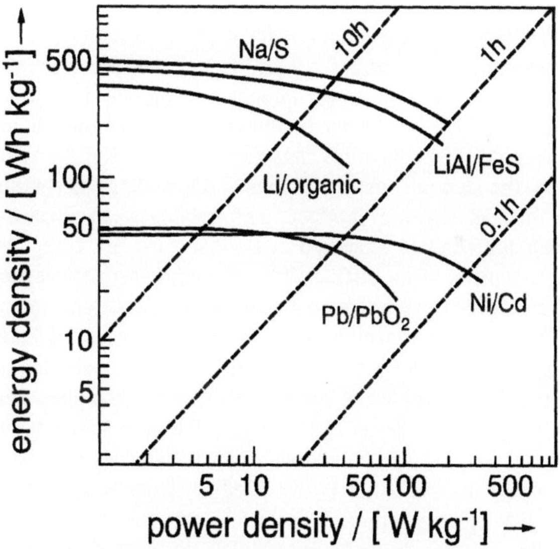

Figure 31. Energy density versus power density for various battery systems. The data may be compared with the typical energy density of gasoline of 3×10^4 Wh kg^{-1}.[63,204] Reprinted from Ch. Julien and G.-A. Nazri, *Solid State Batteries: Materials, Design and Optimization*, Kluwer Academic Publishers, New York (1994). Copyright © 1994 with permission from Springer.

The related "Zebra cell"[205] uses the same solid electrolyte as in the Na–S battery, and operates at similar temperatures, but the cell reaction comprises the reaction of Na with $NiCl_2$ to Ni and NaCl, which is contained in an $NaAlCl_4$ melt (open circuit voltage per cell $\simeq 2.6$ V). Its damage potential is much less: Should the electrolyte ceramic break, Na would gently react with $NaAlCl_4$ to yield Al. The short circuit via Al would even ensure that the cell chain would not lose its function even if 5% of the cell has been destroyed in this manner.

Besides Li-accumulators also Ni metal hydride systems and metal–air systems (like the zinc–air batteries already described‡) come into consideration for electrotraction (Figure 10). Because metal air systems such as the Zn–air system, even though not proper secondary systems (nor are they fuel cells), can be "chemically recharged," they are relevant for fleet operators (e.g., taxi enterprises).

(*iii*) *Other Storage Devices: Supercapacitors and Photobatteries*

In Figure 10a comparison has been made between battery systems and electrochemical capacitors (if the capacity is high, also termed supercapacitors) concerning power density and energy density. While for batteries the stored amount of materials is higher, the speed at which the energy can be released is typically much greater in the case of electrochemical capacitors (see e.g., Ref.[85]). While in a battery the electrons discharge ions and the energy is chemically stored in the reaction products and while in an electrostatic capacitor the charges separated in the dielectric are compensated by the electrons of the metal plate, in electrochemical capacitors the charging process refers to the electrolyte/electrode interface. Even though some capacitors involve Faradaic or partial Faradaic processes, in the latter case the "reaction" remains restricted to the interface. This involves a lower capacity, but a higher charge/discharge rate and a better cyclability. While in batteries typically already the second cycle exhibits slight differences compared to the first and the cycle life is restricted to typically one thousand cycles, the electrochemical capacitors show a high degree of cyclability, viz. on the order of 10^5–10^6 times. Both applications are hence complementing each other. Even though local capacitances can be quite high (cf. double layer capacitance) and even though the high surface area can lead to quite substantial overall values, the capacities of the electrochemical capacitors are normally distinctly smaller than those of batteries. The key of the design is—while striving for an increased capacity—not to lose then the advantage of the fast charging and discharging rates.

Table 7 gives an overview of various types of nonbattery storage capacitors and the modes of energy storage. It is also interesting to compare the two limiting cases, the ideal battery ($C = \partial$ (stored charge)/∂ (voltage) $= \infty$) and the ideal capacitor ($C =$ const). In the first case the discharge curve (U versus charge) is horizontal, in the second a straight line with the slope $1/C$. Hence, if we charge a capacitor from $U = 0$ to a value U_0, the energy needed ($\int U \mathrm{d}Q$) is obviously half the value needed for charging an ideal battery with the same charge and voltage U_0(viz. $\frac{1}{2}U_0 Q_0$). For more details on materials issues the reader is referred to a recent monograph by Conway.[206]

Table 7.
Types of capacitors and mode of energy storage after Ref.[206] Reprinted from B.E. Conway, *Electrochemical supercapacitors. Scientific Fundamentals and Technological Applications*, Kluwer Academic/Plenum Publishers, New York (1999). Copyright © 1999 with permission from Kluwer Academic Publishers.

Type	Basis of charge or energy storage	Examples
(a) Vacuum	Electrostatic	–
(b) Dielectric	Electrostatic	Mica, Mylar, paper
(c) Oxide electrolytic (thin film)	Electrostatic	Ta_2O_5, Al_2O_3
(d) Double-layer	Electrostatic (charge separation at double-layer at electrode)	C preparations, powders, fibers
(e) Colloidal electrolyte	Electrostatic (special double-layer system)	Undeveloped
(f) Redox oxide films	Faradaic charge transfer (pseudocapacitance)	RuO_x, IrO_x, Co_3O_4
(g) Redox polymer film	Faradaic charge transfer (pseudocapacitance)	Polyaniline, polythiophenes
(h) Soluble redox system	Faradaic charge transfer (pseudocapacitance)	$Fe(CN)_6^{4-}/Fe(CN)_6^{3-}$, $V^{2+}/V^{3+}/VO^{2+}$

Here we will just mention a special case that is interesting in the case of nanocrystalline electrodes for Li-batteries mentioned above. A sloped regime (in the voltage versus charge curves) indicating the occurrence of interfacial capacitors is seen in the process of Li storage in oxides MO (or fluorides MeF_2) after the reduction to the metal M at the expense of forming Li_2O. Obviously excess lithium can be stored interfacially. In Ref.[194] the following model was

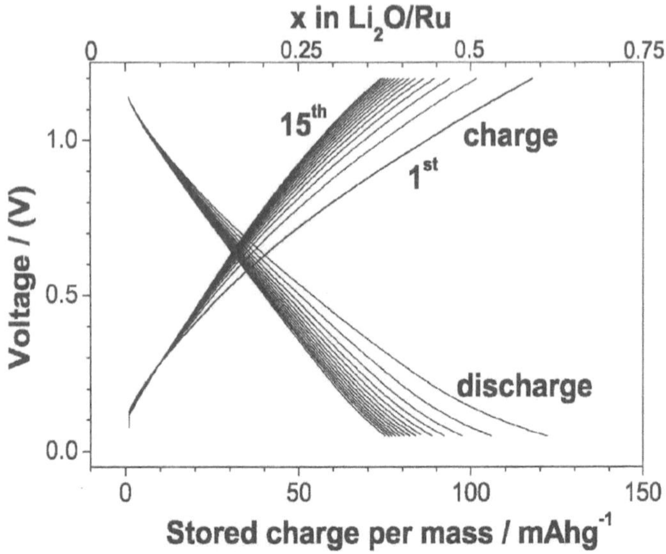

Figure 32. Pseudocapacitive charging of an Li_2O/Ru composite in an Li cell (cf. Figure 28c).

considered for the excess storage at such a metal/Li_2O interface: Li_2O cannot perceptibly take up Li, i.e., the homogeneity range towards Li is very small, not because Li^+ cannot be accommodated, rather it is the e^- that cannot find a low energy accommodation in Li_2O. At the metal contact, however, it is possible that Li^+ is stored on the Li_2O side, while the electrons are accumulated at the metal surface, a mechanism that bears similarity with underpotential deposition of metals. Especially, if the Li_2O particles are very small with a Debye length exceeding the particle size, this storage enables a substantial storage via a mechanism on the borderline of supercapacitor and battery. Figure 32 compares the three storage modes of Li discussed here and in the previous section.

A further promising device is a photobattery which is able to electrochemically store optical energy. In a semiconductor the impact of photons of energies larger than the bandgap can produce electron–hole pairs. At an appropriate interface of the semiconductor to an electrolyte, the electron–hole pairs are separated and can do electrochemical work while recombining otherwise. To be specific, let us consider TiO_2 in contact with an electrolyte that contains a redox pair (ox, red). The hole separated from the electron which is taken up by the outer circuit, acts, if transferred to the electrolyte, oxidizing and increases the ratio [ox]/[red], while the electrons flowing to the other side do the reverse. The thus

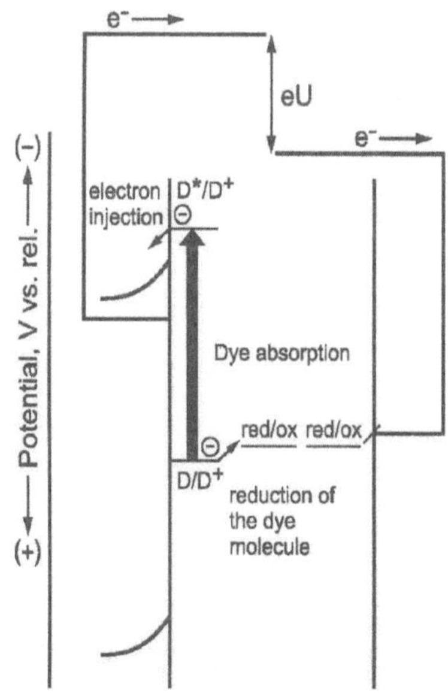

Figure 33. Dye (*D*) sensitized electrochemical photovoltaic cell based on a wide bandgap semiconductor. Reprinted from R. Memming, *Semiconductor Electrochemistry*, Wiley-VCH Verlag GmbH, Weinheim (2001). Copyright © 2001 with permission from Wiley-VCH Verlag GmbH.

achieved chemical potential gradient can in turn be used to release electrical energy. A major problem is to find a photoactive material that exhibits high enough an efficiency and is simultaneously stable enough (see Table 8). Most oxides that come into consideration have too high band gaps; sulfides are more appropriate in this respect but easily undergo self-decomposition (e.g., hole oxidizes S^{2-}). A promising way out is to sensitize the photoelectrodes (see Figure 33) by dyes to obtain a better efficiency. Also here the strategy is to look for solid electrolytes and composite electrode materials.[207,208]

Instead of storing energy, photoelectrochemical cells can also be used for producing chemicals, such as photoelectrolyzers which split water into hydrogen and oxygen. If sufficiently efficient, such a device could be of key importance for a future hydrogen economy.

In spite of its fascination, we have to leave this field and now turn to the second half of this contribution, describing electrochemical techniques which enable us to determine materials parameters.

Table 8.
Electrochemical photovoltaic cell systems. Photovoltage (U_{ph}), photocurrent (j_{ph}), fill factor FF, efficiency (here η) Reprinted from R. Memming, *Semiconductor Electrochemistry*, Wiley-VCH Verlag GmbH, Weinheim (2001).

Cell	E_g (eV)	Solvent	U_{ph} (V)	j_{ph} (mA cm^{-2})	FF	η (%)
(1) n-CdSe/(S^{2-}/S_n^{2-})	1.7	NaOH	0.75	12	–	8
(2) n-CdSe, Te/(S^{2-}/S_n^{2-})	1.7	H_2O	0.78	22	0.65	12.5
(3) n-GaAs/(Se^{2-}/Se_n^{2-})	1.4	NaOH	0.65	20	–	12
(4) n-CdS/(I^-/I_3^-)	2.5	CH_3CN	0.95	0.03	–	009.5
(5) n-CuInSe$_2$/(I^-/I_3^-)	1.01	H_2O	0.64	21	–	009.7

(6) *n*-$MoSe_2/(I^-/I_3^-)$	1.1	H_2O	0.55	9	–	–
(7) *n*-$WSe_2/(I^-/I_3^-)$	1.2	H_2O	0.63	28	–	>14
(8) *n*-$FeS_2/(I^-/I_3^-)$	0.95	H_2O	0.25	10	–	2.8
(9) *n*-WSe_2/(Fephen$^{2+/3+}$)	1.2	H_2SO_4	0.65	10	–	–
(10) *n*-Si/(Br^-/Br_2)	1.1	H_2O	0.68	22	–	>10
(11) *n*-Si/(Fc^{1+}/Fc)[a]	1.1	CH_3OH	0.67	20	>0.7	>10
(12) *n*-GaAs/(Fc^{1+}/Fc)[a]	1.4	CH_3CN	0.7	20	–	11
(13) *n*-GaAs/(Cu^{2+}/Cu^+)	1.4	HCl	0.65	0.5	–	–
(14) p-InP(V^{3+}/V^{2+})	1.3	HCl	0.65	25	0.65	11.5

[a] Ferrocene

III. ELECTROCHEMICAL TECHNIQUES

In the following, we study the response of electrochemical cells to electrical driving forces. We emphasize techniques that allow studying characteristic solid state properties, i.e., techniques by the help of which one can deconvolute ionic and electronic conductivities, establish precise stoichiometries, determine rate coefficients of such compositional changes, extract interfacial properties, or analyze inhomogeneities and heterogeneities (see also Table 10). As long as the information provided by Poisson's equation and linear irreversible thermodynamics is sufficient, the problem is, on a linear response level in principle solved by equivalent circuits of the type given in Figure 69 of Part I, if we introduce the respective boundary and initial conditions. However, here we are more interested in an intelligible understanding.

We will discuss the response of essentially seven cell types which are designed to measure a mixed conductor which conducts e^- and O^{2-} (as regards the systematics we follow Ref.[3,15,11,209,210]):

$$\left(O^{2-},e^-\rangle\middle|MO\middle|\langle e^-,O^{2-}\right) \tag{1}$$

$$\left(O^{2-},e^-\rangle\middle|MO\middle|\langle e^-,O^{2-}\right)' \tag{2}$$

$$\left(O^{2-},e^-\rangle\middle|MO\middle|\langle e^-\right) \tag{3}$$

$$\left(e^-\rangle\middle|MO\middle|\langle e^-\right) \tag{4}$$

$$\left(O^{2-},e^-\rangle\middle|MO\middle|\langle O^{2-}\right) \tag{5}$$

$$\left(O^{2-}\rangle\middle|MO\middle|\langle O^{2-}\right) \tag{6}$$

$$\left(O^{2-}\rangle\middle|MO\middle|\langle e^-\right). \tag{7}$$

They can be discussed according to whether the sample MO is contacted by reversible—designated by (O^{2-},e^-)—or selectively blocking electrodes. The latter ones can be blocking for ions (but reversible for electrons) or blocking for electrons (but reversible for ions); they are designated by (e^-) or (O^{2-}), respectively. The angle bracket in the above list points towards the interface at which the blocking occurs. If we are concerned with an oxide and sufficiently high temperatures, reversible electrodes can be porous Pt electrodes, while compact graphite may act blocking for O^{2-}. The combination $Pt,O_2|YSZ$ acts blocking for electrons (at the rhs) if the electronic resistance YSZ (YSZ ≡ Y-doped ZrO_2) is greater than the ionic resistance of the sample (otherwise the roles of blocker and sample are exchanged) (see Ref.[211]). Hence, we specifically deal with

$$\mathrm{Pt,O_2(P)|MO|O_2(P),Pt} \tag{1'}$$

$$\mathrm{Pt,O_2(P)|MO|O_2(P'),Pt} \tag{2'}$$

$$\mathrm{Pt,O_2(P)|MO|C} \tag{3'}$$

$$\mathrm{C|MO|C} \tag{4'}$$

$$\mathrm{Pt,O_2(P)|MO|YSZ|O_2(P),Pt} \tag{5'}$$

$$\mathrm{Pt,O_2(P)\,|\,YSZ\,|\,MO\,|\,YSZ\,|\,O_2(P),Pt} \tag{6'}$$

$$\mathrm{Pt,O_2(P)\,|\,YSZ\,|\,MO\,|\,C.} \tag{7'}$$

Analogous cells for a cationic mixed conductor (conductivity via electrons and cations, e.g., Ag_2S) can be formulated (Ag|AgI instead of O_2, Pt|YSZ; Ag or an Ag-alloy instead of O_2, Pt). In this case sealing is not necessary; however, the component activity cannot be tuned so simply.

We will consider these cells, primarily the oxygen cells, under open circuit conditions and under load (or even short-circuit condition). In the transient and in the steady state it is not necessary to treat them all in detail, since (as outlined below) cells with one selectively blocking electrode and those with two of the same kind show far-reaching similarities (compare cell 3 with cell 4 and cell 5 with cell 6). The same is true if we compare cells with electrodes that are selectively blocking for electrons with cells that are specifically blocking for ions (compare cell 3 with cell 5 and cell 4 with cell 6): it is easy to show that the relations are symmetrical as regards the indices e^- and O^{2-} (see below and Appendix 1).[210,11]

We will see that in the steady state of the blocking cells, we can extract partial conductivities, and from the transients chemical diffusion coefficients (and/or interfacial rate constants). Cell 7 combines electronic with ionic electrodes: here a steady state does not occur but the cell can be used to titrate the sample, i.e., to precisely tune stoichiometry. Cell 1 is an equilibrium cell which allows the determination of total conductivity, dielectric constant or boundary parameters as a function of state parameters. In contrast to cell 1, cell 2 exhibits a chemical gradient, and can be used to e.g., derive partial conductivities. If these oxygen potentials are made of phase mixtures[212] (e.g., AO, A or ABO_3, B_2O_3, A) and if MO is a solid electrolyte, thermodynamic formation data can be extracted for the electrode phases.

Furthermore, it is also not necessary to discuss different excitations in detail as long as we restrict ourselves to the linear response regime. There it holds that the response to any excitation allows the calculation of the response to other excitations via the convolution theorem of cybernetics.[213] In the galvanostatic mode, e.g., we switch the current on from zero to I_p (or switch it off from I_p to zero) and follow $U(t)$ as a response to the current step. The response to a sinusoidal excitation then is determined through the complex impedance which is given by the Laplace transform of the response to the step function multiplied with $j\omega\,(j \equiv \sqrt{-1}, \omega \equiv \text{angular frequency})$.

1. Determination of Bulk Parameters

Here, we concentrate on cell 1 and assume negligible electrode effects. If a constant current is switched on, both a faradaic as well as a displacement current flows (cf. Section I). Hence the actual current can be ionic/electronic or capacitive, the relative proportions depending on the electronic (σ_{eon}) and ionic (σ_{ion}) conductivities and the dielectric constant. Correspondingly, the elements are, as long as σ_{eon} and σ_{ion} are summed locally, in parallel (∞ denotes the bulk and $R_\infty = R_{eon}R_{ion}/(R_{eon}+R_{ion})$) and the equivalent circuit is given by (cf. also Eq. (5))

$$\text{Equivalent Circuit} = \text{Par}\left(R_{eon}, R_{ion}, C_\infty\right) = \text{Par}\left(R_\infty, C_\infty\right). \tag{60}$$

(the operators Par and Ser, respectively, connect the elements in parallel and in series, the parameters of which are given in the brackets.) For small excitations and not too high frequencies (i.e., we do not deal with ultra-short time resolution) the circuit elements are invariant, and the d.c. response is an exponential voltage time dependence characterized by the relaxation time $\tau_\infty = R_\infty C_\infty$ (cf. Eq. (5)) with the steady state value IR_∞, whilst the a.c. measurement gives the complex impedance as a semicircle in the Gaussian plane with the diameters R_∞ and the peak frequency $\omega = \tau_\infty^{-1}$. Anomalies in the high-frequency behavior are discussed later (Section III.8). In this way, one can easily obtain the bulk conductivity as a function of temperature (T), partial pressure (P) and dopant's concentration (C), which allows for the construction of defect chemical models as discussed in Part I. In the case of simple defect models, the conductivity of the carrier k in a binary compound is given as sectional solution of the form (u: mobility with prefactor u_0 and migration enthalpy $\Delta H^{\neq}$, z: charge number, K: mass action constants of the defect chemical problem with prefactor K_0 and reaction enthalpy ΔH°)

$$\sigma_k = |z_k| \mathrm{F} u_k(T) c_k(T,P,C) = |z_k| \mathrm{F} u_{0k} \exp - \frac{\Delta H_k^{\neq}}{\mathrm{R}T} \Sigma_r K_{0r}^{\gamma_{rk}} \exp - \frac{\Sigma_r \gamma_{rk} \Delta_r H^{\circ}}{\mathrm{R}T} P^{N_k} C^{\mathrm{M}_k} \tag{61}$$

where the simple rational numbers $\gamma_{rk}, \mathrm{N}_k, \mathrm{M}_k$ are characteristic for the defect model under consideration.[214–216] Consequently the variation of P and C provides worthwhile information on the carrier-type and the defect model, whilst the T-dependence gives insight into the thermochemistry ($\Delta_r H^{\circ}$) and the transport mechanism ($\Delta H_k^{\neq}$). In ternary and multinary systems further P-terms appear or, under quenched conditions, the C-term contains also native constituents. This is all set out in detail in Part I[2] and will not be considered further.

If the amount of mass transferred is analyzed, further conclusions on the carrier type (ionic/electronic) can be drawn according to "Faraday's law," simply because of the fact that the pure stationary electronic flux will not cause mass displacements. If also the direction of mass transport is analyzed, e.g., by recording the position variation of the boundaries, cationic and anionic conductivities may be distinguished (Tubandt–Hittorf experiments).

The bulk capacitance is (besides geometric parameters) determined by the bulk dielectric constant ε, like the mobility, ε is usually quite insensitive with respect to P, C, and—unlike the mobility—also to T. Only at high carrier concentrations (e.g., formation of polar associates) or in special compounds (e.g., ferroelectrica) strong variations are to be expected.

Cell 2 differs from cell 1 by the fact that the chemical component potential is varied which in particular enables the separation of ionic and electronic conductivities. These effects are to be discussed later (Section III.6).

2. Determination of Boundary Parameters

Boundary impedances refer either to heterocontacts, such as the contacts of the electrode to the electrolyte or contacts between the grains of different phases in composites, or to homocontacts, i.e., grain boundaries. In this chapter, we restrict ourselves to resistive boundaries which are then perceived as serial contributions. Effects of highly conductive boundaries will be discussed in Section III.9. While such grain boundary or electrode processes can be usually easily distinguished from bulk processes by exploiting the different relaxation times, e.g., by impedance spectroscopy (see e.g., Figure 34), it can be more complicated to distinguish between electrode and grain boundary effects. One possibility is the separation by multi-point measurements using cell 1 (Section III.9), another is grain size variation, a third one the evaluation of the capacitances, as considered in more detail below.

A reasonable approximation of the impact of such a boundary in series to the current flow is provided by a parallel *R–C* circuit $(R_\perp, C_\perp)$ in series to the bulk impedance

$$\text{Equivalent Circuit} = \text{Ser}\left(\text{Par}\left(R_\perp, C_\perp\right), \text{Par}\left(R_\infty, C_\infty\right)\right). \tag{62}$$

Eq. (62) can take account of the dielectric response of a boundary (electrode, grain boundary) but would not be sufficient to describe bulk polarization phenomena (appearing at longer times or lower frequencies) induced by strongly selectively blocking electrodes or grain boundaries (see below). The latter effect will be touched upon in the next section.

The boundary resistance $R_\perp$ is dominated either by the space charge resistance in the current direction or the proper charge transfer resistance through the boundary core (Part I, Section VI.6). (Resistance effects due to laterally inhomogeneous contact conditions are discussed in Section III.9).

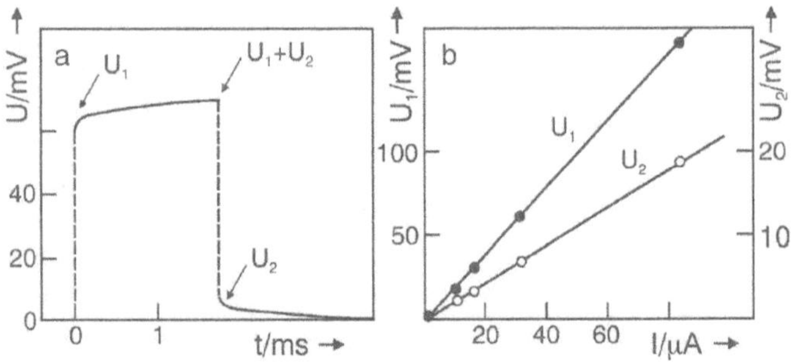

Figure 34. Galvanostatic step experiment in the case of PbO (orh.) at 550°C and 0.05 kPa O_2. With the help of the detailed time dependence the parameters $R_1 (\hat{=} R_\infty)$, $R_2 (\hat{=} R_\perp)$, $C_1 (\hat{=} C_\infty)$, $C_2 (\hat{=} C_\perp)$ can be extracted.[15,217] Reprinted from J. Maier, *Solid State Phenom.*, **39/40** (1994) 35–60. Copyright © 1994 with permission from Trans Tech Publications Ltd.

The inverse boundary capacitance $C_\perp$ is essentially the sum of the inverse contribution of the space charge layer[218] and the rigid double layer[219–221] and usually dominated by the first contribution. Since the effective dielectric permeabilities of bulk and boundary are comparable, the ratio $C_\perp/C_\infty$ significantly differs from unity (it is typically of the order of L/λ, L being the sample thickness, and λ the Debye length). This explains (i) why for comparable $R_\perp$ and R_∞ values the relaxation times $\tau_\perp$ are typically drastically different, and (ii) why the uncertainty whether C_∞ is in parallel to the whole MX-part of the cell or only to the proper bulk part as assumed in Eq. (62), corresponds to a quantité négligeable. Note that in a polycrystalline material $1/C_\perp$ as well as $R_\perp$ are—if they refer to the impact of the internal boundaries—not only proportional to the inverse capacitance or conductance of a single interface but also to the number of interfaces (given by the ratio of sample size and grain size), whereas $\tau_\perp$ is invariant with respect to geometrical parameters. Hence, the evaluation of the capacitances allows usually a clear assignment of impedance arcs either to electrode interfaces or grain boundary interfaces, while the τ-values are quite similar. In other words, electrode and (series) grain boundary impedances are likely to occur at similar frequencies, but characterized by completely different capacitance values; a typical figure may be 10 pF for bulk, 10 nF for grain boundary, 10 μF for electrode corresponding to a ratio of (grain or electrode) boundary thickness to grain size to sample size of 1 nm : 1 μm : 1 mm (ε: 10, electrode area: 1 cm^2). (Hence they are usually easily identified but it is difficult to separate them if they both occur with comparable resistances.) The same thickness effect is responsible for the fact that for comparable voltages drops, bias dependencies are usually negligible for the bulk response where the drop is partitioned over $\sim 10^6$ elementary distances (L = 1 mm), while for the electrode response a moderate drop can make the current voltage relation nonlinear (see Part I,[2] Section VI.2). The same occurs for a grain boundary impedance in the bicrystal experiment shown in Figure 35, while for fine-grained thick samples nonlinearity requires high voltages.

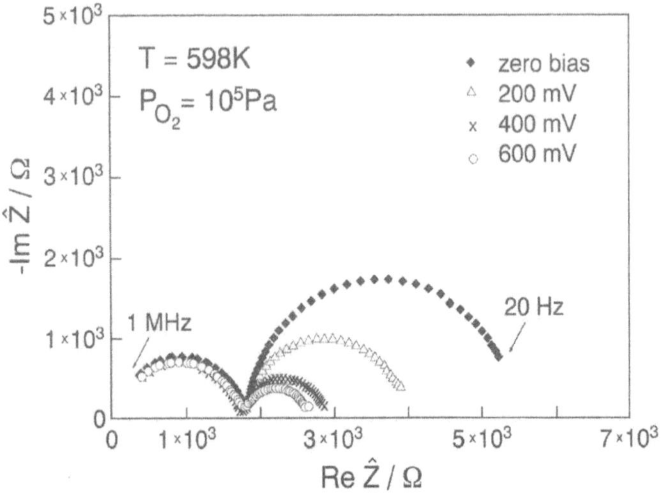

Figure 35. Impedance spectra of the cell O_2, Pt|$SrTiO_3$|$SrTiO_3$|Pt, O_2 as function of d.c. bias. Electrodes are parallel to the bicrystal boundary (Σ5 tilt grain boundary, iron content: 2×10^{18} cm^{-3}). Both bulk and boundary resistances are predominantly electronic resistances.[222] Reprinted from I. Denk, J. Claus and J. Maier, *J. Electrochem. Soc.*, **144** (1997) 3526–3536. Copyright © 1997 with permission from The Electrochemical Society.

As far as Eq. (62) is concerned, in the galvanostatic switching-on experiment the $U(t)$ function is obviously the sum of two exponentials. On the time scale of τ_∞ the boundary process is not of interest and on the time scale of $\tau_\perp$ the boundary impedance is what remains in series with a constant IR_∞ drop corresponding to the blocking bulk capacitor. Figure 34 shows the response to a step function of the current signal for PbO displayed on the time scale of $\tau_\perp$. As already explained, the linearity of $U_2(t)$ presupposes small excitations. Accordingly two semicircles are observed in the impedance plot, with the low frequency semicircle corresponding to the boundary part (See Figures 35 and 36). Figure 36 shows besides the complex impedance in the Gaussian plane—analogous plots also of other conveniently defined system functions.[3,15] The dielectric modulus $\hat{M} = j\omega\hat{Z}$ deserves special attention, as in the $\hat{M}$-plot the capacitances are as conveniently extracted as the resistances from the $\hat{Z}$ presentation.

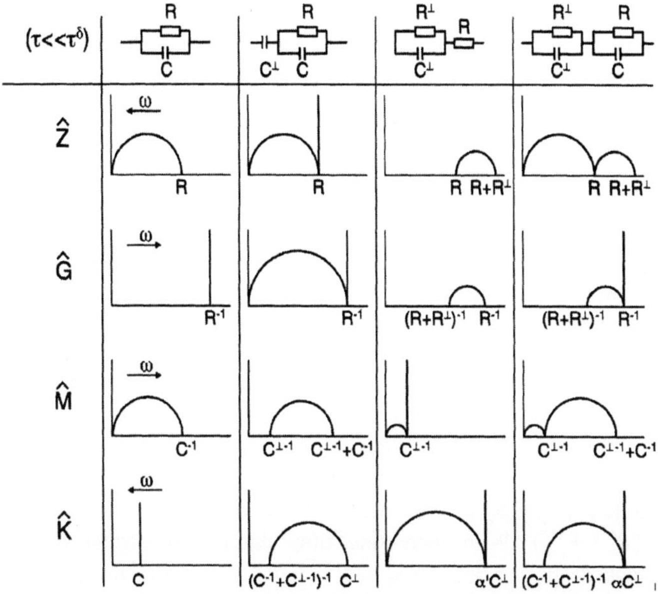

Figure 36. Plots of different system functions for the Equivalent Circuit given by Eq. (62) and special cases of it.[3,15] Reprinted from J. Maier, "Evaluation of Electrochemical Methods in Solid State Research and Their Generalization for Defects with Variable Charges", *Z. Phys. Chem. NF*, (1984) 191–215. Copyright © 1984 with permission from Oldenbourg Verlagsgruppe.[223]

Let us consider in more detail Figure 35 which shows the impedance spectrum ($\hat{Z}$ -plot) of an $SrTiO_3$ bicrystal in which the (serial) boundary effect is determined by Mott–Schottky depletion layers. Resistances and capacitances can be evaluated as a function of temperature, oxygen partial pressure, doping content, and bias as extensively described in Part I (Sections IV and V). In nanocrystalline $SrTiO_3$ ceramics (in fact already below a grain size of 100 nm because of the high dielectric constant) the high frequency semicircle disappears. Here depletion layers start to overlap and the electroneutral bulk to disappear.[224] For a further interpretation of space charge impedances or impedances stemming from the electrode reaction the reader is also referred to Part I.[2]

3. Electrochemical Polarization—The Effect of Selectively Blocking Electrodes

(*i*) *Heuristic Considerations*

In the following text (Section III.3) we discuss the electrochemical response of the cells (3–6) with a mixed conductor as the central phase, extending from $x = 0$ to $x = L$. The treatment is based on work by Wagner and Yokota[225,210] and neglects further boundary effects as well as the presence of defect chemical reactions within the solid MX. The generalization to internal reversible reactions treated by Maier in Ref.[3,15,226] will be separately considered in Section III.7. As before we restrict our considerations to the case of small signals (except Section III.3.ii). The common feature of these cells is the use of selectively blocking electrodes: In cells 3 and 4 $i_{\rm ion} = 0$ at $x = 0$ and $x = L$ (ionically blocking electrodes = electronic electrodes), whilst in the case of the cells 5 and 6 $i_{\rm eon}$ disappears at the boundaries $x = 0$ and $x = L$ (electronically blocking electrodes = ionic electrodes). Before we turn to the detailed consideration of the transport theory, we discuss the matter on the heuristic level of simple equivalent circuit elements.[3]

Immediately after switching-on a constant current both ions and electrons do flow according to their transference numbers while the voltage increases from zero to IR_∞ (or more precisely to $I(R_\infty + R_\perp)$). With increasing time the partial current of the blocked species decreases, and eventually vanishes leading to a steady state in which the total current is carried only by the nonblocked species, i.e., the electrons in cells 3 and 4 or the ions in cells 5 and 6. Hence the steady state reveals the conductivity of the nonblocked species. The comparison with the total conductivity (e.g., obtained from the IR-drop at the beginning of the experiment) also yields the conductivity of the blocked species.

If we switch-off the current after the steady state has been reached, the voltage relaxes to the initial zero-level. The electrical behavior can be taken into account by introducing in the simple equivalent circuit (eqc) (Eq. (60)) a capacitor C_δ in series to $R_{\rm ion}$ or to $R_{\rm eon}$ (if ions or electrons are blocked): In the language of system theory the equivalent circuit of the bulk represents a PDT$_1$-element and reads[3,15] e.g., for cells 3 and 4 $\mathrm{Par}\left(C_\infty, \mathrm{Par}\left(R_{\rm eon}, \mathrm{Ser}\left(R_{\rm ion}, C_\delta\right)\right)\right)$.

This additional capacitive element (C_δ)[3,15] represents a chemical capacitance in the sense of Part I,[2] Section VI.7 (see Figure 37).

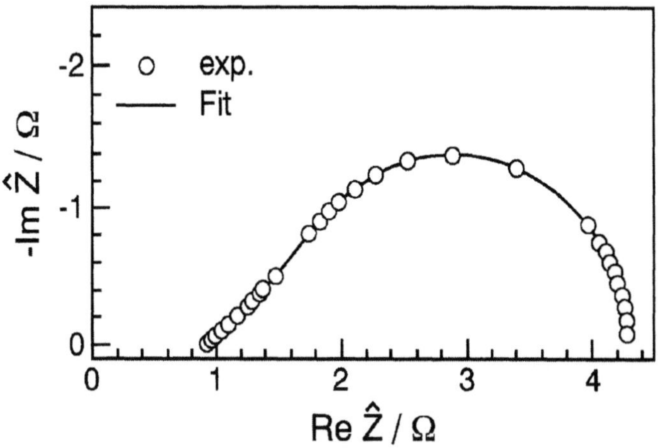

Figure 37. The impedance spectrum of the cell Pt|Ag_2Te|Pt displays the stoichiometry polarization (200°C, fine composition set up by Coulometric titration). The transition from a straight 45° *line to a semicircle*, before the maximum frequency, corresponds to the theoretical treatment given in Section III.3.[227] Reprinted from R. Andreaus and W. Sitte, *J. Electrochem. Soc.*, **144** (1997) 1040–1044. Copyright © 1997 with permission from The Electrochemical Society, Inc.

To take into account boundary impedances at the nonblocked side (or of the nonblocked carrier on the other side) the circuit has to be complemented by an additional $\mathrm{Par}(R_{\perp}, C_{\perp})$. The same circuit element may also represent the response of a serial grain boundary (Figure 35).

The simple equivalent circuit (eqc) which reads for ion blocking as

$$\mathrm{eqc} = \mathrm{Ser}\Big(\mathrm{Par}\big(R_{\perp}, C_{\perp}\big), \mathrm{Par}\Big(C_{\infty}, \mathrm{Par}\big(R_{\mathrm{eon}}, \mathrm{Ser}\big(R_{\mathrm{ion}}, C_{\delta}\big)\big)\Big)\Big) \quad (63)$$

reproduces the correct transient behavior to a surprising degree (see Figures 38 and 39). It relies on the fact that the capacitance C_{δ} is characterized by a very high value and hence the relaxation time of the diffusional process $\tau_{\delta} = (R_{\mathrm{eon}} + R_{\mathrm{ion}})C_{\delta}$ being normally large compared to both τ_{∞} and $\tau_{\perp}$.

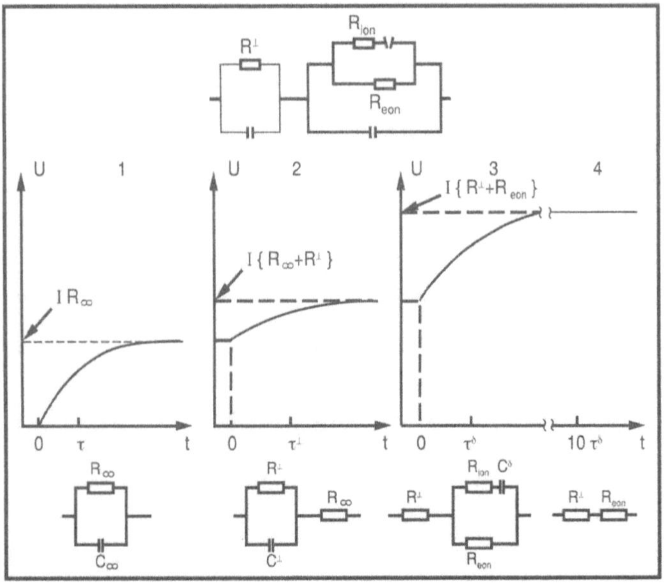

Figure 38. The voltage response on a galvanostatic polarization of a mixed conductor with ion-blocking electrodes (~ eqc. (63)[3,15]). Reprinted from J. Maier, "Evaluation of Electrochemical Methods in Solid State Research and Their Generalization for Defects with Variable Charges", Z. Physik. Chemie N.F., 191–215, Copyright © 1984 with permission from Oldenbourg Verlagsgruppe.

In the case of electron blocking the indices eon and ion have to be exchanged, but also the total equivalent circuit has to be complemented by circuit elements that take account of the bulk impedance of the blocking electrodes (e.g., YSZ).

It is instructive to explicitly consider the charging process of eqc. Eq. (63) by a constant current I_p (Figure 38). According to Kirchhoff's laws, the voltage response is given by:

$$U(t) = I_\mathrm{p} R_\infty \left(1 - \exp\left(-t/\tau_\infty\right)\right) + I_\mathrm{p} R_\perp \left(1 - \exp\left(-t/\tau_\perp\right)\right) + I_\mathrm{p} \frac{R_\mathrm{eon}^2}{R_\mathrm{eon} + R_\mathrm{ion}} \left(1 - \exp\left(-t/\tau_\delta\right)\right). \tag{64}$$

In the short-time regime of the order of τ_∞, the capacitances $C_\perp$ and C_δ are completely permeable and the voltage response consists of an exponential increase from zero to $I_\mathrm{p} R_\infty$ corresponding to the first term in Eq. (64); terms 2 and 3 are zero.

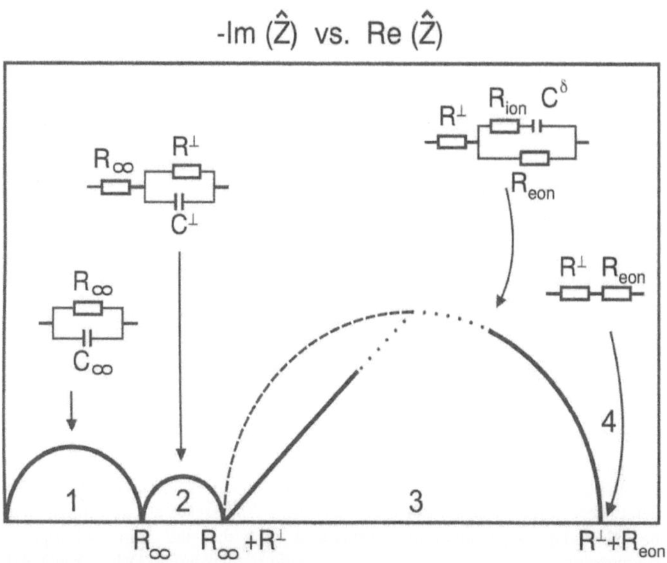

Figure 39. The complete impedance spectrum of a mixed conductor contacted by ion-blocking electrodes.[3,15] Regarding signal 3: The *dashed curve* corresponds to the heuristic approach (Eq. (64)), the *straight line* to the solution of the diffusion law; with respect to the detailed behavior around the maximum see Figure 37. Figure 39 is the translation of Figure 38 into the frequency domain. Reprinted from J. Maier, *Z. Phys. Chem.* NF, (1984) 191–215. Copyright © 1984 with permission from Oldenbourg Verlagsgruppe.

For intermediate time resolutions (of the order of $\tau_\perp$) the bulk capacitor has become impermeable, and the boundary circuit is relevant for the time dependence (second term in Eq. (64)); term 1 is constant, while term 3 is still zero. Finally in the long-time regime, at $t \sim \tau_\delta$, the stoichiometric polarization occurs while both bulk and boundary responses constitute the initial voltage jump from $U = 0$ to $U = I_p(R_\infty + R_\perp)$; note that both corresponding capacitors are completely impermeable, i.e., terms 1 and 2 are constant. In the steady state ($t \gg \tau_\delta$) all the capacitors block, and $R_\perp + R_{eon}$ are obtained as the stationary resistance value. Obviously time-resolved dc experiments allow the partial conductivities and the capacitances to be measured together with the chemical diffusion coefficient ($\tau_\delta \propto 1/D^\delta$). The switching-off behavior is analogous.

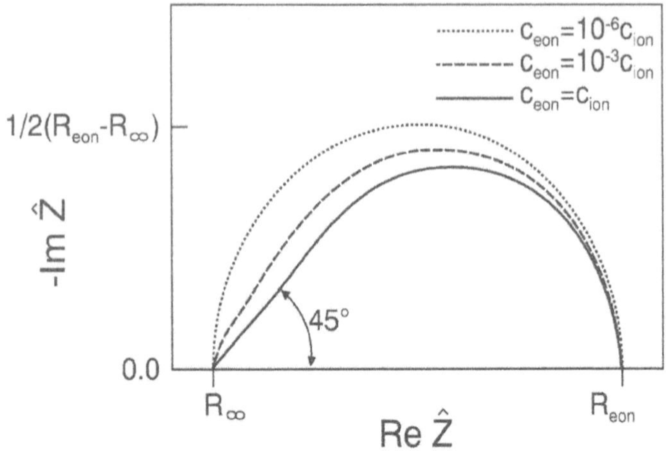

Figure 40. Numerically calculated impedance for ion blockage. The variation of the defect concentration leads to the transition from Warburg to a pure semicircular behavior. The impedances are normalized such that the points of highest frequency coincide. (For the mobility ratio, ratio of thickness to Debye length and charge numbers the following values are assumed: $u_{eon} = 10u_{ion}, L/\lambda = 10^4, z_{ion} = 1 = -z_{eon}$).[228] Reprinted from J. Jamnik, J. Maier, S. Pejovnik, *Electrochim. Acta*, **44** (1999) 4139–4145. Copyright © 1999 with permission from Elsevier.

According to the superposition theorem of system theory for linear responses, this response to a step-function in the current can be employed to deduce the impedance behavior. As regards a qualitative discussion, one can adopt the above description by just replacing short/long times by high/small frequencies. Quantitatively the impedance is given by a Laplace transformation of Eq. (64) (or equivalently by applying Kirchhoff's laws to the equivalent circuit (Eq. (63))) with the result

$$\hat{Z} = R_\infty \frac{1 - j\omega\tau_\infty}{1 + \omega^2\tau_\infty^2} + R_\perp \frac{1 - j\omega\tau_\perp}{1 + \omega^2\tau_\perp^2} + \frac{R_{eon}^2}{R_{eon} + R_{ion}} \frac{1 - j\omega\tau_\delta}{1 + \omega^2\tau_\delta^2}. \tag{65}$$

The dc and ac responses are compared in Figures 38 and 39.[229]

The more accurate relationships obtained from the underlying microscopic current equations are discussed below (see Section III.3.iii). There we will see that the impedance that describes the stoichiometry polarization is composed of a so-called Warburg behavior (linear increase with a slope of 45° followed by a semicircular behavior, see Figure 39). Figure 37 shows an experimental example that comes close to the ideal situation. The detailed behavior around the maximum is also in accord with the precise treatment (see Section III.3.iii).

In many examples, however, impedance features have been observed that deviate from the description in Figures 38 and 39; at low frequencies, e.g., slopes larger than 45° can often be found (cf. Figure 40). The deficiency of the above considerations becomes obvious if we proceed from a mixed conductor to a pure ionic or pure electronic conductor and compare this with the result which we would expect by restricting to a single carrier right from the beginning. If we start with the above approach (Eq. (65)) and let the concentration of the nonblocked species approach zero, an infinitely large semicircle is predicted that initially exhibits a 45° slope but not a steeper one, whereas a 90° slope is expected by completely neglecting the counter carrier (pure capacitive behavior). The solution of this dilemma lies in the fact that space charge regions have to be taken account of in a more precise analysis, and it is the ratio of chemical and electrical capacitance that determines the crossover from a pure capacitive blocking to a stoichiometric blocking effect. A recent quantitative and pertinent treatment is given in Ref.[228]

What has been ignored so far and will only be briefly mentioned is that a stoichiometric polarization is also caused by grain boundaries if the ratio of ionic and electronic conductivities differs from the bulk value, as it is usually the case.[230] Figure 41 gives a clear example of this. In the general case of blocking electrodes and grain boundaries we expect even two stoichiometry polarization processes.

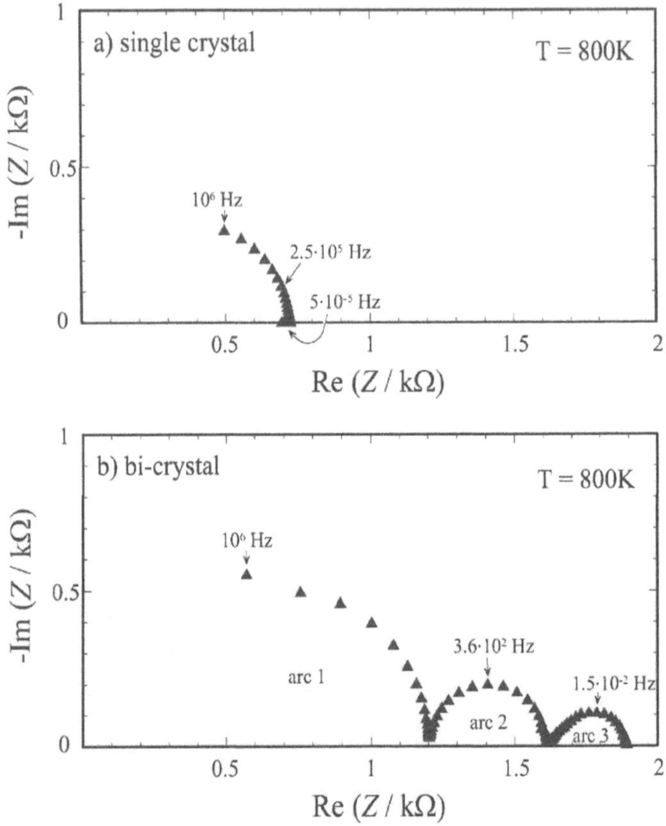

Figure 41. Impedance of an $SrTiO_3$ single crystal (a) and an $SrTiO_3$ bicrystal (b). In both cases reversible $YBa_2Cu_3O_{6+x}$ electrodes were used,[230] such that electrode impedances are absent. Reprinted from J. Jamnik, X. Guo, and J. Maier, *Appl. Phys. Lett.*, **82** (2003) 2820–2822. Copyright © 2003 with permission from the American Institute of Physics.

(ii) The Steady-State Response: The Evaluation of Partial Conductivities

Now we wish to consider the electrochemical polarization with the help of selectively blocking electrodes (connected with the neutral phase at $x = L$, while $x = 0$ is the position of the reversible electrode contact) on a more fundamental level[3,15,210,225,231] and refer, to be specific, to a galvanostatic experiment on cells 3 and 4. We start with the steady state.

There the ionic current vanishes, that is (for a nonzero σ_{ion}) the driving force $\nabla\tilde{\mu}_{\text{ion}}$ must be zero. Because of

$\nabla\mu_{\mathrm{O}} = \nabla\tilde{\mu}_{\mathrm{O}^{2-}} - 2\nabla\tilde{\mu}_{\mathrm{e}^-} = -2\nabla\tilde{\mu}_{\mathrm{e}^-} \propto i/\sigma_{\mathrm{eon}}$ the local electronic conductivity is determined by the total current divided by the local change of the electrons' electrochemical potential and simultaneously by the local variation in stoichiometry ($\nabla\tilde{\mu}_{\mathrm{e}^-} \propto \nabla\mu_{\mathrm{O}} = 1/2\nabla\mu_{\mathrm{O}_2}$). If the conductivity is approximately constant, the electronic conductivity is obtained from the ratio of current density and voltage, which is directly obvious from the equivalent circuit. More accurately, the partial conductivity is derived from the slope of the current–voltage relationship.

In more detail: As $i = \mathrm{const}$, the explicit current function follows by spatial integration as

$$i = -\frac{1}{L}\int_{\tilde{\mu}_{\mathrm{e}^-}(0)}^{\tilde{\mu}_{\mathrm{e}^-}(L)} \frac{\sigma_{\mathrm{e}^-}}{\mathrm{F}}\mathrm{d}\tilde{\mu}_{\mathrm{e}^-} = \frac{1}{4L}\int_{\mu_{\mathrm{O}_2}(0)}^{\mu_{\mathrm{O}_2}(L)} \frac{\sigma_{\mathrm{e}^-}}{\mathrm{F}}\mathrm{d}\mu_{\mathrm{O}_2}. \tag{66}$$

Herefrom the electronic conductivity is precisely arrived at by differentiating the current with respect to the oxygen potential at the contact of the blocking electrode ($x = L$). As $\partial\mu_{\mathrm{O}_2}(L) = \partial(\mu_{\mathrm{O}_2}(L) - \mu_{\mathrm{O}_2}(0)) = -4\partial(\tilde{\mu}_{\mathrm{e}^-}(L) - \tilde{\mu}_{\mathrm{e}^-}(0)) = -4\mathrm{F}\partial U$, it follows that:

$$\partial i/\partial U = \sigma_{\mathrm{e}^-}/L. \tag{67}$$

In the case of the asymmetrical cell considered, the voltage variation leads to an one-sided extension of the P_{O_2} range, and the detailed evaluation of the correct voltage relation contains easily extractable information on partial conductivities. This is obtained as follows: If the P_{O_2} dependence is a power law ($\sigma_{\mathrm{h}^{\cdot};\mathrm{e}'} \propto P^{\pm N}$), Eq. (66) directly leads to

$$i_{\mathrm{eon}} = \frac{\mathrm{R}T}{4N\mathrm{F}L}\left\{\sigma_{\mathrm{e}'}(0)\left[\exp\left(+\frac{4N\mathrm{F}U}{\mathrm{R}T}\right)-1\right]+\sigma_{\mathrm{h}^{\cdot}}(0)\left[1-\exp\left(-\frac{4N\mathrm{F}U}{\mathrm{R}T}\right)\right]\right\} \tag{68}$$

from which the partial conductivities can be obtained and which takes the Wagner–Hebb form for $N = 1/4$, i.e., for predominant ionic disorder. The behavior for $N \neq 1/4$ is studied in Ref.[58,234–237] where it is also described how to proceed if power laws are not valid.

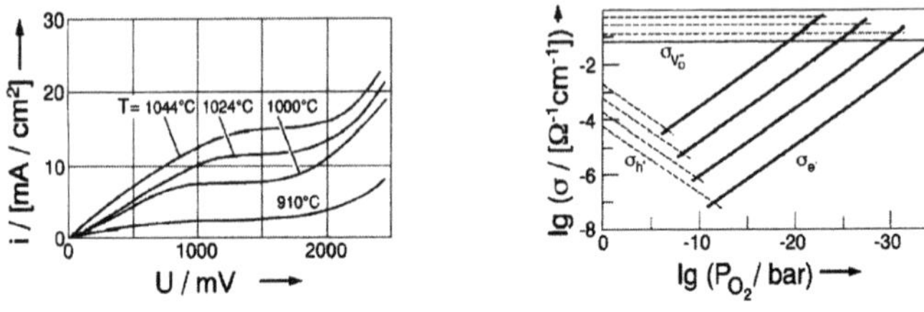

Figure 42. Steady-state results of a Wagner–Hebb polarization of Y-doped ZrO_2 or ThO_2 with the aid of the cell ⊖N_2, Pt|YSZ or YST|air, Pt ⊕ (*left*: current–voltage curve for $Th_{0.9}Y_{0.1}O_{1.95}$ (YST), *right*: partial conductivities for $Zr_{0.9}Y_{0.1}O_{1.95}$ (YSZ)).[232] Reprinted from L.D. Burke, H. Rickert, and R. Steiner, *Z. Phys. Chem. N.F.*, **74** (1971) 146–167. Copyright © 1971 with permission from Oldenbourg Verlagsgruppe.

Figure 42 depicts steady-state results of a Wagner–Hebb polarization of (Y_2O_3-doped) ZrO_2 and ThO_2 for which the ideal Wagner–Hebb relation is valid.[232]

The analogous conclusions are arrived at for the ionic conductivity in the case of cell 5. The blocking technique can be extended in order to de-convolute ionic and electronic conductivities for a grain boundary of a bicrystal (see Figure 43)[236] or even in composites.[240] In both the cases, the relations are substantially more complicated because of local inhomogeneities. Some of the pitfalls of the Wagner–Hebb technique have been described by Riess.[241]

In contrast to such asymmetrical cells, in a symmetrical polarization cell (cells 4 and 6) variation of the voltage is not of great worth, as this only affects the magnitude of the P_{O_2} variation around the mean value. (The corresponding current voltage relations are discussed in Ref.[242,243]). Nonetheless a mean value of σ_{eon} can be referred to the P_{O_2} in the unpolarized state. Repeating the polarization experiment with different initial states, allows one to extract reliable information about the stoichiometry dependence provided the polarization voltage is sufficiently small. This is treated in greater detail in Ref.[244] Figures 44 and 45 show the evaluation of the ionic conductivity of the high temperature superconductor $YBa_2Cu_3O_{6+x}$ and the evaluation of the electronic conductivity in the mixed conductor Fe-doped $SrTiO_3$ obtained from symmetrical polarization cells.

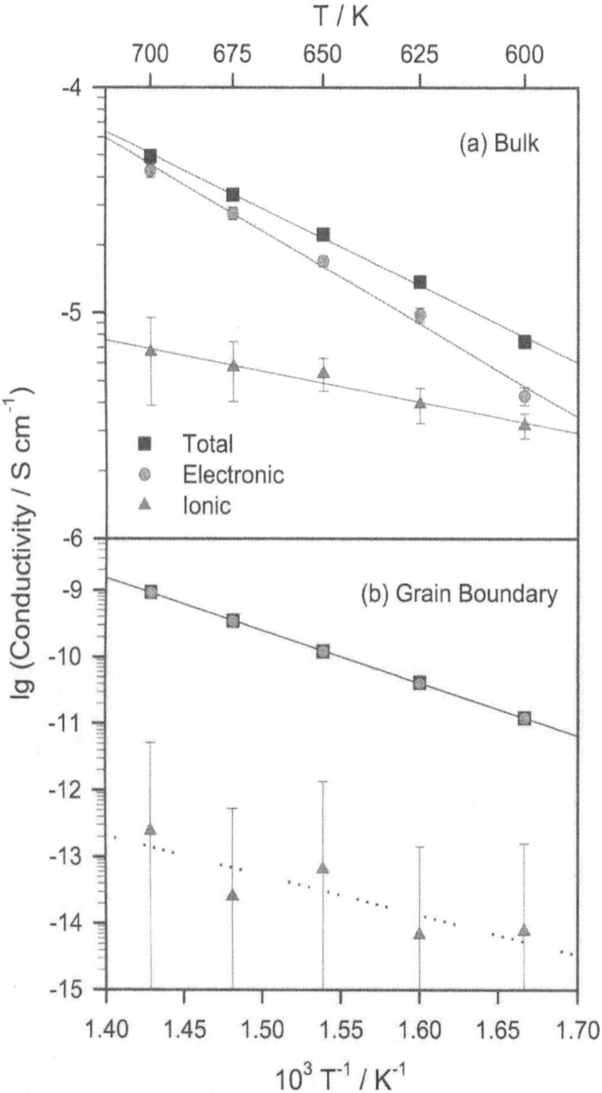

Figure 43. Temperature dependence of total and partial conductivities of (a) bulk and (b) grain boundary[111] in a 0.016 wt% Fe-doped $SrTiO_3$ bicrystal at about 2.0 Pa oxygen. The lower electronic and the much lower ionic grain boundary values are due to Mott–Schottky layer (see Part I, Section V).[233] Reprinted from X. Guo, J. Fleig and J. Maier, *J. Electrochem. Soc.*, **148** (2001) J50–J53. Copyright © 2001 with permission from The Electrochemical Society, Inc.

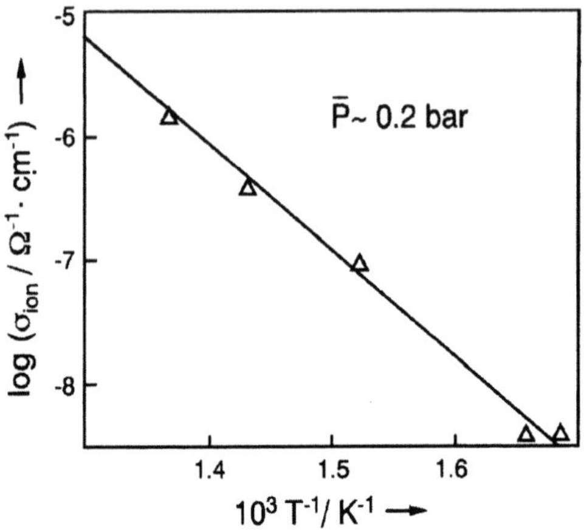

Figure 44. The steady-state voltage of an electron-blocking experiment on $YBa_2Cu_3O_{6+x}$ allows the determination of its ion conductivity as a function of temperature and the assignment to an oxygen partial pressure.[11,238] Reprinted from J. Maier, P. Murugaraj, G. Pfundtner, W. Sitte, *Ber. Bunsenges. Phys. Chem.*, 1350–1356, (1989) **93**, © 1989 with permission from Deutsche Bunsen-Gesellschaft für Physikalische Chemie.

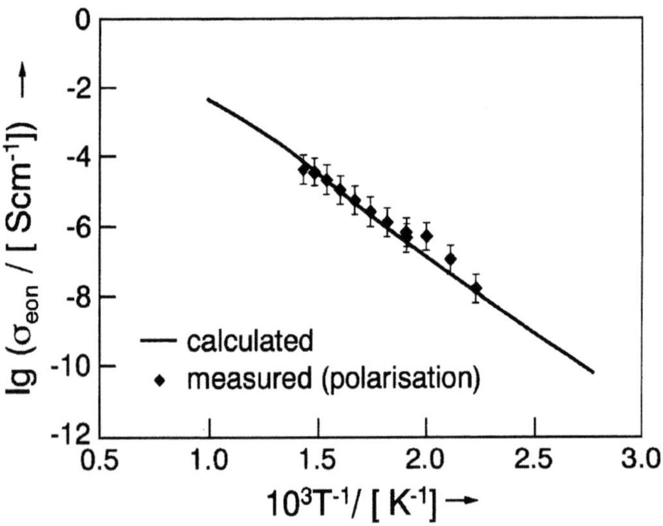

Figure 45. The steady-state voltage of the ion blocking experiment on $SrTiO_3$ allows one to determine the partial electronic conductivity. The experiments succeed without encapsulation, simply on the basis of inhibited exchange kinetics (in contrast to Figures 44 and 46).[11,239] Reprinted from I. Denk, W. Münch, and J. Maier, *J. Am. Ceram. Soc.*, **78** (1995) 3265–3272. Copyright © 1995 with permission from "The American Ceramic Society".

The polarization methods are generally more tricky to apply for conducting oxides than it is, e.g., for silver conducting silver halides, as the blocking of oxygen exchange requires a careful cell design. As in the case of asymmetric Wagner–Hebb cells the use of a small exhaustible gas volume at the minus pole (flushed with Ar) instead of a blocking electrode results in a suppression of an oxygen ion current, too, since in the steady-state oxygen cannot be supplied[232] (see figure above). As, however, now the exchange processes at the boundaries are reversible, the boundary condition of a zero concentration gradient, which usually allows for a straightforward evaluation of D^{δ} from the instationary behavior, is not fulfilled here and the evaluation more complicated. For that reason the application of two-sided blocking cells have been more popular in the case of oxides.[217,238,245] A serious experimental complication consists in the following point: Unlike in the case of metal ion conductors, the lateral parts of the sample have to be protected against oxygen access as well, unless one uses extreme aspect ratios. Otherwise too high apparent steady-state conductivities are measured as verified in the case of $Bi_2Sr_2CaCu_2O_{8+\delta}$ (Figure 46).[246]

An elegant variant of the method has been used for $SrTiO_3$ where all these problems have been overcome by working in the kinetic regime in which the electrode and surface reactions are completely hindered.[247] Special electrodes as well as special protecting coatings are not necessary (see Figure 45). For more details the reader is referred to the Ref.[248]

As already mentioned, the formal transition to the frequency domain does not offer problems, yet in most cases the experimental application of impedance spectroscopy for evaluating the stoichiometric polarization is not of great worth as the frequencies used in the impedance experiments (as shown in Figure 44) are too high to observe the diffusion. Nonetheless, in such cases a combination of dc and ac methods is helpful, as the impedance branches which include bulk and boundary effects of ZrO_2 can be used to correct the steady-state value of the dc experiment (steady state: $R_{ion} + \Sigma R_{ZrO_2}$; initial value: $R_{\infty} + \Sigma R_{ZrO_2}$).[249] Only in cases of very short stoichiometric relaxation times (see Section III.3.iii) the polarization can be followed by impedance or even a dc limit extracted from it; Figure 37 gave such an example.

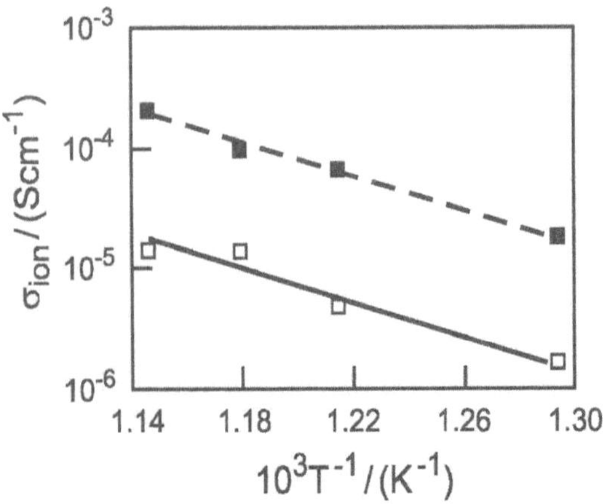

Figure 46. The steady-state voltage of an electron-blocking experiment on $Bi_2Sr_2CaCu_2O_{8+\delta}$ (*upper curve*: nonsealed cell; *lower curve*: sealed cell) shows the necessity of a lateral coating (only the *lower curve* corresponds to true ionic conductivities).[246] Reprinted from M. Quilitz and J. Maier, *J. Superconductivity*, **9** (1996) 121–127. Copyright © 1996 with permission from Kluwer Academic Publishers (Springer).

(iii) The Instationary Behavior: The Evaluation of the Chemical Diffusion Coefficient

Now let us discuss the evaluation of the chemical diffusion constant D^δ from the transient behavior.[3,58,210] For this discussion we assume sufficiently small signals, as to be able to neglect stoichiometry dependencies and thus positional dependencies of the transport parameters $\sigma_{\rm eon}, \sigma_{\rm ion}$, and D^δ.

As basic ambipolar theory[210,225] shows (cf. Part I[2], Section VI.3.iv), in the interior of the sample the flux densities $j_{\rm k}$ are a combination of an ohmic term and a stoichiometry term. In the case of a mixed (O^{2-}/e^-) conductor (in the absence of internal defect chemical reactions), $j_{\rm k}({\rm k = eon, ion})$ is given by

$$j_{\rm k} = -\frac{\sigma_{\rm k}}{\sigma}\frac{i}{2{\rm F}} - D^\delta \nabla c_{\rm k} \tag{69}$$

where

$$D^\delta = \frac{1}{4{\rm F}^2}\frac{\sigma_{\rm eon}\sigma_{\rm ion}}{\sigma}\frac{{\rm d}\mu_{\rm O}}{{\rm d}c_{\rm O}} \tag{70}$$

is composed of an inverse chemical resistance ($\propto \frac{\sigma_{\rm eon}\sigma_{\rm ion}}{\sigma}$) and an inverse chemical capacitance ($\propto {\rm d}\mu/{\rm d}c_{\rm O}$). The total current density is then

$$i = \frac{\sigma}{2\mathrm{F}}\nabla\tilde{\mu}_{\mathrm{O}^{2-}} - \frac{\sigma_{\mathrm{eon}}}{4\mathrm{F}}\nabla\mu_{\mathrm{O}_2} = \frac{\sigma}{\mathrm{F}}\nabla\tilde{\mu}_{\mathrm{e}^-} + \frac{\sigma_{\mathrm{O}^{2-}}}{4\mathrm{F}}\nabla\mu_{\mathrm{O}_2}. \tag{71}$$

According to Appendix 1 the cell voltage is determined by the electrochemical potentials of the electrons at the terminal leads under the above assumptions. Their difference is proportional to the difference of the $\tilde{\mu}_{\mathrm{e}}$ - values at $x = 0$ and $x = L$ up to constant resistance contributions stemming from processes in or at the electrode. This internal $\tilde{\mu}_{\mathrm{e}}$ difference is given by (cf. Eq. (71))

$$-\frac{\Delta\tilde{\mu}_{\mathrm{eon}}}{\mathrm{F}} = U_{\mathrm{eon}} = \int_0^L \mathrm{d}x\left\{\frac{i}{\sigma} - \frac{1}{4\mathrm{F}}\frac{\sigma_{\mathrm{ion}}}{\sigma}\frac{\mathrm{d}\mu_{\mathrm{O}_2}}{\mathrm{d}x}\right\}. \tag{72}$$

In the cases of the cells 5 and 6 the cell voltage is analogously related to U_{ion} (see Appendix 1), which is determined by the difference in the electrochemical potentials of the mobile ions at $x = 0$ and $x = L$

$$U_{\mathrm{ion}} = \int_0^L \mathrm{d}x\left\{\frac{1}{\sigma} + \frac{1}{4\mathrm{F}}\frac{\sigma_{\mathrm{eon}}}{\sigma}\frac{\mathrm{d}\mu_{\mathrm{O}_2}}{\mathrm{d}x}\right\}. \tag{73}$$

By substitution we can, for small signals, express the voltage in terms of concentration differences as

$$U_{\mathrm{eon,\,ion}}(t) = \frac{iL}{\sigma} \mp 2\mathrm{F}D_0^{\delta}\,\frac{c_{\mathrm{O}}(x = L, t) - c_{\mathrm{O}}(x = 0)}{\sigma_{\mathrm{eon,\,ion}}}. \tag{74}$$

This concentration difference can be obtained by solving Fick's second law under the appropriate boundary conditions of constant concentrations or concentration gradients. The initial values are determined by the constant concentration profile in the case of the polarization, and by the linear steady-state profile (linear in this simplest approximation) in the case of the depolarization. The relevant solutions are given in Appendix 2. The desired *U*(*t*) functions of the polarization cells 3 and 4 simplify for long and short terms to

$$U_{\mathrm{eon}} = \frac{i_{\mathrm{p}}L}{\sigma} + \begin{cases} \dfrac{\sigma_{\mathrm{ion}}}{\sigma}\dfrac{i_{\mathrm{p}}L}{\sigma_{\mathrm{eon}}}\left\{1 - \dfrac{8}{\pi^2}\exp\left[-\dfrac{t}{\tau_\delta}\right]\right\} & t \geq \tau_\delta/2 \\ \dfrac{\sigma_{\mathrm{ion}}}{\sigma}\dfrac{i_{\mathrm{p}}L}{\sigma_{\mathrm{eon}}}\dfrac{4}{\pi^{3/2}}\sqrt{\dfrac{t}{\tau_\delta}} & t \leq \tau_\delta/2 \end{cases} \tag{75}$$

with the steady-state value $V_{\mathrm{eon}} = i_{\mathrm{p}}L/\sigma_{\mathrm{eon}}$. The results for cells 5 and 6 are symmetrical to Eq. (75) in exchanging eon by ion and vice versa. The relaxation time τ_δ for cell 3 is $4L^2/(\pi^2 D^\delta)$ and $L^2/(\pi^2 D^\delta)$ for cell 4. In other words: concerning the instationary behavior the most essential difference of applying two blocking electrodes instead of one, is that the polarization occurs four times faster. The comparison with Eq. (64) reveals that—besides the appearance of $8/\pi^2$ instead of 1—for long times (i.e., finite boundary conditions) the $U(t)$ function indeed corresponds to the naive capacitor picture given above. As $\tau_\delta = C_\delta(R_{\text{eon}} + R_{\text{ion}})$, the chemical capacitance C_δ follows from Eq. (70) as

$$\frac{C_\delta}{A} = 4\frac{(4 \text{ or } 1 \times \mathrm{F}^2)}{\pi^2} L \left(\frac{\mathrm{d}\mu_\mathrm{O}}{\mathrm{d}c_\mathrm{O}}\right)^{-1}. \quad (76)$$

This materials-specific term is proportional to the inverse of the thermodynamic factor and measures the increase of particle number density with chemical potential (while the electrical capacitance measures the increase of charge with electrical potential). For short times at which the profile near one electrode does not yet perceive the influence of the second one, the result is a $\sqrt{t}$ -law, and obviously differs from the heuristic approach. Thus more correctly one has to replace C_δ by a Warburg-type capacitance as already discussed above (for a more exact description cf. Part I[2], Section VI.7). Figure 45 shows a kinetic analysis for $YBa_2Cu_3O_{6+x}$ for the short- and the long-time behavior in the time domain yielding identical D^δ values. (Note that in these figures different symbols have been used for D^δ.)

The steady-state voltage $V_{\text{eon,ion}} \left(= \frac{i_\mathrm{p} L}{\sigma_{\text{eon, ion}}}\right)$ refers to the steady-state voltage at the chemical capacitor (C_δ) and corresponds to the Nernst-value $(\Delta\tilde{\mu}_{\mathrm{e}^-} \propto \Delta\mu_{\mathrm{O}_2})$.

The depolarization behavior is derived in the same way (Appendix 2). Figure 47 shows both the polarization and depolarization transients of a blocking experiment on PbO (orh.) characterized by consistent diffusion values. As already stated, only

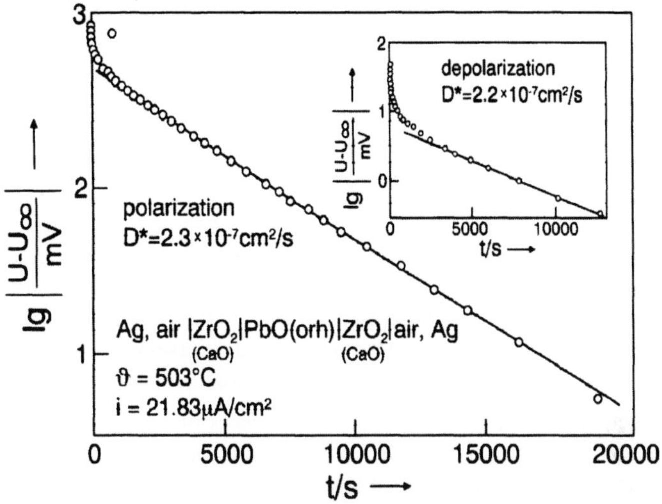

Figure 47. The time-dependence of both the polarization and the depolarization voltage (galvanostatic mode) of a symmetrical blocking cell applied to PbO (orh.) allows the determination of the chemical diffusion coefficient (here denoted by D^{*}).[15,217] Reprinted from J. Maier, Solid State Phenom., **39/40** (1994), 35–60. Copyright © 1994 with permission from Trans Tech Publications Ltd.

for samples with very short τ_δ (large D^δ, small L) the stoichiometric polarization can be conveniently observed by impedance spectroscopy. The Ag_2Te example in Figure 37 discussed in Section III.3.i fulfilled this requirement and also revealed the theoretical behavior for negligible space charge effects.

4. Chemically Imposed Gradients

(*i*) *Chemical Polarization and Concentration Cell Experiment*

The same stoichiometric gradients can also be generated by varying the gas composition in cell 1. Let us assume that two electrode chambers are isolated from each other and the oxygen partial pressure in one chamber (say at $x = L$) is suddenly changed. We also assume the electrode reactions and the surface exchange at the electrode contact to be very fast, while the uncontacted surfaces are sealed or inactive. The concentration profiles in the interior can then be calculated from Fick's second law with fixed concentrations on both sides as boundary conditions. The solutions are given in Appendix 3 (see also Ref.[250]). The transient may be followed, e.g., by conductance experiments (given appropriate mobilities) what again enables the determination of D^δ.

As the cell voltage is determined by the boundary values only, it is at all times given by the zero-current value of Eqs. (72) and (73). Using electronic probes the result is[209]

$$U = -\frac{\mathrm{R}T}{4\mathrm{F}}\langle\frac{\sigma_{\mathrm{ion}}}{\sigma}\rangle \ln\frac{P(x=L)}{P(x=0)}. \tag{77}$$

(Deviations from Eq. (77) that are due to surface effects should allow conclusions on effective surface rate constants). The more detailed derivation has been given in Section II.2 (Eq. (35)). In Eq. (77) the ionic transference number has been averaged over the P_{O_2}-range covered. A more precise evaluation involves the explicit calculation of the average value on the basis of a detailed defect model[251] (see Appendix 4). Then the domain boundaries of the P, I, N regime can be determined (see Section IV, Part I[2]). A further possibility of a more thorough analysis is to consider the change of the emf with changing P_{O_2} at $x = L$.

The steady-state result (Eq. (77)) can be directly used to separate ionic and electronic conductivities; the disadvantage of the technique is that it presupposes gas-separation. If not special measures are taken, it becomes unreliable for the ionic transport number less than ~1%. Thus, this method well complements the Wagner–Hebb method which is very sensitive to small transference numbers. The partial conductivities of PbO shown in Figure 48 have been de-convoluted by the emf technique just described.[3]

The heuristic equivalent to Eq. (77) can also be directly derived from the equivalent circuit $\mathrm{Par}(\mathrm{Ser}(R_{\mathrm{ion}}, C_\delta), R_{\mathrm{eon}})$ if the voltage at C_δ is identified with the Nernst-voltage.[3,15] Evidently

$$U = E - I_{\mathrm{ion}}R_{\mathrm{ion}} = E - \frac{ER_{\mathrm{ion}}}{R_{\mathrm{ion}} + R_{\mathrm{eon}}} = t_{\mathrm{ion}}E \tag{78}$$

with $E = \frac{\mathrm{R}T}{4\mathrm{F}}\ln\frac{P(L)}{P(0)}$. If the cell voltage is measured by ionic probes, analogously $U/E = t_{\mathrm{eon}}$ follows.

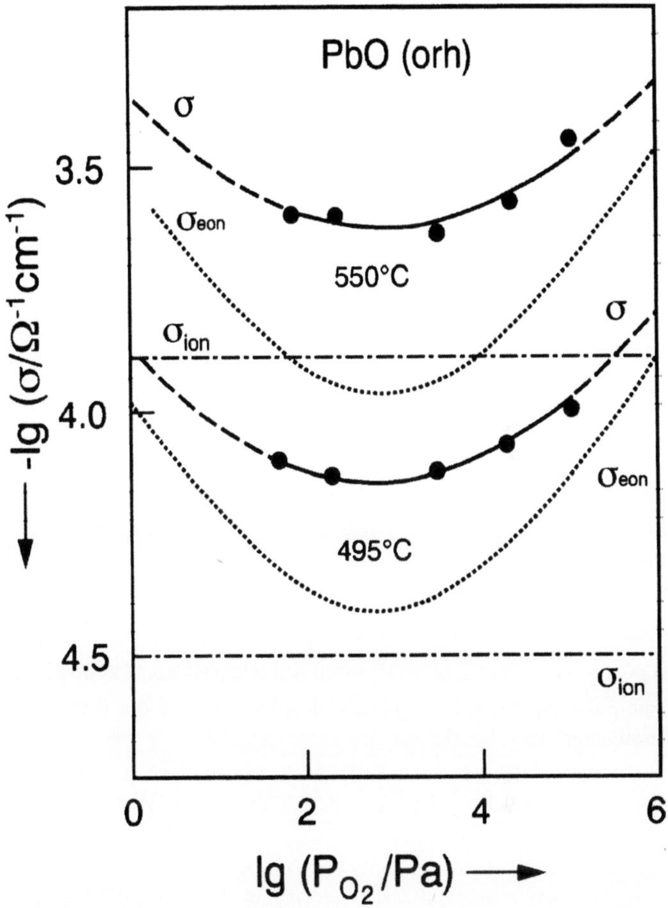

Figure 48. The concentration cell experiment together with impedance spectroscopy allows one to separate ionic (σ(O)) and electronic (σ(e)) partial conductivities of PbO (orh.).[252] If YSZ is used to separate the gas chambers (e.g., with glass seals) both U and E (i.e., $\frac{\mathrm{R}T}{4\mathrm{F}}\ln\frac{P(x=L)}{P(x=0)}$) can be measured directly.

Reprinted from J. Maier and G. Schwitzgebel, *Mater. Res. Bull.*, **17** (1982) 1061–1069. Copyright © 1982 with permission from Elsevier.

The symmetrical reversal of the above chemical polarization experiment is the sudden replacement of the oxygen partial pressure on $x = L$ by the original one: Then the concentration profiles decay by chemical depolarization (allowing again for a measurement of D^δ). If, however, the gas flow at that gas chamber is just stopped, the gas pressure equilibration takes more time in that it not only affects the sample volume but also the volume of the gas chamber. This directly connects with the permeation technique described in the next section.

(*ii*) *Oxygen Permeation*

The permeation technique as discussed in the context of permeation membranes in Section II.2, can also be used to determine diffusion coefficients and minority conductivities (in fact $\sigma^\delta = \sigma_{\text{ion}}\sigma_{\text{eon}}/\sigma$), $D^\delta \propto \sigma^\delta / C_\delta$ by evaluating the transients, σ^δ by measuring the steady-state oxygen flux that is given by (see Eq. (34))

$$j_{\text{O}} \propto \frac{1}{L}\int_{\text{cell}} D^\delta \mathrm{d}c_{\text{O}} \propto \frac{1}{L}\int_{\text{cell}} \frac{\sigma_{\text{eon}}\sigma_{\text{ion}}}{\sigma}\mathrm{d}\mu_{\text{O}}. \tag{79}$$

As the oxygen partial pressure ratio, and hence $\Delta\mu_{\text{O}}$, is known, the ambipolar conductivity is readily determined from the flux. This knowledge can be further used to calculate the partial conductivities, and by knowing D^δ from the transient (i.e., by also evaluating the delay time[231]) to derive the thermodynamic factor (i.e., the chemical capacitance).

As described in Section II.2 variants of the permeation techniques use systems in which electronic and ionic pathways are locally separated, e.g., an ion conductor which is internally short-circuited by percolating metallic inclusions (or materials in which grain boundary and bulk possess different conductivity types (e.g., nano-CeO_2)). In all these cases the permeation flux is determined by the lowest partial conductivity.

(*iii*) *Zero-Driving Force Method*

A related method to separate electronic and ionic conductivities has been extensively discussed by Riess.[253] As in the previous section a stationary chemical gradient is imposed, here, however, the electronic current flows via external leads. The two electrodes are now short-circuited by the outer circuit and the short-circuit current is measured. The very high electronic conductivity of the outside part of the circuit, nullifies the electrochemical potential difference of the electrons $(i = i_{\text{eon}} \propto \sigma_{\text{eon}} \nabla \tilde{\mu}_{\text{eon}})$. Then, there is also no driving force for the internal electronic current. Hence, i_{eon} in the sample is zero and only an ionic current flows through it. This ionic current is proportional to both σ_{ion} and the chemical potential gradient $(\nabla \mu_{\text{O}} \simeq \nabla \tilde{\mu}_{\text{O}^{2-}} - 2\nabla \tilde{\mu}_{\text{eon}} = \nabla \tilde{\mu}_{\text{O}^{2-}} \propto 1/2 \nabla \ln P_{\text{O}_2})$ allowing σ_{ion} to be determined from i and $P_{\text{O}_2}(x=0)/P_{\text{O}_2}(x=L)$. It is instructive to compare this to a blocking experiment for which i_{eon} disappears due to the boundary conditions (but $\sigma_{\text{eon}} \neq 0$) or to a solid electrolyte cell in which the electronic current is zero as a consequence of $\sigma_{\text{eon}} = 0$ (but $\nabla \tilde{\mu}_{\text{eon}} \neq 0$).

(iv) *Chemical Relaxation*

Even though D_0^δ can be determined from the transients of a chemical or electrochemical (de-)polarization (see Section III.4.i), it is usually simpler to employ a chemical relaxation,[254] i.e., to chemically change from one homogeneous situation $(P = P_1)$ to the next homogeneous situation $(P = P_2)$. Consider again cell 1 in equilibrium with P_1 where σ has been determined $(\sigma(P_1))$, and change suddenly the oxygen partial pressure to P_2. The transient into the new equilibrium state can be followed by a signal that is an unambiguous function of the defect concentration, usually it is proportional to it. Examples are the time-change of the overall conductance parallel or in series to the mass-flow

$$R^{-1\,\parallel} \propto \int c(x)\mathrm{d}x \text{ or} \tag{80a}$$

$$R^{\perp} \propto \int c^{-1}(\mathrm{x})\mathrm{d}x. \tag{80b}$$

The analytical solution follows from Fick's second law for the boundary conditions $c(x=0,t) = c(x=L,t) = c_2$ and the initial

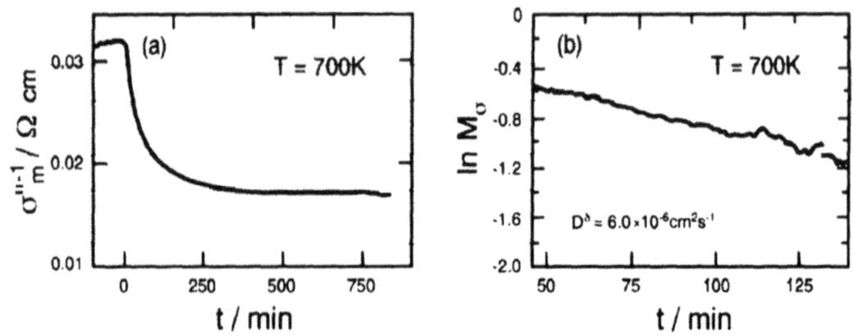

Figure 49. The mean resistivity response on a sudden partial pressure step yields the chemical diffusion coefficient of $YBa_2Cu_3O_{6+x}$ over the corresponding P_{O_2} range. For a more defined activity correlation see Ref.[15,244,255] Reprinted from J. Maier, Solid State Phenom., **39/40** (1994) 35–60. Copyright © 1994 with permission from Trans Tech Publications Ltd.

condition $c(x,t=0)=c_1$. Appendix 3 gives the detailed solution also for the case of surface hindrance.[250] The value of the chemical relaxation consists in its simplicity and the possibility to use a variety of detection methods such as optical absorption (note: the integral absorbance is proportional to $\int c dx$ if the light beam is perpendicular to the interface), EPR, NMR, IR, gravimetry, etc.

An example of an electrically followed chemical relaxation is displayed by Figure 49. According to the diffusion laws (Appendix 3), for long times the mean resistivity follows an exponential behavior, the time constant of which (slope of the logarithmic representation) is given by D^δ and L. Such an integral technique can also be evaluated in the mixed kinetic regime where both surface and bulk effects constitute. However, the deconvolution of surface and bulk processes is not always unambiguous, while in the case of a spatial resolution, as achieved by the optical technique (see Figure 50), this distinction is very clear.

Figure 50 shows three examples for (different) kinetic situations in which bulk diffusion, surface reaction, and transport across a grain boundary are the sluggish steps. Nonetheless, the other parameters can also be evaluated. This becomes especially clear from the top figure, where the nonunity intercepts reveal surface effects. Similarly, the nonzero bending of the profiles in the other two figures indicates transport resistances.

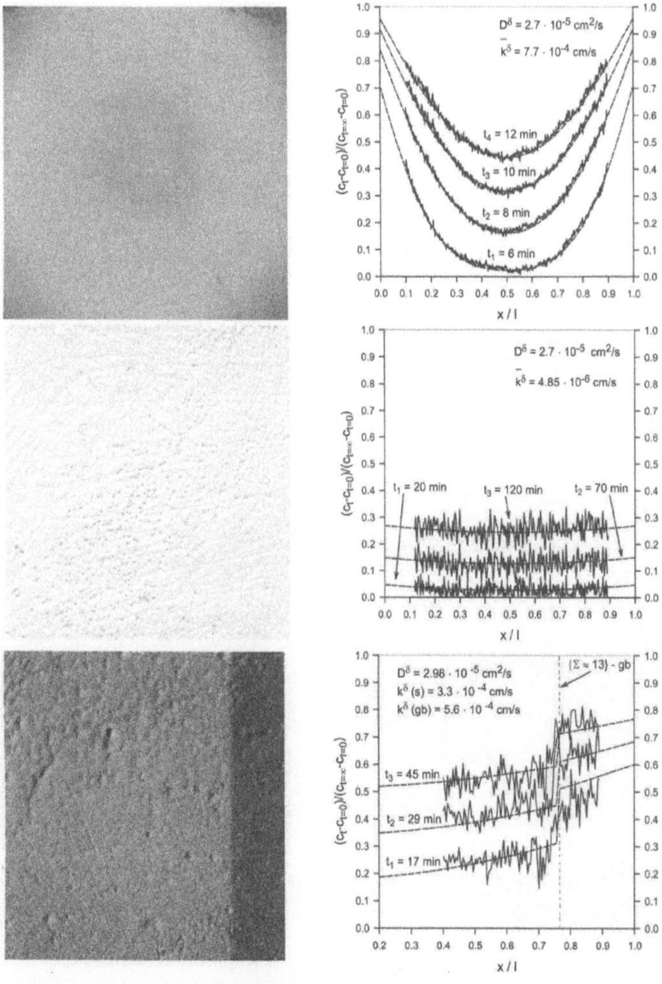

Figure 50. Snapshots of oxygen incorporation experiments in Fe-doped $SrTiO_3$, recorded by in situ time and space resolved optical absorption spectroscopy.[256] Rhs column refers to the corresponding oxygen concentration profiles, in a normalized representation. *Top row* refers a predominantly diffusion controlled case (single crystal), *center row* to a predominantly surface reaction controlled case (single crystal), *bottom row* to transport across depletion layers at a bicrystal interface.[257,258] For more details on temperature, partial pressure, doping content, structure see Part I and Ref.[257–259] Reprinted from J. Maier, *Solid State Ionics*, **135** (2000) 575–588. Copyright © 2000 with permission from Elsevier.

A major advantage of the chemical relaxation method is that a successive change of the partial pressure allows the detection of D^{δ} as well as of the properties measured in the homogeneous states (e.g., $\sigma_{eon}/\sigma_{ion}$) as a function of composition (or P_{O_2}). Such a step-by-step variation of the composition can also be achieved electrically by Coulometric titration.

5. Coulometric Titration

In this technique the stoichiometry is changed by a current flow through cell 7. The current through the oxygen ion conductor is necessarily an O^{2-} flow which is converted into an electronic flow already within MO or at the latest at the MO/C boundary. This is connected with the incorporation or excorporation of neutral oxygen. There is of course no stationary case and the corresponding impedance spectrum corresponds to a capacitive behavior (no dc intercept). By interrupting the current after a given time period, the stoichiometry of MO[209,260] can be very sensitively varied, and measured by recording the open circuit potential. The emf is zero if the O-content corresponds to phase equilibrium with the outer O_2 partial pressure (maximum O content):

$$\begin{aligned}&\mathrm{emf}\left\{\mathrm{C}\middle|\mathrm{MO}_{1+\delta_{\max}}\middle|\mathrm{YSZ}\middle|\mathrm{Pt},\mathrm{O}_2(P_1)\right\}\\&=\mathrm{emf}\left\{\mathrm{O}_2(P_1),\mathrm{Pt}\middle|\mathrm{MO}_{1+\delta_{\max}}\middle|\mathrm{YSZ}\middle|\mathrm{Pt},\mathrm{O}_2(P_1)\right\}=0\end{aligned} \tag{81}$$

It is tacitly assumed that there is no higher oxide stable at this temperature. If the minimum O-content is reached, M is about to be formed (or, what we ignore here, alternatively a lower oxide) and the emf is given by the free formation enthalpy of the oxide (see Chapter 2)

$$\begin{aligned}&\mathrm{emf}\left\{\mathrm{C}\middle|\mathrm{MO}_{1+\delta_{\min}}\middle|\mathrm{YSZ}\,|\,\mathrm{Pt},\mathrm{O}_2(P_1)\right\}\\&=\mathrm{emf}\left\{\mathrm{M}\middle|\mathrm{MO}_{1+\delta_{\min}}\middle|\mathrm{YSZ}\middle|\mathrm{Pt},\mathrm{O}_2(P_1)\right\}\propto\Delta_f\mathrm{G}(P_1)\end{aligned} \tag{82}$$

The technique obviously allows one to measure the stoichiometry range and also the free formation enthalpy. The actual non-stoichiometry within the phase-width can be directly calculated from the defect model. As discussed in Part I, for a simple ionic disorder the nonstoichiometry δ is a sinh-function in the difference of the chemical component potential to the value of the stoichiometric point ($\delta = 0$) (see Part I,[2] Section IV). Owing to

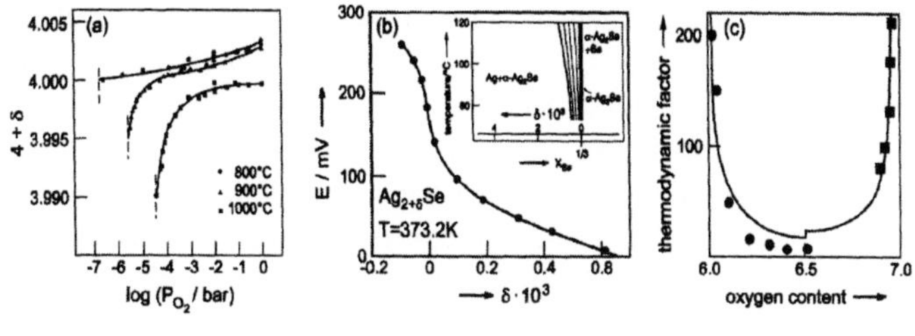

Figure 51. (a) Dependence of oxygen content in $La_{1.95}Sr_{0.05}CuO_{4+\delta}$ on $\log P_{O_2}$ at 800 … 1000°C. The *broken lines* represent the decomposition partial pressures.[261] Reprinted from H. Kanai, J. Mizusaki, H. Tagawa, S. Hoshiyama, K. Hirano, K. Fujita, M. Tezuka and T. Hashimoto, *J. Solid State Chem.*, **131** (1997) 150–159. Copyright © 1997 with permission from Elsevier. (b) Coulometric titration of α-Ag_2Se and derived phase width.[262] Reprinted from U. von Oehsen and H. Schmalzried, *Ber. Bunsenges. Phys. Chem.*, **85** (1981), 7–14. Copyright © 1981 with permission from Deutsche Bunsen-Gesellschaft für Physikalische Chemie. (c) A simple statistic model for the thermodynamic factor (w) of $YBa_2Cu_3O_{6+x}$ (*solid line*) partly obtained by Coulometric titration.[263,264] The results are in good agreement with predictions from a simple statistical model.[265,266] In the tetragonal phase ($x<6.5$)$w=2(6+x)/(x(2-x))$ and in the orthorhombic phase ($x\geq 6.5$)$w=(6+x)/x(1-x)$. From Ref.[266]

Nernst equation this then also holds for the variation in the emf. The stoichiometric point can therefore be calculated from the inflection point of the emf versus δ curve. Since current and time can be measured very precisely, at least ppm-resolution is easily accessible. According to Nernst's equation the slope of the emf versus δ curve represents the thermodynamic factor which together with the ambipolar conductivity (available from σ_{eon} and σ_{ion} or directly from the permeation technique) constitutes the chemical diffusion coefficient. Figure 51 refers to results obtained for $La_2CuO_4, YBa_2Cu_3O_{6+x}$ and Ag_2Se. In all cases the fine-structure of the phase diagram can be evaluated and so can the activity dependence of the composition within the phase. In the case of $YBa_2Cu_3O_{6+x}$ (Figure 51c) the thermodynamic factor[263,264] could even be explained by a simple statistical model[265] as in this compound the oxygen excess is so high that oxygen interstitials and holes can be considered to be largely associated.[266]

If the boundary conditions are well-defined—which is not always the case here—the transient behavior of the Coulometric titration, i.e., emf(t) allows one to determine D^{δ} directly. Appendix 5 details the time behavior of the voltage depending on the transference numbers of ions and electrons in the central phase as given by Preis und Sitte[267] (Appendix 5).

In multinary compounds not only the chemical potential of an elementary component,[268] but also the chemical potential of a binary component can be varied. One example is the variation of the Na_2O activity in $Na_2Ti_6O_{13}$ as described in Ref.[269]

As already mentioned in the beginning of this text (almost) every experimental technique described here has its technological counterpart. In the case of Coulometric Titration, this is the intercalation process in secondary electrodes treated in Section II.3.ii. The technological counterpart of what remains to be discussed in the next section, are the emf sensors. Since we dealt with general aspects on equilibrium cells quite extensively already in the application part (Section II.3.i), we will restrict ourselves to only very few remarks.

6. Thermodynamic Data from Electrochemical Cells Involving Solid Electrolytes

As pointed out in Section II.3.ii, the key principle of the galvanic cells relies on chemical asymmetry and a zero σ_{eon} in the electrolyte rendering possible the fact that there can be an electrochemical potential gradient, i.e., a nonzero cell voltage without having a current.

Emf cells are standard tools to elucidate thermodynamic formation data. Table 9 gives various examples (the underlined compounds are the ones for which thermodynamic data have been measured).

In order for an emf to be completely defined, the phase rule must be obeyed for each electrode side. Cells involving ideal solid electrolytes ($t_{ion} = 1$) can be usually measured over a wider temperature range and are hence convenient means to deduce thermodynamic data such as formation enthalpies and formation entropies. The cell given in Eq. (82)

$$\mathrm{M}|\mathrm{MO}|\mathrm{YSZ}|\mathrm{Pt}, \mathrm{O}_2\left(P_{\mathrm{O}_2}\right) \tag{83}$$

Table 9.
Examples of equilibrium cells which provided thermodynamic data of the compounds underlined.[16,270,271]

$Mg, MgF_2 \mid CaF_2 \mid \underline{ThF_4}, Th,$	$\underline{(Ni, Mn)O}, Ni \mid ThO_2 (+Y_2O_3) \mid Ni, NiO$
$Th, ThF_4 \mid CaF_2 \mid \underline{AlF_3}, Al,$	$Co, SiO_2, \underline{Co_2SiO_4} \mid ZrO_2 (+CaO) \mid Co, CoO$
$U, \underline{UF_3} \mid CaF_2 \mid AlF_3, Al,$	$Co, Al_2O_3, \underline{(Co, Mg)Al_2O_4} \mid ZrO_2 (+CaO) \mid Co, Al_2O_3, CoAl_2O_4$
$Th, ThF_4 \mid CaF_2 \mid \underline{NiF_2}, Ni,$	$Ni, NiO \mid ZrO_2 (+CaO) \mid \underline{(Cu, Ni)}, NiO$
$Al, AlF_3 \mid CaF_2 \mid \underline{PbF_2}, Pb,$	$Ag \mid AgI \mid \underline{(Ag, Te)}$

By an independent measurement of the specific heat over a wide T-range down to very low temperatures, a complete data set can be obtained for all the temperatures of interest. If the sample for which the thermodynamics is to be studied is a pure ionic conductor itself, a most simple cell can be used, such as

$$Ag|AgCl|Pt, Cl_2 \tag{84}$$

in the case of AgCl. In Section II.1.iii we considered a CO_2 sensor cell, which—when P_{CO_2} is known—can be used to elucidate the thermodynamic data of ternary oxides such as Na_2ZrO_3 (cf. Ref.[42]). The cell reads

$$NaZrO_3, ZrO_2, Au|\text{Na-}\beta\text{-alumina}|Au, CO_2, Na_2CO_3 \tag{85}$$

with the cell reaction

$$Na_2CO_3 + ZrO_2 \rightleftharpoons Na_2ZrO_3 + CO_2 \tag{86}$$

from which $\Delta_f G^\circ_{Na_2ZrO_3}$ can be obtained. If reversibility of the electrode reaction is guaranteed, also more complex conditions can be analyzed such as the mixture thermodynamics of a highly multinary system such as Nasicon ($Na_{1+x}Zr_2P_{3-x}Si_xO_{12}$, cf. Figure 52), or the excess free energy of nanocrystalline constituents.[272–275]

Exploiting the reversibility of the lhs electrode, the cells

$$O_2, Na_2Ti_6O_{13}, TiO_2(\text{rutile})|\beta\text{-alumina}|TiO_2(\text{anatase}), Na_2Ti_6O_{13}, O_2 \tag{87}$$

$$O_2, Na_2Ti_6O_{13}, TiO_2(\text{rutile}, r=\infty)|\beta\text{-alumina}|TiO_2(\text{rutile}, r), Na_2Ti_6O_{13}, O_2 \tag{88}$$

can be used to study the stability rutile versus anatase as a function of grain size (r) and thus, e.g., to evaluate the interfacial energies and entropies.

Also the thermodynamics of nanocrystalline silver have been studied by cells of the type

$$Ag(\text{nano})|AgCl|Ag\ (\text{micro}). \tag{89}$$

Here, however, unlike binaries or multinaries, the emf is inherently unstable[273,274] as the same processes allowing for an excess emf to be measured are also allowing for sintering and grain growth (electrochemical Ostwald ripening[275,273]). In addition, there has to be paid attention to the fact that the emf refers to the excess chemical potential of silver directly at the electrode/electrolyte interface.[273]

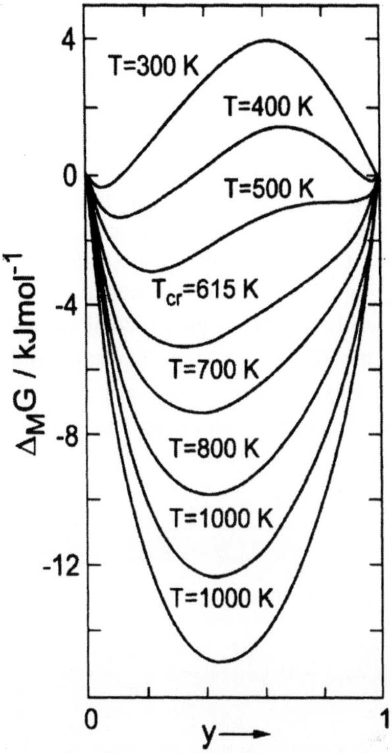

Figure 52. The free enthalpy of mixing of the fast ion conductor Nasicon ($Na_{1+x}Zr_2P_{3-x}Si_xO_{12}$) ($y = x/3$) obtained from emf measurements as a function of composition.[272] The shape of the curves indicates the stability of a thermodynamic two phase system at $T < 615$ K. Reprinted from U. Warhus, J. Maier, and A. Rabenau, *J. Solid St. Chem.*, **72** (1988) 113–125. Copyright © 1988 with permission from Elsevier.

In extension of the previous sections dealing with transport properties, the following chapters will consider more complex cases, in particular complications in defect chemistry, geometry, microstructure, and spatial distribution.

7. Modifications in the Evaluation of Electrochemical Measurements Due to Internal Defect Reactions

The above sections assumed the presence of one electronic and one ionic carrier. Ref.[3,15,59,226] considered explicitly the implications of a more general defect chemistry in which several types of carriers are in local annihilation and generation equilibrium such as the change of the valence states of ionic carriers by association with electronic carriers. (The case that the internal reactions are slow is not considered; in these cases the rate constants of these reactions would appear explicitly.) In such a qualitative case, i.e., structural, changes of the evaluation formulas occur. This is by far not only an academic exercise, as many solids have been shown to be seriously affected by such complications. One example is the occurrence of O_i'', O_i', O_i^x, all with different mobilities, in $YBa_2Cu_3O_{6+x}$, another is the occurrence of different mobile hydrogen species $H_i^{\cdot}, H_i^x, H_i'$ in proton conductors. Let us consider the presence of mobile O^{2-}, O^- and O° centers and concentrate on the time-domain (the frequency behavior follows from Laplace transformation).[3,15] Summing up the individual flux–force relations, the total oxygen flux reads[3,59,226]

$$j_O = -\frac{\sigma_{O^{2-}} + 2\sigma_{O^-}}{\sigma}\frac{i}{2F} - D_O^\delta \frac{\partial}{\partial x} c_O \tag{90}$$

with D^δ being now given as

$$D_O^\delta = \frac{1}{4F^2}\left[2\sigma_{O^-} + 4s + \frac{\left(\sigma_{eon} - \sigma_{O^-}\right)\left(\sigma_{O^{2-}} + 2\sigma_{O^-}\right)}{\sigma}\right]\frac{d\mu_O}{dc_O}. \tag{91}$$

In Eqs. (90) and (91), $\sigma_{O^{2-}} = \sigma_{V_O^{\cdot\cdot}} + \sigma_{O_i''}; \sigma_{O^-} = \sigma_{V_O^{\cdot}} + \sigma_{O_i'}$, $s \propto D_{V_O^x} + D_{O_i^x}$; the thermodynamic factor ($\propto d\mu_O/dc_O$), too, is different from the simple case and is given by

$$\frac{d\mu_O}{dc_O} = RT\left(\frac{\chi_{O_i''}}{c_{O_i''}} + 4\frac{\chi_{h^\cdot}}{c_{h^\cdot}}\right) = RT\left(\frac{\chi_{V_O^{\cdot\cdot}}}{c_{V_O^{\cdot\cdot}}} + 4\frac{\chi_{e'}}{c_{e'}}\right). \tag{92}$$

The χ_k-terms refer to differential defect fractions defined via the corresponding conservative ensembles according to

$$\chi_k = \frac{\partial c_k}{\partial c_{\{k\}}} \tag{93}$$

($c_{\{k\}} = c_{\{O\}}$ or $c_{\{e\}}$ in the case of the excess particles (O_i'', e'), or $-c_{\{O\}}$ and $-c_{\{e\}}$ in the case of missing $(V_O^{\cdot\cdot}, h^\cdot)$ particles; $c_{\{O\}} = -c_{\{e\}}/2$ is proportional to the oxygen excess). These χ-parameters can be calculated from the defect chemistry.

If the carriers used for the explicit formulation in Eq. (92), (O_i'', e'), approximately represent the total ionic and electronic carrier concentrations (i.e., weak association), the χ-terms can obviously be considered to be activity corrections.

If we exclude associates totally, the result

$$\chi_{\mathrm{O}_i''}^{-1} = \frac{\partial c_{\mathrm{O}_i''} - \partial c_{\mathrm{V_{\ddot{O}}}}}{\partial c_{\mathrm{O}_i''}} = 1 - \frac{\partial c_{\mathrm{V_{\ddot{O}}}}}{\partial c_{\mathrm{O}_i''}} = 1 - \frac{c_{\mathrm{V_{\ddot{O}}}}}{c_{\mathrm{O}_i''}} \frac{\partial \ln c_{\mathrm{V_{\ddot{O}}}}}{\partial \ln c_{\mathrm{O}_i''}}, \tag{94}$$

leads, because of $\partial \ln c_{\mathrm{V_{\ddot{O}}}} = -\partial \ln c_{\mathrm{O}_i''}$ (on account of the Frenkel equilibrium), to the simple relation (but already more general than Eq. (70))

$$\chi_{\mathrm{O}_i''} / c_{\mathrm{O}_i''} = \frac{1}{\left(c_{\mathrm{O}_i''} + c_{\mathrm{V_{\ddot{O}}}}\right)}. \tag{95}$$

In the presence of reversible trapping, according to the altered conductivity expression in Eq. (91), there can now be a chemical diffusion of oxygen even if $\sigma_{\mathrm{eon}} = 0$, simply by a neutral oxygen flux but also by a counterflux of O^{2-} and $2\mathrm{O}^-$. The modifications referring to the $\mathrm{d}\mu_\mathrm{O}/\mathrm{d}c_\mathrm{O}$ term have been shown to have extremely strong influence on D^δ [59,276] (see also Part I). In $\mathrm{SrTiO_3}$ and YSZ contaminated with redox-active impurities the relevant internal interactions are dynamical trapping effects due to these impurities which lead to strong increase of the chemical capacitance and hence decrease of D^δ corresponding to this internal buffer effect.[277] Note that this internal buffer effect, which is shown in $\mathrm{d}\mu_\mathrm{O}/\mathrm{d}c_\mathrm{O}$, is not a trivial association or doping effect. It is even present if σ^δ is not affected (as for YSZ). While these points have already been considered in Part I,[2] we now turn to the implications for the evaluation formulas of electrochemical techniques (see Section VI.4.ii in Part I and Ref.[3,15,226,276] for more details).

It can be easily shown that the analysis of transients differs in the short-term behavior but not in the case of long-term approximation, where D^δ can be obtained from the slope as before but of course has a different meaning.

As regards the steady state of the electrochemical polarization, the internal cell voltage (cf. Eqs. (72) and (73)) for cells 3, 4 and cells 5, 6, respectively, has to be generalized to

$$U_{\mathrm{eon}} = \int_0^L \mathrm{d}x \left\{ \frac{i}{\sigma} - \frac{1}{4\mathrm{F}} \frac{\sigma_{\mathrm{O}^{2-}} + 2\sigma_{\mathrm{O}^-}}{\sigma} \frac{\mathrm{d}\mu_{\mathrm{O}_2}}{\mathrm{d}x} \right\} \tag{96a}$$

$$U_{\text{ion}} = \int_0^L \mathrm{d}x \left\{ \frac{i}{\sigma} + \frac{1}{4\mathrm{F}} \frac{\sigma_{\mathrm{e}^-} - \sigma_{\mathrm{O}^-}}{\sigma} \frac{\mathrm{d}\mu_{\mathrm{O}_2}}{\mathrm{d}x} \right\}. \tag{96b}$$

The decisive result for a steady-state stoichiometry polarization, instead of Eq. (67) and its ionic anologue,[3,59] is now

$$\partial U/\partial i = \left\{ \begin{matrix} L\dfrac{1-2Bh_{\{\text{eon}\}}}{\sigma_{\{\text{eon}\}}} & \text{(cells 3,4)} \\ L\dfrac{1-2Bh_{\{\text{ion}\}}}{\sigma_{\{\text{ion}\}}} & \text{(cells 5,6)} \end{matrix} \right\} \tag{97}$$

with $\sigma_{\{e\}} \equiv \sigma_{\text{eon}} - \sigma_{\mathrm{O}^-}, \sigma_{\text{ion}} \equiv \sigma_{\text{ion}} + 2\sigma_{\mathrm{O}^-}$; $B \equiv (1 + 2h_{\{\text{eon}\}} + 2h_{\{\text{ion}\}})^{-1}$, $h_{\{\text{ion}\}} \equiv (\sigma_{\mathrm{O}^-} + 2s)/\sigma_{\{\text{ion}\}}, h_{\{\text{eon}\}} \equiv (\sigma_{\mathrm{O}^-} + 2s)/\sigma_{\{\text{eon}\}}$. One obvious consequence of Eq. (97) is the disappearance of the polarization for large concentrations and diffusivities of neutral oxygen defects (if $s \gg \sigma_{\mathrm{O}^-}, \sigma_{\mathrm{O}^{2-}}$, then $\partial U/\partial i = L/\sigma$); another is that an O^{2-} flux accompanied by a counterflux of O^- or neutral oxygen is equivalent to an effective electronic flow, and an internal short circuit becomes possible without an explicit electronic current.

8. Dynamic Interactions

In Part I (Section VI.4.i), the consequences of a perceptible relaxation response[278–282] of the charge carriers on the jump of a central carrier have been discussed. This obviously leads to a frequency-dependent bulk conductivity. The higher the frequency the more jumps are counted successfully while for dc conditions only the long-range transport is important. Figure 53 gives the frequency-dependent conductivity of $\mathrm{RbAg_4I_5}$. In the $\hat{Z}$-plot the bulk semicircle appearing at high frequencies (see Section III.1) would be significantly depressed. Such dynamic inhomogeneities typically occur in disordered crystals or in glasses. In addition to dynamic inhomogeneities static inhomogeneities occur there and lead to similar effects.[283] In the latter case, and in particular in the case of heterogeneities, percolation effects are very important.[284]

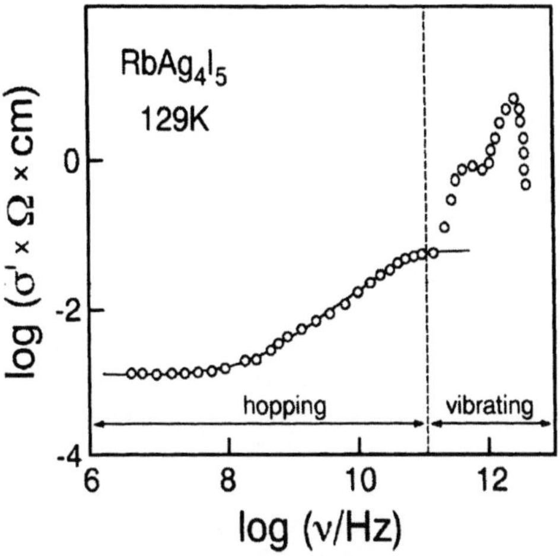

Figure 53. Frequency-dependent conductivity of $RbAg_4I_5$ at 129 K. (More precisely σ' represents the real part of the complex conductivity.) As the *continuous line* shows, the jump–relaxation model in Ref.[278–280] can well describe the behavior in the hopping regime.[278] Reprinted from K. Funke, *Prog. Solid St. Chem.*, **22** (1993) 111–195. Copyright © 1993 with permission from Elsevier.

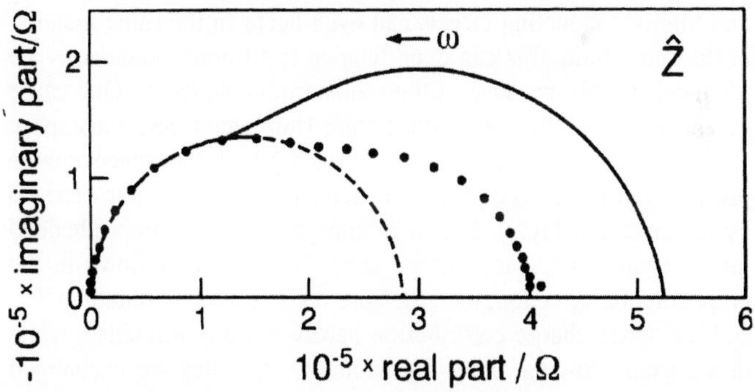

Figure 54. Impedance spectra of an AgBr bicrystal[285] with the current direction normal to the boundary. Reprinted from J. Maier, *Ber. Bunsenges. Phys. Chem.*, **90** (1986) 26–33. Copyright © 1986 with permission from Deutsche Bunsen-Gesellschaft für Physikalische Chemie.

9. Transport in Inhomogeneous, Heterogeneous, and Composite Systems

We mentioned the effect of atomistic inhomogeneities (corresponding to inhomogeneities in the local relaxation times) on impedance. In this section, we shall take a serious look at the practically important case of inhomogeneities and heterogeneities on a larger scale (e.g., microstructural effects). While a continuous distribution of relaxation time is reflected by a depression of impedance arcs, regions of distinctly different relaxation times being connected in series lead to more or less separated arcs, such as in the case of a bicrystal with a highly resistive boundary perpendicular to the flux density.

Figure 54 shows a two-point experiment on an AgBr bicrystal using Ag electrodes. The low-frequency response is solely due to the internal interface. (Figure 35 shows another example for $SrTiO_3$.) The distinction with regard to electrode impedance is possible via four point measurements but also via an evaluation of the capacitance as described in Section III.2.

If grain boundaries constitute high conduction pathways their deconvolution is more difficult. An elegant possibility is offered by the microelectrode impedance technique described below; another is the grain size variation. In polycrystalline materials, blocking and highly conducting effects can even occur in the same material at the same time; this can even happen if all grain boundaries are identical, simply because of their anisotropic character. One cause of anisotropy is the composite nature (interfacial core plus space charge regions), another is the anisotropy of the space charge profile itself (in particular if the space charge zone is characterized by an inversion layer). Let us assume a resistive core embedded into a conductive space charge zone. If the current flows in the perpendicular direction, then the core resistance is dominating $\hat{Z}^{\perp}$, and the space charge contribution hardly has any influence, while in a current flowing along the boundary the roles are exchanged ($\hat{Z}^{\parallel}$ is dominated by space charge effects). Note also that the space charge contributions are different as in the first case, the more resistive parts of the profile prevail while in the last case the least resistive parts dominate. In polycrystalline materials "serial boundaries" are most important for the resistive effect while "parallel boundaries" are decisive for the case of enhanced conductivity. The "brick layer model" (Figure 55) offers a first approximation of a superposition of these effects. Figure 55 shows the corresponding equivalent circuit. (Whether the resistor representing the highly conducting boundary zones is only parallel to the impedance of the grain interior, as is assumed here, or to the whole grain impedance, depends on the behavior of the intersections of the grain interfaces.)

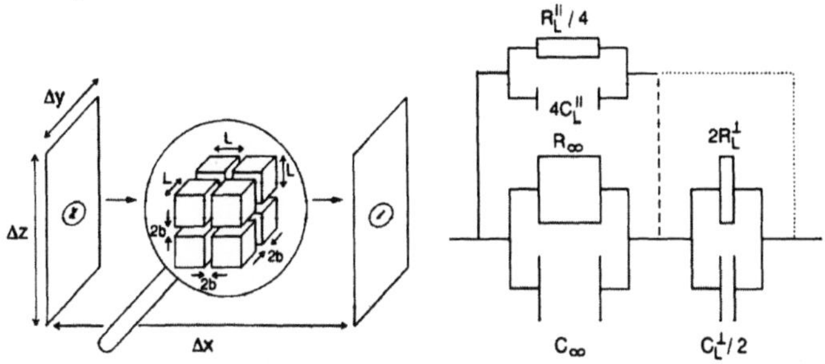

Figure 55. "Brick layer model" as a simple microstructural model and corresponding equivalent circuit. The validity of *dashed* or *dotted curves* is dependent on the local behavior of the intersections of the grain boundaries. The approximation used here refers to the *broken connecting line* corresponding to a blockage of the highly conducting grain boundary paths (e.g., space charge zones) by the perpendicular ones (e.g., core region). The factors 4 and 1/2 refer to the number of grain boundary contributions per grain (4 parallel and 2 serial half grain boundaries) according to Ref.[285] Reprinted from J. Maier, *Ber. Bunsenges. Phys. Chem.*, **90** (1986) 26–33 © 1986 with permission from Wiley-VCH.

If we calculate the impedance of the brick layer polycrystal by taking account of all these pathways[112] we approximately (*dashed lines* in Figure 55) obtain for the effective complex specific conductivity ($\hat{\sigma}_m \equiv \hat{Z}^{-1} L/a$)

$$\hat{\sigma}_m = \frac{\hat{\sigma}_\infty \hat{\sigma}_L^{\perp} + \beta_L^{\|}\varphi_L \hat{\sigma}_L^{\|}\hat{\sigma}_L^{\perp}}{\hat{\sigma}_L^{\perp} + \beta_L^{\perp}\varphi_L \hat{\sigma}_\infty}. \tag{98}$$

The effective terms $\hat{\sigma}^{\|}$ and $\hat{\sigma}^{\perp}$ may be broken down into core (co) and space charge (sc) contributions,[285] φ_L is the volume fraction of the grain boundaries (core and space charge zone), $\beta_L^{\perp} = 1/3$ and $\beta_L^{\|} = 2/3$ in the ideal brick layer model.

If $\tau_L \gg \tau_\infty$, Eq. (98) approximately results in at least two semicircles in the impedance spectrum. The high-frequency semicircle is described by $\hat{\sigma}_m = \hat{\sigma}_\infty + (2/3)\varphi_L \hat{\sigma}_L^{\|}$ (perpendicular boundary capacitances ($C_L^{\perp}$ being permeable)) and includes the values of

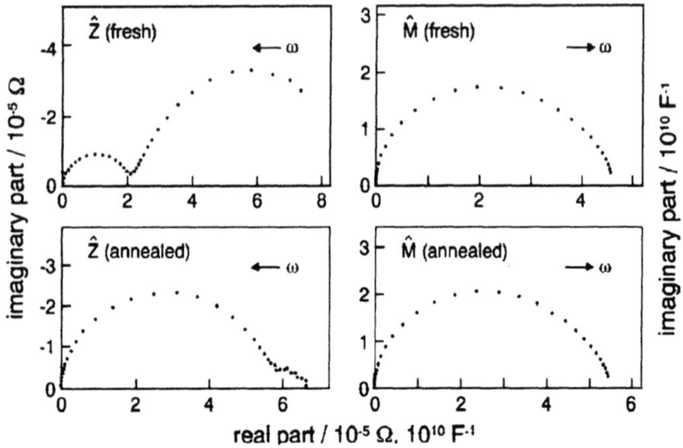

Figure 56. Impedance ($\hat{Z}$) and modulus ($\widehat{M}$) spectra of an AgCl polycrystalline sample before and after being annealed.[15,285] Reprinted from J. Maier, *Solid State Phenom.*, **39/40** (1994), 35–60. Copyright © 1994 with permission from Trans Tech Publications Ltd.

the parallel grain boundaries. The $\perp$-contributions appear as a low-frequency semicircle for which $\hat{\sigma}_{\mathrm{m}}^{-1} = [\sigma_\infty + \beta_{\mathrm{L}}^{\|}\varphi_{\mathrm{L}}\sigma_{\mathrm{L}}^{\|}]^{-1} + \beta_{\mathrm{L}}^{\perp}\varphi_{\mathrm{L}}\hat{\sigma}_{\mathrm{L}}^{\perp -1}$. The real axis intercept represents the dc limit $(\omega \to 0)\hat{\sigma}_{\mathrm{m}}^{-1} = (\sigma_\infty + \beta_{\mathrm{L}}^{\|}\varphi_{\mathrm{L}}\sigma_{\mathrm{L}}^{\|})^{-1} + \beta_{\mathrm{L}}^{\perp}\varphi_{\mathrm{L}}\sigma_{\mathrm{L}}^{\perp -1}$.

It is instructive to consider Figures 54 and 56 in this context both referring to silver halides. Annealing of the bicrystal (Figure 54) (perpendicular grain boundary) results in the diminishing of the second semicircle while the bulk semicircle remains invariant. The polycrystal in Figure 56 also shows two semicircles. Again the low-frequency semicircle shrinks during annealing. In strict contrast to Figure 54, however, now the high-frequency semicircle grows, which is due to the parallel boundary contributions that are reduced on annealing. The detailed analysis of the frequency distribution or more directly the analysis of the intercepts of the $\widehat{M}$-plot (Figure 56, see also Figure 36) confirms this. Assuming the dielectric constant to be positionally constant, the sample thickness turns out to be the effective thickness for the left-hand semicircle; the sample thickness multiplied by the ratio of grain boundary thickness to grain thickness is the one for the right-hand semicircle. The high-frequency semicircle in the $\hat{Z}$-plot varies significantly, whereas the high-frequency semicircle of the

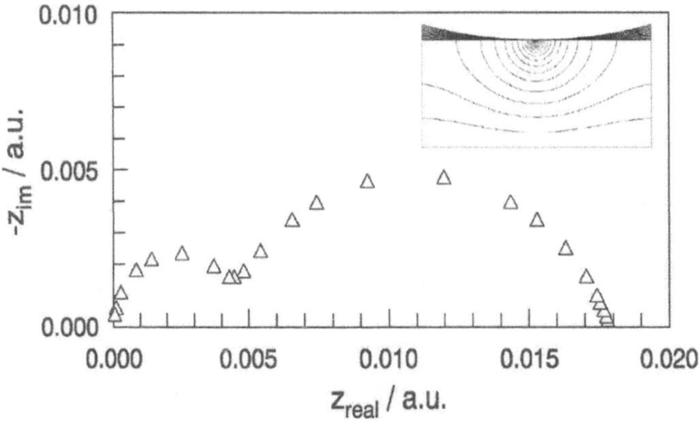

Figure 57. In the case of current constriction induced by a partially contacted metal electrode (shown by electrical potential lines in the *inset*; contact in the *center*, separation by an air gap otherwise) the impedance response ideally consists of two semicircles. At high frequencies the "air gap" (cf. distance between curved electrode and plane surface) becomes dielectrically permeable.[286] Reprinted from J. Fleig and J. Maier, *Electrochim. Acta*, **41** (1996), 1003–1009. Copyright © 1996 with permission from Elsevier.

$\widehat{M}$-plot is invariant. This is because $C^{\|}$ is negligible with respect to C_{∞}, while $R^{\|-1}$ contributes significantly. (Another example of the interplay of parallel and perpendicular boundaries in which even different carriers contribute, is given by nanocrystalline CeO_2.[55])

The fact that in the case of the bicrystal in Figure 54 the effective thickness calculated from the low-frequency semicircle is very much greater than expected from the width of the boundary, while the activation energy is almost equal to that of the bulk, points towards a frequently overlooked complication, namely to current-constriction effects. Such constriction effects occur[287] when the crystal grains are not ideally sintered together, if pores or second phases are included, and interrupt the lateral conductivity of boundaries, as is the case for inhomogeneous electrode contacts.

If the homogeneous direct current distribution meets an array of insulating sites, it becomes channeled and only widens to a homogeneous distribution after some distance, what formally introduces further capacitative elements and, hence, new effective relaxation times. Numerical calculations[286] reveal that often, even though the current distribution is nontrivial, two semicircles are formed (Figure 57 as in the case of a homogeneous blocking). In the frequency range of the high-frequency semicircle the insulating parts are permeable dielectrically: accordingly, the diameter of this semicircle is determined by the ideal bulk resistance ($\propto 1/\sigma$). However, the dc resistance (i.e., the value at $x \rightarrow 0$, i.e., the sum of the two semicircle diameters) is different, but is (as shown by Eq. (99)), proportional to $1/\sigma$ (this is then also true for the low-frequency arc) and hence exhibits the activation energy of the bulk.[‡]

The constriction effect can also be used purposefully to measure inhomogeneities. If point electrodes (of radius b) are employed instead of extended ones (Appendix 6), constriction effects occur on current flow, which lead to the fact that the resistance is determined by the immediate surroundings and is independent of sample size L (if $b \ll L$). The result of the calculation (Appendix 6) is[288]

$$R = (\text{const}\, b\sigma)^{-1} \tag{99}$$

whereby const = 4 for a circular flat applied electrode; const = 2π for an embedded hemispherical electrode. The technique is extremely well suited to measure surface conductance,[§] to record conductivity profiles (e.g., in polycrystalline materials or in inhomogeneous crystals) or to even study electrode kinetics.[290–293] Also polarization experiments (see Section III.3) using microelectrodes have been reported.[294]

Table 10 compiles multipoint methods named after van der Pauw and Valdes, which allow conductivity detection of complex geometries. (See specialist Ref.[295] for more details.)

Local resolution can also be obtained by frequency variation, if a "higher-dimensional" electrochemical chain is constructed (analogous to the construction of a field effect transistor)[298] such as

$$\text{Cell 2D} = \frac{\dfrac{\text{Ag} \mid \text{AgCl}}{\text{SiO}_2}}{\text{Si}}$$

For further developments to obtain spatial resolution such as scanning "impedance spectroscopy"[299] the reader is referred to the current literature.

The above considerations referred to the practically important examples of more or less ordered heterogeneities. If we face random distribution, usually effective medium and percolation theory have to be referred to in order to evaluate the inhomogeneous situations properly. However, attention has to be paid to the fact that they often require nonrealistic approximations. For more details see Ref.[300] In such cases numerical calculations, e.g., via finite element methods are more reliable.

‡ Provided that morphology does not change. The dependences on component potential and doping have to be discussed analogously.
§ An alternative technique uses guard rings.[289]

Table 10.
Experimental configuration and evaluation formulae of the methods named after van der Pauw and Valdes.[296,297] Reprinted from J. Maier, *Physical Chemistry of Ionic Materials. Ions and Electrons in Solids*, (2004). Copyright © 2004 with permission from John Wiley and Sons, Ltd.

van der Pauw method for plane parallel sheets of thickness d	Any form 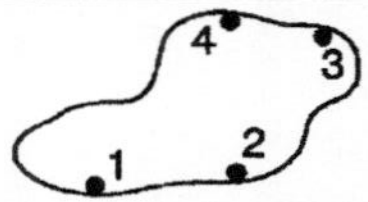	$\exp\left(-\pi d\sigma \frac{U_{43}}{I_{12}}\right) + \exp\left(-\pi d\sigma \frac{U_{14}}{I_{23}}\right) = 1$
	Mirror symmetry 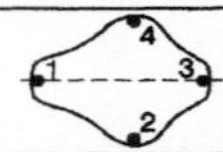	$\frac{U_{43}}{I_{12}} = \frac{U_{14}}{I_{23}} = \frac{\ln 2}{\pi d\sigma}$
Valdes method[a] for sheets of thickness d	Linear array of 4 probes 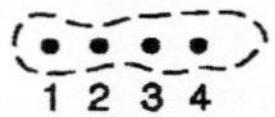	$\frac{U_{32}}{I_{14}} = \frac{\ln 2}{\pi d\sigma}$ (distance (1,4) $\ll$ size)

[a] If the distance between the probes is not great compared with d, other correction factors have to be used.[295]

10. Related Techniques

At the end of the discussion of electrochemical measurement techniques, let us, however, briefly mention that there are other techniques that are not exclusively electrochemical in nature but related to the above methods such as thermoelectric measurements and Hall-effect measurements. Both techniques are extremely helpful in combination with conductivity experiments as they then allow the splitting of the conductivities into carrier concentration and mobilities. The first method relies on the emf formed as a sheer consequence of temperature differences (cross-effects in the thermal and chemical flux–force relations), while the second technique refers to concentration changes upon application of magnetic fields. Both techniques are particularly worked out for electronic carriers but are more tricky and much less straightforward for ionic carriers. For more details the reader is referred to Ref.[16,301,302]

Both techniques in principle bear the potential to distinguish between p- and n-type conductors and—if applicable to ion conductors—to distinguish between interstitial and vacancy contribution provided the nature of the mobile ion is known.

In electronics, a well-established procedure to make statements on the sign of the electronic carriers is establishing the appropriate junctions (cf. diodes). The transformability of the semiconductor experiments to ion conductors suffers from the fact that the situation in ion conductors is more related to the situation in relaxation type semiconductors than to lifetime semiconductors; note that only the latter shows the typical significant electronic effects such as in diodes or transistors. Nonetheless setting up ionic diodes and ionic transistors may be a worthwhile task for the future. (One such attempt to find out the nature of the ionic carriers (O_i'' or $V_{\ddot{O}}$) in PbO by diode effects, viz. by a contact to the vacancy conductive YSZ, has been reported in Ref.[217])

IV. CONCLUSIONS

In Part II we have seen that electrochemical cells based on solid materials are useful for a variety of electrochemical applications involving energy and information transformation. Many of these applications cannot be replaced by electronic devices and will play an increasingly important role in our future life. The same principles and cells being operative for electrochemical devices can also be used for the determination of decisive thermodynamic and kinetic parameters. The treatment should have shown that an understanding of solid state physics and chemistry is important to design and critically evaluate the measurements. They unfold their power in particular if combined with each other and with alternative solid state techniques such as optical spectroscopy.

ACKNOWLEDGMENT

The author is very much indebted to Dr. J. Fleig for helpful discussions and critically reading the manuscript.

A APPENDIX 1—TERMINAL POTENTIAL DIFFERENCE

Let us prove the relations between the voltage drop over the central phase and the difference in the respective electrochemical potentials given in Section III.3.iii on page 94. First, we consider the symmetrical cell 6 (on page 75) and assume transfer equilibrium for the non blocked charge carrier. The contact metal, O_2/ion conductor is, on the metal-side, indicated with 1 (on the other side of the cell with 8) on the ion conductor side the contact is indicated with 2 and 7, respectively. The ion conductor/sample contact is denoted by 3 and 6 on the ion conductor side, and on the sample-side by 4 and 5. Hence

$$FU = \tilde{\mu}_{e^-}^{(1)} - \tilde{\mu}_{e^-}^{(8)} = \frac{1}{2}\tilde{\mu}_{O^{2-}}^{(2)} - \frac{1}{2}\tilde{\mu}_{O^{2-}}^{(7)} - \left(\frac{1}{4}\mu_{O_2}^{(1)} - \frac{1}{4}\mu_{O_2}^{(8)}\right)$$

$$= \frac{1}{2}\mu_{O^{2-}}^{(2)} - \frac{1}{2}\mu_{O^{2-}}^{(7)} - \frac{1}{4}\left(\mu_{O_2}^{(1)} - \mu_{O_2}^{(8)}\right) - \left(\phi^{(2)} - \phi^{(7)}\right).$$

Owing to $\mu_{O^{2-}} \simeq \text{const}$ in the ion conductor it follows that

$$\mu_{O^{2-}}^{(2)} - \mu_{O^{2-}}^{(7)} = \mu_{O^{2-}}^{(3)} - \mu_{O^{2-}}^{(6)} = \tilde{\mu}_{O^{2-}}^{(3)} - \tilde{\mu}_{O^{2-}}^{(6)} + \phi^{(3)} - \phi^{(6)} = \tilde{\mu}_{O^{2-}}^{(4)}$$

$$-\tilde{\mu}_{O^{2-}}^{(5)} + \phi^{(3)} - \phi^{(6)}.$$

If the partial pressures on both sides are the same it finally results because $\phi^{(3)} - \phi^{(2)} \equiv \Delta\phi_{I}$ and $\phi^{(7)} - \phi^{(6)} \equiv \Delta\phi_{II}$ that

$U = \left\{\frac{1}{2F}\tilde{\mu}_{O^{2-}}^{(4)} - \frac{1}{2F}\tilde{\mu}_{O^{2-}}^{(5)}\right\} + \left(\Delta\phi_{I} + \Delta\phi_{II}\right)$. The term in the braces we define as $U_{O^{2-}}$; it refers exclusively to the sample, while $\Delta\phi_{I} + \Delta\phi_{II}$ are the electrical potential drops over the two ion conductor samples. Contact resistances occurring at other phase boundaries lead to drops in electrochemical potential, in the stationary state being proportional to the product of current and contact resistance. On the time scale of the stoichiometry polarization, electrical bulk processes and charge transfer processes behave in a quasistationary manner, with the result that $U = \{\ldots\} + \Sigma_i \mathrm{IR}_i$, whereby i refers to all parts outside the sample. In the case of electronic electrodes we have to exchange $\tilde{\mu}_{O^{2-}}$ by $\tilde{\mu}_{e^-}$ and 2F by F.

B APPENDIX 2—ELECTROCHEMICAL POLARIZATION

This appendix refers to Ref.[210] as far as the interaction-free situation is concerned and to Ref.[3] if fast internal defect interactions are to be included.

B.1 Switching-On Experiment

The flux equations for ions, eons read

$$j^{*}_{\mathrm{eon,ion}} = \frac{\sigma^{*}_{\mathrm{eon,ion}}}{\sigma}\frac{i}{z^{*}_{\mathrm{eon,ion}}\mathrm{F}} - D^{\delta *}(\partial/\partial \mathrm{x})c^{*}_{\mathrm{eon,ion}}$$

where $z^{*}_{\mathrm{eon}}(\partial/\partial \mathrm{x})c^{*}_{\mathrm{eon}} = -z^{*}_{\mathrm{ion}}(\partial/\partial \mathrm{x})c^{*}_{\mathrm{ion}}$. As we allow for internal reactions, we have to formulate these equations in terms of conservative ensembles as referred to in Section III.7 (otherwise source terms have to be considered) explicitly. Here stars are used to mark the quantities that have to be generalized (see also Ref.[3]). If internal reaction coupling is absent, the stars are to be ignored and the h-parameters set to zero.

With the initial condition $c^{*}(t=0,x) = c^{*0}$, and the boundary conditions $i^{*}_{\mathrm{ion}} = 0$ (for an electronic electrode), $i^{*}_{\mathrm{eon}} = 0$ for an ionic electrode or, $\partial c^{*}/\partial t = 0$ for a reversible electrode one obtains with the flux equations and the continuity equation the following as solution for the concentration profile:

$$c^* - c^{*0} = \begin{cases} -t^*_{\text{ion}}\zeta^*\left\{\dfrac{x}{L} - \frac{1}{2} + \phi^*(t/\tau^*_\delta, x/L)\right\}, & \text{(cell 4)} \\ +t^*_{\text{eon}}\zeta^*\left\{\dfrac{x}{L} - \frac{1}{2} + \phi^*(t/\tau^*_\delta, x/L)\right\}, & \text{(cell 6)} \\ -t^*_{\text{ion}}\zeta^*\left\{\dfrac{x}{L} + \Theta^*(t/\tau^*_\delta, x/L)\right\}, & \text{(cell 3)} \\ +t^*_{\text{eon}}\zeta^*\left\{\dfrac{x}{L} + \Theta(t/\tau^*_\delta, x/L)\right\}, & \text{(cell 5)} \end{cases} \quad \text{(B1)}$$

with the abbreviations

$$\zeta^* = \left(i_{\text{p}}L/(2\text{F}D^{\delta*})\right)$$

and

$$\phi^* = \frac{4}{\pi^2}\sum_{m=0}^{\infty}(2m+1)^{-2}\exp\left[-(2m+1)^2\cdot\frac{t}{\tau^*_\delta}\right]\cos\left[(2m+1)\pi\frac{x}{L}\right]$$

$$\Theta^* = -\frac{8}{\pi^2}\sum_{m=0}^{\infty}(-1)^m(2m+1)^{-2}\exp\left[-(2m+1)^2\frac{t}{\tau^*_\delta}\right]\sin\left[(2m+1)\frac{\pi}{2}\frac{x}{L}\right],$$

where the time constants τ^*_δ are given by $4L^2/(\pi^2 D^{\delta*})$ for cells 3 and 5 and $L^2/(\pi^2 D^{\delta*})$ for cells 4 and 6. Transformation into voltages uses Appendix 1, expresses the electrochemical potential difference by the integral $\int(\partial\tilde{\mu}/\partial x)\mathrm{d}x$, and replaces the integrand by the ratio of flux and conductivity of the respective carrier. By using the connection between flux and concentration gradient (see above) and by linearizing we obtain the connection between voltage and boundary concentrations[3,15] and hence

$$U_{\text{eon,ion}}(t) = \frac{iL}{\sigma} \mp 2\text{F}D^{\delta*}\frac{c^*(t,L) - c^*(t,0)}{\sigma^*_{\text{eon,ion}} + 2h_{\text{ion,eon}}\sigma},$$

$$U_{\text{eon}} = \frac{i_{\text{p}}L}{\sigma} + t^*_{\text{ion}}\frac{i_{\text{p}}L}{\sigma^*_{\text{eon}} + 2h_{\text{ion}}\sigma}\{1 - 2\phi^*(t/\tau^*_\delta, 0)\} \text{ for cells 3, 4,} \quad \text{(B2)}$$

$$U_{\text{ion}} = \frac{i_{\text{p}}L}{\sigma} + t^*_{\text{eon}}\frac{i_{\text{p}}L}{\sigma^*_{\text{ion}} + 2h_{\text{eon}}\sigma}\{1 - 2\phi^*(t/\tau^*_\delta, 0)\} \text{ for cells 5, 6.} \quad \text{(B3)}$$

For long times ($t \geq \tau^*_\delta/2$) the sum can be approximated by the first term and for short times ($t \leq \tau^*_\delta/2$) the following approximation may be used:

$$1-\frac{8}{\pi^2}\sum_{m=0}^{\infty}(2m+1)^{-2}\exp\left[-(2m+1)^2\frac{t}{\tau_\delta^*}\right]\simeq\frac{4}{\pi^{3/2}}\sqrt{\frac{t}{\tau_\delta^*}}.$$

The result is the generalization of Eq. (75) for internal defect equilibria.

B.2 Switching-off experiment

The boundary conditions for the depolarization are the same as in Appendix B.1, but $i=0$. The initial condition here is the final linear profile in Appendix B.1 with i being the polarization current. The result for the concentration is

$$c^*-c^{*0}=\begin{cases}\pm t^*_{\text{ion;eon}}\dfrac{i_\text{p}L}{2\text{F}D^{\delta*}}\phi^*(\text{t}/\tau_\delta^*,x/L) & \text{for cells } 3;4,\\ \pm t^*_{\text{ion;eon}}\dfrac{i_\text{p}L}{2\text{F}D^{\delta*}}\Theta^*(\text{t}/\tau_\delta^*,x/L) & \text{for cells } 5;6,\end{cases}\tag{B4}$$

and for the sample voltage

$$U_{\text{eon;ion}}=t^*_{\text{ion;eon}}\left\{\frac{i_\text{p}L}{\sigma^*_{\text{eon;ion}}+2h_{\text{ion;eon}}\sigma}\right\}2\phi^*(t/\tau_\delta^*,0)\tag{B5}$$

for (cell 3) and (cell 4); (cell 5) and (cell 6).

C APPENDIX 3—CHEMICAL POLARIZATION AND RELAXATION

The solution of Fick's second law with the following initial and boundary conditions:[303]

$$c^*(0,x)=c^{*0}=\text{const};\quad c^*(t,0)=c_1^*=k\text{onst};\quad c^*(t,L)=c_2^*=K\text{onst}$$

is given by

$$c^*(x,t)=c_1^*+(c_2^*-c_1^*)\frac{x}{L}+\psi^*\left(\frac{x}{L};\frac{t}{\tau_\delta^*}\right)$$

with

$$\begin{aligned}\psi^*&=\frac{2}{\pi}\sum_{n=1}^{\infty}\left[(-1)^n c_2^*-c_1^*\right]n^{-1}\sin\left[n\pi\frac{x}{L}\right]\exp\left[-n^2\frac{t}{\tau_\delta^*}\right]\\&+\frac{4}{\pi}c^{*0}\sum_{m=0}^{\infty}(2m+1)^{-1}\sin\left[(2m+1)\pi\frac{x}{L}\right]\exp\left[-(2m+1)^2\frac{t}{\tau_\delta^*}\right].\end{aligned}\tag{C1}$$

For the case of a normal chemical polarization, $c^{*0} = c_1^* \neq c_2^*$, the above-given expression can be simplified and converted into a voltage function as in Appendix 2.

For a chemical relaxation, $c^{*0} \neq c_1^* = c_2^* = c^{*1}, \psi^*$ simplifies to

$$\Xi^* = \frac{4}{\pi}\sum_{m=0}^{\infty}(2m+1)^{-1}\sin\left[(2m+1)\pi\frac{x}{L}\right]\exp\left[-(2m+1)^2\frac{t}{\tau_\delta^*}\right] \quad \text{(C2)}$$

with the result $c^*(x, t) = c^{*1} + (c^{*0} - c^{*1})\Xi^*(x, t)$.

If the rate is not infinite at both surfaces the solution is more difficult. For simplicity we neglect internal reactions here. The boundary conditions have to be expressed as $-D^\delta \partial c/\partial x = \bar{k}^\delta$ $(c^1 - c(0 \text{ or } L))$ at $x = 0$ or L; and the solution is given by

$$\frac{c(x,t)-c^1}{c^0-c^1} = \sum_{n=1}^{\infty}\frac{2\lambda\cos(\beta_n\xi)}{(\beta_n^2+\lambda^2+\lambda)\cos\beta_n}\exp\left(-\frac{4D^\delta\beta_n^2 t}{L^2}\right)$$

where the β_n-values represent the set of positive roots of $\beta\tan\beta = \lambda$ with $\lambda = \frac{\bar{k}^\delta L}{2D^\delta}$ and $\xi \equiv 2\frac{x}{L} - 1$. Finally one obtains for the integrated mean value the limiting relations

$$\frac{\bar{c}(t)-c^1}{c^0-c^1} = \sum_{n=1}^{\infty}\frac{8}{(2n-1)^2\pi^2}\exp\left(-\frac{D^\delta(2n-1)^2\pi^2 t}{L^2}\right)$$

for diffusion control, and

$$\frac{\bar{c}(t)-c^1}{c^0-c^1} = \exp\left(-2\frac{\bar{k}^\delta t}{L}\right)$$

for reaction control.

D Appendix 4—Electrolytic Domain Boundaries

The expression $t_{\text{ion}} = (1+\sigma_{\text{eon}}/\sigma_{\text{ion}})^{-1}$ can be written in the form $t_{\text{ion}} = [1+(P_\ominus/P)^N + (P/P_\oplus)^N]^{-1}$ provided $\sigma_{e'} \propto P^{-N}, \sigma_{h^\cdot} \propto P^N$, and σ_{ion} is independent of the partial pressure. In this term the partial pressures $P_\ominus$ and $P_\oplus$ represent the domain boundaries between the N- and I- and the I- and P-regimes, respectively, i.e., $\sigma_{e'}(P_\ominus) = \sigma_{\text{ion}}(P_\ominus)$ and $\sigma_{h^\cdot}(P_\oplus) = \sigma_{\text{ion}}(P_\oplus)$. For $P_\oplus \gg P_\ominus$ it follows[251] that:

$$E = \frac{\mathrm{RT}}{\mathrm{F}}\left(\ln\frac{P_{\oplus}^{N}+P_{1}^{N}}{P_{\oplus}^{N}+P_{2}^{N}}+\ln\frac{P_{\ominus}^{N}+P_{2}^{N}}{P_{\ominus}^{N}+P_{1}^{N}}\right)$$

on integration. Hence the knowledge of the defect chemistry allows for an analytical relation $E(P_1,P_2)$, where P_1 and P_2 are the partial pressures of the two sides of the concentration cell.

E Appendix 5—Coulometric Titration

The galvanostatic solution for a coulometric cell has been given in Ref.[267] When switching on the current, the boundary conditions are $\partial c/\partial x = -t_{\mathrm{eon}} i/z\mathrm{F}D^{\delta}$ at $x=0$ and $\partial c/\partial x = t_{\mathrm{ion}} i/z\mathrm{F}D^{\delta}$ at $x=L$. The initial condition is $\Delta c \equiv c - c(t=0) = 0$ at $t=0$. The result is (z: charge number of ion, internal reactions neglected)

$$\Delta c = \frac{t_{\mathrm{eon}} iL^{\delta}}{\mathrm{z}\mathrm{F}D^{\delta}}\left[\frac{D^{\delta}\mathrm{t}}{L^2}+\frac{(L-x)^2}{2L^2}-\frac{1}{6}-\frac{2}{\pi^2}\times\sum_{n=1}^{\infty}\frac{\cos(n\pi x/L)}{n^2}\exp\left(-\frac{D^{\delta}tn^2\pi^2}{L^2}\right)\right]$$
$$+\frac{t_{\mathrm{ion}} iL}{\mathrm{z}\mathrm{F}D^{\delta}}\left[\frac{D^{\delta}\mathrm{t}}{L^2}+\frac{x^2}{2L^2}-\frac{1}{6}-\frac{2}{\pi^2}\times\sum_{n=1}^{\infty}\frac{(-1)^n\cos(n\pi x/L)}{n^2}\times\exp\left(-\frac{D^{\delta}tn^2\pi^2}{L^2}\right)\right]$$

for c, and

$$U = -\frac{iL\mathrm{RT}\vartheta}{(\mathrm{zF})^2 D^{\delta} c(t=0)}\left[\frac{D^{\delta}\mathrm{t}}{L^2}+\frac{1}{3}-\frac{2}{\pi^2}\times\sum_{n=1}^{\infty}\frac{\exp(-D^{\delta}tn^2\pi^2/L^2)}{n^2}\times\left(t_{\mathrm{eon}}^2+t_{\mathrm{ion}}^2\right)-\frac{2}{\pi^2}\right.$$
$$\left.\times\sum_{n=1}^{\infty}\frac{(-1)^{\mathrm{n}}\exp(-D^{\delta}tn^2\pi^2/L^2)}{n^2}\times 2t_{\mathrm{eon}}t_{\mathrm{ion}}\right]$$

for U, where $\vartheta \equiv c(t=0)(\partial\mu(t=0)/\partial c)/RT$.

After having polarized the cell for $\Delta t \equiv t_{\mathrm{p}}$, the current is switched off, i.e., $\partial c/\partial x$ at $x=0$ and L is valid. The initial condition

$$\Delta c(t=0) = \frac{t_{\mathrm{eon}} iL}{\mathrm{z}\mathrm{F}D^{\delta}}\left[\frac{D^{\delta}t_{\mathrm{p}}}{L^2}+\frac{(L-x)^2}{2L^2}-\frac{1}{6}-\frac{2}{\pi^2}\sum_{n=1}^{\infty}\frac{\cos(n\pi x/L)}{n^2}\times\exp\left(-\frac{D^{\delta}t_{\mathrm{p}}n^2\pi^2}{L^2}\right)\right]$$
$$+\frac{t_{\mathrm{ion}} iL}{\mathrm{z}\mathrm{F}D^{\delta}}\left[\frac{D^{\delta}t_{\mathrm{p}}}{L^2}+\frac{x^2}{2L^2}-\frac{1}{6}-\frac{2}{\pi^2}\sum_{n=1}^{\infty}\frac{(-1)^n\cos(n\pi x/L)}{n^2}\times\exp\left(-\frac{D^{\delta}t_{\mathrm{p}}n^2\pi^2}{L^2}\right)\right]$$

leads to the voltage solution

$$U = -\frac{iLRT\vartheta}{(zF)^2 D^\delta c(t=0)} \left\{ \frac{D^\delta t_p}{L^2} + \frac{2}{\pi^2} \sum_{n=1}^{\infty} \frac{\exp(-D^\delta t n^2 \pi^2 / L^2)[1 - \exp(-D^\delta t_p n^2 \pi^2 / L^2)]}{n^2} \right\} \times (t_{eon}^2 + t_{ion}^2)$$

$$- \frac{iLRT\vartheta}{(zF)^2 D^\delta c(t=0)} \left\{ \frac{D^\delta t_p}{L^2} + \frac{2}{\pi^2} \sum_{n=1}^{\infty} \frac{(-1)^n \exp(-D^\delta t n^2 \pi^2 / L^2)[1 - \exp(-D^\delta t_p n^2 \pi^2 / L^2)]}{n^2} \right\} \times 2 t_{eon} t_{ion},$$

F Appendix 6—Point Electrode Resistance

We approximate the electrode by a sphere embedded to one half in the sample and having a radius of b. The extended "normal" counter electrode is applied at a distance $L \gg b$. On account of the potential distribution (see Figure 57) it can be assumed[288] that the problem can be treated by considering a homogeneously conducting body of spherical symmetry hemispherically extending from the contact point. It then follows for the resistance of an infinitesimal section at distance r, that $\mathrm{d}R = \frac{2\mathrm{d}r}{4\pi r^2 \sigma}$. By integration between b and L the result $R = \frac{1}{2\pi\sigma b} - \frac{1}{2\pi\sigma L} \simeq \frac{1}{2\pi b}\frac{1}{\sigma}$ is obtained. On account of the convergence of the integral for large L neither the exact sample thickness nor the exact geometry of the extended counter electrode is important.

SYMBOLS

C	capacitance; dopant concentration
I	current
D	diffusion coefficient
E	cell potential; energy level; electrical field
G	Gibbs energy
H	enthalpy
K	mass action constant
L	thickness
P	partial pressure
Q	charge
R	resistance
$\mathcal{R}$	reaction rate
T	temperature
U	voltage
Z	impedance
a	area, affinity
c	concentration

d	distance
i	current density; interstitial site (as index)
j	flux density
t	time; transfer number
w	efficiency
x	positional coordinate
z	charge number
β	percolation factor
δ	nonstoichiometry; indicates chemical diffusion or permeation (as index)
ε	dielectric permeability
η	overvoltage
μ	chemical potential
$\tilde{\mu}$	electrochemical potential
ν	stoichiometric coefficient
ρ	charge density
σ	conductivity
τ	relaxation time
ϕ	electrical potential
φ	volume fraction
χ	trapping factor
$\vee$	vacancy
$\cdot$	effective positive charge (as index)
$'$	effective negative charge (as index)
$\times$	effective neutral charge (as index)
$\hat{}$	denotes complex number
$\{\}$	denotes conservative ensembles
Par (i, j)	operator that connects circuit elements denoted by the quantities i, j in parallel
Ser (i, j)	operator that connects circuit elements denoted by the quantities i, j in series

Further abbreviations are introduced and explained locally.

REFERENCES

[1]E. A. Guggenheim, *Mixtures*, Oxford University Press, Oxford, 1952.

[2]J. Maier, Solid state electrochemistry I: Thermodynamics and kinetics of charge carriers in solids. in: *Modern Aspects of Electrochemistry* Vol. 38, Ed. by B. E. Conway, C. G. Vayenas, R. E. White, and M. E. Gamboa-Adelco Kluwer Academic/Plenum Pubishers, New York, 2005, pp. 1–173.

[3]J. Maier, *Z. Phys. Chem.* N.F. **140** (1984) 191.

[4] $\partial/\partial\mathbf{a}$ symbolically stands for the vector with the components $(\partial/\partial a_x, \partial/\partial a_y, \partial/\partial a_z) = \nabla_a$.

[5] L. D. Landau and E. M. Lifshitz, *Electrodynamics of Continuous Media (Course of Theoretical Physics)* Vol. 8, Pergamon Press, Oxford, 1963.

[6] S. z. B. M. Henzler and W. Göpel, *Oberflächenphysik des Festkörpers*, B. G. Teubner, Stuttgart, 1991;

[7] G. Kortüm, *Treatise on Electrochemistry*, Elsevier, Amsterdam, 1965.

[8] J. O'M. Bockris and A. K. N. Reddy, *Modern Electrochemistry*, Plenum Press, New York, 1977.

[9] J. S. Newman, *Electrochemical Systems*, Prentice-Hall, Englewood Cliffs, NJ, 1991; A. J. Bard and L. R. Faulkner, *Electrochemical Methods. Fundamentals and Applications*, John Wiley and Sons, New York, 1980; J. O'M. Bockris and S. Srinivasan, *Fuel Cells: Their Electrochemistry*, McGraw-Hill Book Company, New York, 1969; J. O'M. Bockris and A. K. V. Reddy, *Modern Electrochemistry*, Plenum Press, New York, 1970; C. Julien, G. A. Nazri, *Solid State Batteries*, Kluwer Academic Publishers, Norwell, 1994; M. Winter, J. O. Besenhard, M. E. Spahr, and P. Novák, *Adv. Mater.* **10** (1998) 725; F. von Sturm, *Elektrochemische Stromerzeugung*, VCH, Weinheim, 1969; K. J. Vetter, *Electrochemical Kinetics*, Academic Press, New York, 1967.

[10] See later sections for special references.

[11] J. Maier, *Physical Chemistry of Ionic Materials, Ions and Electrons in Solids*, John Wiley and Sons, Ltd., Chichester, 2004.

[12] I. Riess, in: *Encyclopedia of Electrochemistry*, Vol. 1, Ed. by A. J. Bard and M. Stratmann, Wiley-VCH, Weinheim, 2002, Vol. 1, *Thermodynamics and Electrified Interfaces*, (E. Gileadi and M. Urbakh (eds.)), pp. 253.

[13] P. Bruce (ed.), *Solid State Electrochemistry*, Cambridge University Press, Cambridge, 1994.

[14] P. J. Gellings and H. J. M. Bouwmeester (eds.), *CRC Handbook of Solid State Electrochemistry*, CRC Press, Boca Raton, 1997.

[15] J. Maier, *Solid State Phenom.* **39–40** (1994) 35.

[16] H. Rickert, *Electrochemistry of Solids*, Springer-Verlag, Berlin, 1982; T. Kudo and K. Fueki, *Solid State Ionics*, VCH, Tokyo-Kodansha, 1990.

[17] J. Maier, *Solid State Ionics* **62(1,2)** (1993) 105.

[18] J. Maier, in: *Oxygen Ion and Mixed Conductors and their Technological Applications*, Vol. 368 Ed. by H. L. Tuller and J. Schoonman and I. Riess NATO Science Series: E Applied Sciences, Kluwer Academic Publishers, (2000) 399.

[19] B. Kamp, R. Merkle, and J. Maier, *Sens. Actuators B* **77** (2001) 534; J. Jamnik, B. Kamp, R. Merkle and J. Maier, *Solid State Ionics* **150** (2002) 157.

[20] One can also define the terms selectivity and specificity as continuous quantities, as done in Ref.[21] The sensitivity may be defined in such cases by the determinant of the response matrix.

[21] K. Dauter, E. Than, and D. Molch, *Analytik*, Wissenschaftliche Verlagsgesellschaft, Stuttgart, 1977.

[22] J. Maier and W. Münch, *J. Chem. Soc., Faraday Trans.* **92** (12) (1996) 2143.

[23] R. Moos, W. Menesklou, H.-J. Schreiner, and K. H. Härtdl, *Sens. Actuat. B* **67** (2000) 178.

[24] J. Maier, and W. Göpel, J. *Solid State Chem.* **72** (1988) 293; C. G. Fonstad and R. H. Rediker, *J. Appl. Phys.* **42** (1971) 2911.

[25] T. Seiyama, A. Kato, K. Fujiishi, and M. Nagatani, *Anal. Chem.* **34** (1962) 1502; **38** (1966) 1069; N. Taguchi, Jpn. Patent 45-38200, 1962.

[26] K. Sasaki and J. Maier, *J. Appl. Phys.* **86** (10) (1999) 5422, 5434; K. Sasaki, M. Haseidl and J. Maier, in: *Proc. EUROSOLID 4* Ed. by A. Negro and L. Montanaro, Politecnico di Torino, Turin, 1997, pp. 123.

[27]J. Jamnik and J. Maier, *Ber. Bunsenges. Phys. Chem.* **101** (1) (1997) 23; J. Jamnik and J. Maier, *J. Phys. Chem. Solids* **59** (9) (1998) 1555.

[28]J. Jamnik, B. Kamp, R. Merkle, and J. Maier, *Solid State Ionics* **150** (2002) 157.

[29]J. Janata, *Principles of Chemical Sensors*, Plenum Press, New York, 1989.

[30]In a usual field effect transistor an appropriate voltage between the gate over an insulating oxide such as SiO_2, to the underlying Silicon (p-doped) causes an inversion layer at the contact oxide/Si which can be detected by measuring the source–drain current. In a CHEMFET the electrical field is provided by a rudiment of a galvanic cell. The metal gate is replaced by an electrolyte which is in contact to a reference electrode. The process on the working side of the cell is the interaction with the insulating oxide.

[31]W. Göpel, J. Hesse, and J. N. Zemel (eds.), *Sensors, A Comprehensive Study*, VCH, Weinheim, 1987.

[32]K. Kiukkola and C. Wagner, *J. Electrochem. Soc.* **104** (1957) 308; 379.

[33]H. Dietz, W. Haecker, and H. Jahnke, in: *Advances in Electrochemistry and Electrochemical Engineering*, Ed. by H. Gerischer, and C. W. Tobias, John Wiley and Sons, Ltd., New York, 1977.

[34]For a given concentration in the gas phase, the current causes a depletion immediately at the boundary where the reaction takes place. Usually the width over which the gradient is established, remains finite for hydrodynamic reasons. If the concentration at the interface has essentially decreased to zero at the boundary, the diffusion controlled current cannot be increased any longer $\left(i_{\mathrm{diff}} \propto \frac{c_\infty - c_0}{\delta}, i_{\mathrm{diff}}(U \rightarrow \infty) \propto c_\infty\right)$. Hence i_{diff} is proportional to the concentration in the gas atmosphere.

[35]H. Iwahara, H. Uchida, and J. Kondo, *J. Appl. Electrochem.* **13** (1983) 365.

[36]K.-D. Kreuer, E. Schönherr, and J. Maier, *Solid State Ionics* **70/71** (1994) 278.

[37]J.-H. Yu, J.-S. Lee, and J. Maier, in preparation.

[38]J. Maier, in: *Recent Trends in Superionic Solids and Solid Electrolytes*. Ed. by S. Chandra, and A. Laskar, Academic Press, New York (1989), pp. 137.

[39]M. Holzinger, J. Fleig, J. Maier, and W. Sitte, *Ber. Bunsenges. Phys. Chem.* **99(11)** (1995) 1427.

[40]M. Bendahan, P. Lauque, J. L. Seguin, K. Aguir, and P. Knauth, *Sen. Actuat. B* **95** (2003) 170.

[41]M. Gauthier, and A. Chamberland, *J. Electrochem. Soc.* **124** (1977) 1579.

[42]J. Maier and U. Warhus, *J. Chem. Thermodyn.* **18** (1986) 309.

[43]T. Maruyama, S. Sasaki, and Y. Saito, *Solid State Ionics* **23** (1987) 107.

[44]M. Holzinger, J. Maier, and W. Sitte, *Solid State Ionics* **86–88** (1996) 1055.

[45]R. Merkle and J. Maier in: *Encyclopedia of Sensors*, C.A. Grimes, E.C. Dickey, M.V. Pishko, eds. American Scientific Publishers, **Vol. 3** (2006) 139–155.

[46]H.-H. Möbius, P. Shuk, and W. Zastrow, *Fresenius' J. Anal. Chem.* **349** (1996) 684.

[47]M. Holzinger, J. Maier, and W. Sitte, *Solid State Ionics* **94** (1997) 217.

[48]M. A. Careem, K. P. Vidanapathirana, S. Skaarup and K. West, *Solid State Ionics* **175** (2004) 725.

[49]H. Iwahara, *Chem. Solid State Mater.* **2** (1992) 122, Cambridge University Press, Cambridge (UK).

[50]H. Iwahara, *Solid State Ionics* **77** (1995) 289, Elsevier, Amsterdam.

[51]G. C. Vayenas, S. Bebelis, and S. Neophytides, *J. Phys. Chem.* **92** (1988) 5085.

[52]S. P. Balameneon and C. G. Vayenas, *J. Electrochem. Soc.* **151** (2004) A 1874; see also A. Hashibon, S. Raz, and I. Riess, *Solid State Ionics* **149** (2002) 117.

[53]J. Fleig, and J. Jamnik, *J. Electrochem. Soc.* **152** (4) (2005) E138.

[54]A. Tschöpe, *Solid State Ionics* **139** (2001) 267.

[55]S. Kim and J. Maier, *J. Electrochem. Soc.* **149** (10) (2002) J73.
[56]C. Wagner, Z. *Phys. Chem.* **B21** (1933) 25; *Corr. Sci.* **9** (1969) 91.
[57]C. Wagner, in *Progress in Solid State Chemistry*, Vol. 6, Ed. by H. Reiss and J.O. McCaldin, Pergamon Press, Oxford, 1972, pp. 1.
[58]J. Maier and G. Schwitzgebel, *Phys. Stat. Sol.* (b) **113** (1982) 535.
[59]J. Maier, *J. Am. Ceram. Soc.* **76(5)** (1993) 1212, 1218, 1223, 1228.
[60]T. Norby, and P. Kofstad, *Solid State Ionics* **20** (1986) 164.
[61]J. Jamnik and J. Maier, *Phys. Chem. Chem. Phys.* **3**(9) (2001) 1668.
[62]See e.g., P. M. S. Mouk, J. A. Duffy, and M. D. Ingram, *Electrochim. Acta* **38** (1993) 2759.
[63]Ch. Julien and G.-A. Nazri, *Solid State Batteries: Materials, Design and Optimization*, Kluwer Academic Publishers, New York, 1994.
[64]M. Wakihara. O. Yamamoto (eds.), *Lithium Ion Batteries*, Wiley-VCH, Weinheim, 1998.
[65]M. Winter, J. O. Besenhard, M. E. Spahr, and P. Novák, *Adv. Mater.* **10** (1998) 725, VCH, Weinheim.
[66]B. C. H. Steele, *Philos. Trans. R. Soc. London A* **354** (1996) 1695; T. Kanada and H. Yokokawa, *Key Eng. Mater.* **125–126** (1997) 187; S. C. Singhal, in: *Solid Oxide Fuel Cells V* Vol. PV 97–40, Ed. by U. Stimming, S. C. Singhal, and H. Tagawa, and W. Lehnert, p. 37, The Electrochemical Society, Pennington (NJ), 1997.
[67]W. Ostwald, *Z. Elektrochem.* **1** (1894/95) 122.
[68]Please note that, unlike in the other parts of the text, here we follow our initial nomenclature: $\Delta_r G$, etc. are not referred to 1 mol, rather $\Delta_r G = n\Delta_r G_m$ etc.
[69]F. von Sturm, *Elektrochemische Stromerzeugung*, VCH, Weinheim, 1969.
[70]This does not contravene the First Law, since the surroundings are cooled in such circumstances. This does not contravene the Second Law, since it is not possible by this process to construct a periodic machine, that carries out work merely by cooling.
[71]K.-D. Kreuer and J. Maier, *Spekt. Wissen.* **7** (1995) 92.
[72]K. J. Vetter, *Electrochemical Kinetics*, Academic Press, New York, 1967,
[73]W. Gajewski, Spekt. Wissen. 7 (1995) 88.
[74]A. Eisenberg, and H. L. Yeager (eds.), *Perfluorinated Ionomer Membranes*, The American Chemical Society, Washington, DC, 1982.
[75]G. G. Scherer, H. P. Brack, F. N. Buchi, B. Gupta, O. Haas, and M. Rota, in: *Proceedings of 11th World Hydrogen Energy Conference*, Vol. 2, Ed. by T. N. Veziroglu, p. 1727, International Association for Hydrogen Energy, Coral Gables (FL), 1996; K.-D. Kreuer, Th. Dippel, and J. Maier, in: *Proton Conducting Membrane Fuel Cells I*, Vol. PV 95-23, p. 241, The Electrochemical Society, Pennington (NJ), 1995.
[76]M. Ise, K. D. Kreuer and J. Maier, *Solid State Ionics* **125** (1999) 213.
[77]M. Rehahn, A. D. Schlüter, and G. Wegner, *Makromol. Chem.* **191** (1990) 1991.
[78]K. D. Kreuer, *Solid State Ionics* **97** (1997) 1.
[79]K. D. Kreuer, A. Fuchs, M. Ise, M. Spaeth, and J. Maier, *Electrochim. Acta* **43** (1998) 1281; K. D. Kreuer, J. Membr. Sci. **185** (2001) 29; M. Schuster, W. H. Meyer, G. Wegner, H. G. Herz, M. Ise, K. D. Kreuer, and J. Maier, *Solid State Ionics* **145** (2001) 85; H. G. Herz, K. D. Kreuer, J. Maier, G. Scharfenberger, M. F. H. Schuster and W. H. Meyer, *Electrochim. Acta* **48** (2003) 2165.
[80]J. H. Park R. N. and Blumenthal, *J. Electrochem. Soc.* **136** (1994) 2867.
[81]R. E. W. Casselton, *Phys. Stat. Sol.* (a) **2** (1970) 571.
[82]S. P. S. Badwal, *Solid State Ionics* **52** (1992) 23.
[83]W. Weppner, *J. Solid State Chem.* **20** (1977) 305.
[84]K. Sasaki and J. Maier, *Solid State Ionics* **134** (2000) 303.

[85]T. Takahashi, in: *Physics of Electrolytes*, **Vol. 2**, Ed. by J. Hladic, Academic Press, London, 1972, p. 980.
[86]J. F. Baumard and P. Abelard, in: *Advances in Ceramics*, **Vol. 12**, Ed. by N. Claussen, M. Rühle and A. H. Heuer, The American Ceramic Society, Columbus, OH, USA, 1984, p. 555.
[87]M. O. Zacate, L. Minervini, D. J. Bradfield, R. W. Grimes, and K. E. Sickafus, *Solid State Ionics* **128** (2000) 243.
[88]Y. Yamamuru, S. Kawasaki, and H. Sakai, *Solid State Ionics* **126** (1999) 181.
[89]M. Meyer, N. Nicoloso, and V. Jaenisch, *Phys. Rev. B* **56** (1997) 5961.
[90]F. M. Spiridonov, L. N. Popova, and R. Ya, *J. Solid State Chem.* **2** (1970) 430.
[91]S. B. S. Badwal, F. T. Ciacchi, S. Rajendran, and J. Drennan, *Solid State Ionics* **109** (1998) 167.
[92]K. Nomura, Y. Mizutani, M. Kawai, Y. Nakamura, and O. Yamamoto, *Solid State Ionics* **132** (2000) 235.
[93]S. B. S. Badwal, F. T. Ciacchi, and D. Milosevic, *Solid State Ionics* **136–137** (2000) 91.
[94]M. Kilo, G. Borchardt, B. Lesage, S. Weber, S. Scherrer, M. Schroeder, and M. Martin, *Key Eng. Mater.* **206–213** (2002) 601; M. Martin, *J. Chem. Thermodyn.* **35** (2002) 1291.
[95]J. E. Bauerle, *J. Phys. Chem. Solids* **30** (1969) 2657.
[96]S. P. S. Badwal, *Solid State Ionics* **76** (1995) 67.
[97]T. van Dijk and A. J. Burggraaf, *Phys. Stat. Sol.* (a) **63** (1981) 229.
[98]M. J. Verkerk, B. J. Middelhuis, and A. J. Burggraaf, *Solid State Ionics* **6** (1982) 159.
[99]S. P. S. Badwal and J. Drennan, *J. Mater. Sci.* **22** (1987) 3231.
[100]M. Kleitz, H. Bernard, E. Fernandez, and E. Schouler, in: *Advances in Ceramics. Science and Technology of Zirconia* **Vol. 3**, Ed. by A. H. Heuer and L. W. Hobbs, American Ceramic Society, (1981) p. 310, Washington, DC.
[101]M. C. Steil, F. Thevenot, and M. Kleitz, *J. Electrochem. Soc.* **144** (1997) 390.
[102]S. P. S. Badwal and S. Rajendran, *Solid State Ionics* **70–71** (1994) 83.
[103]M. Aoki, Y.-M. Chiang, I. Kosacki, L. Jong-Ren Lee, H. Tuller, and Y. Liu, *J. Am. Ceram. Soc.* **79** (1996) 1169.
[104]X. Guo and J. Maier, *J. Electrochem. Soc.* **148** (3) (2001) E121; X. Guo, W. Sigle, J. Fleig and J. Maier, *Solid State Ionics* **154–155** (2002) 555; X. Guo, W. Sigle, and J. Maier. *J. Am. Ceram. Soc.* **86** (1) (2003) 77.
[105]M. Gödickemeier, B. Michel, A. Orliukas, P. Bohac, K. Sasaki, L. Gauckler, H. Heinrich, P. Schwander, K. Kostorz, H. Hofmann, and O. Frei, *J. Mater. Res.* **9** (1994) 1228.
[106]B. C. H. Steele, *Solid State Ionics* **129** (2000) 95.
[107]R. Gerhardt and A. S. Nowick, *J. Am. Ceram. Soc.* **69** (1986) 641.
[108]G. M. Christie and F. P. F. van Berkel, *Solid State Ionics* **83** (1996) 17.
[109]K. E. Adham and H. Hammou, *Solid State Ionics* **9–10** (1983) 905.
[110]J. Fleig, K. D. Kreuer, and J. Maier, in: *Handbook of Advanced Ceramics. Volume II: Processing and Their Applications*, Ed. by S. Sōmiya, F. Aldinger, N. Claussen, R. M. Spriggs, K. Uchino, K. Koumoto and M. Kaneno, Elsevier Academic Press, Amsterdam, 2003, p. 59.
[111]X. Guo, W. Sigle, J. Fleig, and J. Maier, *Solid State Ionics* **154–155** (2002) 555.
[112]J. Maier, Ber. Bunsenges. *Phys. Chem.* **90** (1986) 26.
[113]J. Fleig, S. Rodewald, and J. Maier, *Solid State Ionics* **136–137** (2000) 905.
[114]E. G. Moya, in: *Science of Ceramics Interfaces II*, Ed. by J. Novotny, p. 277, Elsevier Science, Amsterdam (1994).
[115]J. Fleig and J. Maier, *J. Electroceram.* **1** (1997) 73.
[116]J. Fleig and J. Maier, in: *Proceedings of the Fifth International Symposium on Solid Oxide Fuel Cells*, **Vol. 97–40**, Ed. by U. Stimming, S. C. Singhal, H. Tagawa, W. Lehnert, The Electrochem. Soc., Pennington (1997), p. 1374.

[117]F. H. van Heuveln, H. J. M. Bouwmeester, and van F. P. F. Berkel, *J. Electrochem. Soc.* **144** (1997) 126.
[118]K. Huang, M. Feng, and J. B. Goodenough, *J. Am. Ceram. Soc.* **81** (1998) 357.
[119]L. Minervini, M. O. Zacate, and R. W. Grimes, *Solid State Ionics* **116** (1999) 339.
[120]M. Mogensen, N. M. Sammes, and G. A. Tompsett, *Solid State Ionics* **129** (2000) 63.
[121]H. L. Tuller and A. S. Nowick, *J. Electrochem. Soc.* **122** (1975) 255.
[122]I. Riess, *Solid State Ionics* **52** (1992) 127.
[123]N. S. Choudhury and J. W. Patterson, *J. Electrochem. Soc.* **118** (1971) 1389.
[124]M. Gödickemeier and L. J. Gauckler, *J. Electrochem. Soc.* **145** (1998) 414.
[125]I. Riess, M. Gödickemeier, and L. J. Gauckler, *Solid State Ionics* **90** (1996) 91.
[126]T. S. Stefanik and H. L. Tuller, *J. Electroceramics* **13** (2004) 799.
[127]J. Maier, *Prog. Solid State Chem.* **23** (3) (1995) 171.
[128]T. Ishihara, H. Matsuda, and Y. Takita, *J. Am. Ceram. Soc.* **116** (1994) 3801.
[129]M. Feng and J. B. Goodenough, *Eur. J. Solid State Inorg. Chem.* **31** (1994) 663.
[130]J. W. Stevenson, T. R. Armstrong, D. E. McCready, L. R. Pederson, and W. J. Weber, *J. Electrochem. Soc.* **144** (1997) 3613.
[131]K. Huang, R. S. Tichy, and J. B. Goodenough, *J. Am. Ceram. Soc.* **81** (1998) 2565.
[132]P. Huang and A. Petric, *J. Electrochem. Soc.* **143** (1996) 1644.
[133]N. Trofimenko and H. Ullmann, *Solid State Ionics* **118** (1999) 215.
[134]K. Yamaji, T. Horita, M. Ishikawa, N. Sakai, and H. Yokokawa, *Solid State Ionics* **121** (1999) 217.
[135]K. Yamaji, H. Negishi, T. Horita, N. Sakai, and H. Yokokawa, *Solid State Ionics* **135** (2000) 389.
[136]S. W. Tao, F. W. Poulsen, G. Y. Meng, and O. T. Sorensen, *J. Mater. Chem.* **10** (2000) 1829.
[137]T. L. Nguyen and M. Dokiya, *Solid State Ionics* **132** (2000) 217.
[138]T. Ishihara, H. Furutani, M. Honda, T. Akbay, T. Sakai, N. Yokokawa, and Y. Takita, Chem. Mater. **11** (1999) 2081; T. Ishihara, T. Shibayama, S. Ishikawa, K. Hosoi, H. Nishiguchi, and Y. Takita, *J. Eur. Ceram. Soc.* **24** (2004) 1329.
[139]Z. Shao and S. M. Haile, *Nature* **431** (2004) 170.
[140]R. A. De Souza, and J. A. Kilner, *Solid State Ionics* **126** (1999) 153.
[141]R. H. E. van Doorn, I. C. Fullarton, R. A. De Souza, J. A. Kilner, H. J. M. Bouwmeester, and A. J. Burggraaf, *Solid State Ionics* **96** (1997) 1.
[142]T. Ishigaki, S. Yamauchi, K. Kishio, J. Mizusaki, and K. Fueki, *J. Solid State Chem.* **73** (1988) 179.
[143]P. S. Manning, J. D. Sirman, R. A. De Souza, and J. A. Kilner, *Solid State Ionics* **100** (1997) 1.
[144]E. Ruiz-Trejo, J. D. Sirman, Y. M. Baikov, and J. A. Kilner, *Solid State Ionics* **113** (1998) 565.
[145]T. Horita, K. Yamaji, N. Sakai, H. Yokokawa, T. Kawada, and T. Kato, *Solid State Ionics* **127** (2000) 55.
[146]V. Brichzin, J. Fleig, H.-U. Habermeier, and J. Maier, *Electrochem. Solid-State Lett.* **3** (9) (2000) 403; V. Brichzin, J. Fleig, H.-U. Habermeier, and J. Maier, in: *Solid Oxide Fuel Cells VII*, **Vol. 2001–16**, Ed. by H. Yokokawa and S. C. Singhal, The Electrochemical Society, Pennington/NJ (2001), p. 555; *J. Fleig, Annu. Rev. Mater. Res.* **33** (2003) 361.
[147]F. Baumann, J. Fleig, and J. Maier, in preparation.
[148]A. Atkinson, S. Barnett, R. J. Gorte, J. T. S. Irvine, A. J. McEvoy, M. Mogensen, S. C. Singhal, and J. Vohs, *Nature Mater.* **3** (2004) 17.
[149]I. Yasuda and T. Hikita, *J. Electrochem. Soc.* **140** (1993) 1699.
[150]S. Tao and J. T. S. Irvine, *Nature Mater.* **2** (5) (2003) 320.
[151]S. C. Singhal and K. Kendall, *Solid Oxide Fuel Cells*, Elsevier, Oxford, 2003.

[152]H. L. Tuller, in: *Proceedings of 17th Risø International Symposium Materials Science: High Temperature Electrochemistry: Ceramics and Metals*, Ed. by F. W. Poulsen, N. Bonanos, S. Linderoth, M. Mogensen, B. Zachau-Christiansen, p. 139, Risø National Laboratory, Roskilde, Denmark, 1996.
[153] T. Ishihara, H. Matsuda, and Y. Takita, *J. Am. Chem. Soc.* **116** (1994) 3801; T. Ishihara, H. Furutani, H. Nishiguchi,Y. Takita, in: *Ionic and Mixed Conducting Ceramics III*, Vol. PV 97–24, Ed. by T. A. Ramanarayanan, W. L. Worrell, H. L. Tuller, M. Mogensen, A. C. Khandkar, p. 834, The Electrochemical Society, Pennington (NJ), 1997.
[154] J. B. Goodenough, J. E. Ruiz-Diaz, and Y. S. Zhen, *Solid State Ionics* **44** (1990) 21.
[155] S. B. Adler, J. A. Reimer, J. Baltisberger, and U. Werner, *J. Am. Ceram. Soc.* **116** (1994) 675; G. B. Zhang, D. M. Smyth, *Solid State Ionics* **82** (1996) 161.
[156] W. van Gool, *Philips Res. Rept.* **20** (1965) 82.
[157] C. D. Dyer, Nature **343** (1990) 542.
[158] T. Hibino, K. Ushiki, and Y. Kuwahara, *Solid State Ionics* **91** (1996) 69; T. Hibino, Sh. Wang, Sh. Kakimoto, and M. Sano, *Electrochem. Solid-State Lett.* **2** (1999) 317; T. Hibino et al., *Science* **288** (2000) 183.
[159] T. Tsai and S. A. Barnet, *J. Electrochem. Soc.* **144** (1997) L131.
[160] T. Tsai and S. A. Barnet, *Solid State Ionics* **98** (1997) 191.
[161] A. Virkar, *J. Electrochem. Soc.* **138** (1991) 1481.
[162] E. D. Wachsman, *Solid State Ionics* **151/152** (2002) 657.
[163] W. G. Coors, *J. Power Sources* **118** (1–2) (2003) 150.
[164] K. D. Kreuer, *Annu. Rev. Mater. Res.* **33** (2003) 333.
[165] K. D. Kreuer, St. Adams, W. Münch, A. Fuchs, U. Klock, and J. Maier, *Solid State Ionics* **145** (2001) 295.
[166] K. D. Kreuer, W. Münch, M. Ise, T. He, A. Fuchs, U. Traub, and J. Maier, *Ber. Bunsenges. Phys. Chem.* **101** (9) (1997) 1344.
[167] S. M. Haile, D. A. Boysen, C. R. I. Chisholm, and R. B. Merle, *Nature* **410** (2001) 910.
[168] B. B. Owens, J. E. Oxley, and A. F. Sammells, in *Solid Electrolytes*, Ed. by S. Geller, p. 67, Springer, Berlin, Heidelberg, New York 1977.
[169] C. C. Liang, A. V. Joshi, and W. E. Hamilton, *J. Electrochem. Soc.* **135** (1988) 2901.
[170] A. J. Hills and N. A. Hampson, *J. Power Sources* **24** (1988) 253.
[171] J. Maier, *Einführung in die Physikalische Festkörperchemie*, Lecture Notes, Tübingen-Stuttgart, 1990.
[172] J.-S. Lee, S. Adams, and J. Maier, *J. Electrochem. Soc.* **147** (2000) 2407; K. Shahi and J. B. Wagner, *J. Electrochem. Soc.* **128** (1981) 6.
[173] M. S. Whittingham, *Prog. Solid State Chem.* **12** (1978) 41.
[174] P. G. Dickens and M. S. Whittingham, *Quart. Rev.* **22** (1968) 30; M. S. Whittingham, R. A. Huggins, J. Chem. Phys. **54** (1971) 414; M. S. Whittingham, *J. Electrochem. Soc.* **125** (1976) 315; M. S. Whittingham, *Prog. Solid State Chem.* **12** (1978) 41.
[175] P. Hagenmüller, *Prog. Solid State Chem.* **5** (1971) 71.
[176] K. Mizushima, P. C. Jones, P. J. Wiseman, and J. B. Goodenough, *J. Electrochem. Soc.* **132** (1995) 1521; M. G. Thomas, W. I. David, J. B. Goodenough, and P. Groves, *Mater. Res. Bull.* **20** (1985) 1137; L. A. de Piciotto, and M. M. Thackereay, *Mater Res. Bull.* **19** (1984) 1497.
[177] M M. Thackeray, P. J. Johnson, L. A. de Piciotto, P. G. Bruce, and J. B. Goodenough, *Mater. Res. Bull.* **19** (1984) 179; J. M. Tarascon, E. Wang, F. Shokoohi, W. R. McKinnon, and S. Colson, *J. Electrochem. Soc.* **198** (1991) 2859.
[178] K. Mitzushima, P. C. Jones, P. J. Wiseman, and J. B. Goodenough, *Mater. Res. Bull.* **17** (1980) 785.
[179] E. Ceder, Y.-M. Chiang, D. R. Sadoway, M. K. Aydinol, Y.-J. Jang, and B. Huang, *Nature* **392** (1998) 694.

[180]G. G. Amatucci, J. M. Tarascon, and L. C. Klein, *J. Electrochem. Soc.* **143** (1996) 1114; T. Ohzuku, A. Ueda, M. Nagayama, *J. Electrochem. Soc.* **140** (1993) 1862.
[181]T. Ohzuku, M. Kitagawa, and T. Hirai, *J. Electrochem. Soc.* **137** (1990) 769.
[182]H. Kawai, N. Nagata, H. Tukamoto, and A. R. West, *J. Mater. Chem.* **8** (1998) 837.
[183]N. Ravet, J. B. Goodenough, S. Besner, M. Simoneau, P. Hovington, and M. Armand, in: *Proceedings of the 196th ECS Meeting*, Honolulu, October 1999.
[184]N. Ravet, Y. Chouinard, J. F. Magnan, S. Besner, M. Gauthier, and M. Armand, *J. Power Sources* **97–98** (2001) 503.
[185]S. Y. Chung, J. T. Bloking, and Y.-M. Chiang, *Nature Mater.*, **2** (2002) 123.
[186]P. S. Herle, B. Ellis, N. Coombs, and L. F. Nazar, *Nature Mater.*, **3** (2004) 147.
[187]R. Dominko, M. M. Gaberšček, J. Drofenik, M. Bele, and J. Jamnik, *Electrochim. Acta* **48** (2003) 3709.
[188]J. A. Bruce and M. D. Ingram, *Solid State Ionics* **9/10** (1983) 717.
[189]N. A. W. Holzwarth, S. G. Lonic, and S. Rabii, Phys. Rev. B **28** (1983) 1013; J. Dahn, R. Frong, M. J. Spoon, *Phys. Rev. B* **42** (1990) 6424.
[190]P. Poizot, S. Laruelle, S. Grugeon, L. Dupont, and J-M. Tarascon, *Nature* **407** (2000) 496.
[191]P. Balaya, H. Li, L. Kienle, and J. Maier, *Adv. Funct. Mater.* **13** (8) (2003) 621.
[192]H. Li, G. Richter, and J. Maier, *Adv. Mater.* **15** (9) (2003) 736.
[193]M. Bervas, F. Badway, L. C. Klein, and G. G. Amatucci, *Electrochem., Solid-State Lett.*, **8** (2005) A179.
[194]J. Jamnik and J. Maier, *Phys. Chem. Chem. Phys.* **5** (23) (2003) 5215.
[195]See e.g., M. Morita, M. Ishikawa, Y. Matsuda in: *Lithium-Ion Batteries*, Ed. by M. Wakihara, O. Yamamoto Wiley-VCH, Weinheim, 1998.
[196]D. Aurbach and Y. Ein-Eli, *J. Electrochem. Soc.* **142** (1995) 1746.
[197]M. Armand, *Solid State Ionics* **69** (1994) 309; G. Feullade, P. Perche, *J. Appl. Electrochem.* **5** (1975) 63.
[198]B. Scrosati, in: *Lithium-Ion Batteries*, Ed. by M. Wakihara, O. Yamamoto, VCH, Weinheim, 1998.
[199]F. Croce, G. B. Appetechi, L. Persi, and B. Scrosati, *Nature* **394** (1998) 456; W. Wieczorek, Z. Florjanczyk, and R. Stevens, *Electrochim. Acta* **40** (13–14) (1995) 2251.
[200]A. J. Bhattacharyya and J. Maier, *Adv. Mater.* **16** (9–10) (2004) 811.
[201]A. J. Bhattacharyya, M. Dollé and J. Maier, *Electrochem. Solid-State Lett.* **7** (7) (2004) A432.
[202]A. J. Bhattacharyya et al., in preparation. The courtesy of Prof. J.-M. Tarascon, Université de Picardie Jules Verne, of providing the $LiFePO_4$ samples is acknowledged.
[203]J. H. Kennedy, in: *Solid Electrolytes*, Ed. by S. Geller, p. 105, Springer, Berlin, Heidelberg New York 1977.
[204]R. Selim and P. Bro, *J. Electrochem. Soc.* **121** (1974) 1457.
[205]A. van Zyl, *Solid State Ionics* **86/88** (1996) 883.
[206]B. E. Conway, *Electrochemical supercapacitors. Scientific Fundamentals and Technological Applications*, Kluwer Academic/Plenum Publishers, New York, 1999.
[207]B. O'Reagan and M. Grätzel, *Nature* **353** (1991) 737.
[208]M. Grätzel, *Nature* **414** (2001) 338.
[209]C. Wagner in: *Proceedings 7th Meeting International Communications on Electrochem. Thermodynamics and Kinetics*, Lindau (1955), Butterworth, London.
[210]I. Yokota, *J. Phys. Soc. Jpn.* **16** (1961) 2213.
[211]I. Riess, *Solid State Ionics* **44** (1991) 199.
[212]Note that the number of phases necessary to fix the oxygen potential unambiguously is given by the phase rule demanding, for given temperature and pressure, as many phases as components.

[213]N. Wiener, *Cybernetics or Control and Communication in the Animal and the Machine*, John Wiley and Sons, Ltd., New York, 1948; G. J. Murphy, *Basic Automatic Control Theory*, van Nostrand, Princeton (NJ), 1957; K. Göldner, *Mathematische Grundlagen der Systemanalyse*, Verlag Harri Deutsch, Thun, 1981; W. Hahn, F. L. Bauer, *Physikalische und elektrotechnische Grundlagen für Information*, Springer, Berlin, Heidelberg New York 1975.

[214]F. A. Kröger, *Chemistry of Imperfect Crystals*, North Holland Publ. Comp., Amsterdam, 1964.

[215]J. Maier, Angew. *Chem. Int. Ed. Engl.* **32** (3) (1993) 313.

[216]J. Maier, Angew. *Chem. Int. Ed. Engl.* **32** (4) (1993) 528.

[217]J. Maier and G. Schwitzgebel, *Mater. Res. Bull.* **18** (1983) 601.

[218]A more general approach to evaluate the space charge resistance in the linear regime is given in Ref.[221] According to this the current-voltage-relation may be more generally written as $\mathcal{A}\delta E = \rho(x)\delta I$ where δE and δI denote the first-order perturbation in the electric field and the current density; ρ is the local resistivity and $\mathcal{A}$ is a Sturm-Lioville operator defined by $\mathcal{A} \equiv 1 - 2\frac{\partial}{\partial x}\frac{\frac{\partial}{\partial x}}{p(x)+n(x)} + i\Omega\rho(x)$ (*p*, *n*: equilibrium densities of the positive and negative carriers, Ω: frequency).

[219]S. H. Liu and T. Kaplan, *Solid State Ionics* **18 and 19** (1986) 65; J. R. Macdonald, D. R. Franceschetti, and A.P. Lehnen, *J. Chem. Phys.* **73** (1980) 5272.

[220]J. R. Macdonald, *Impedance Spectroscopy*, Wiley and Sons, New York, 1987.

[221]J. Jamnik, S. Pejovnik, and J. Maier, *Electrochim. Acta* **14** (1993) 1975; J. Jamnik, Ph D-Thesis, Ljubljana 1994.

[222]I. Denk, J. Claus, and J. Maier, *J. Electrochem. Soc.* **144** (10) (1997) 3526.

[223]In the original reference (Ref.[3]) the last row was incorrectly presented (α as well as α' were missing).

[224]P. Balaya, J. Jamnik, and J. Maier, *Appl. Phys.* 44. **88** (2006) 062109.

[225]C. Wagner, in: *Progress in Solid State Chemistry*, Vol. 6 Ed. by H. Reiss, Pergamon Press, Oxford, 1971.

[226]J. Maier and G. Schwitzgebel, *Phys. Stat. Sol.* (b) **113** (1982) 535; J. Maier, in: *Ionic and Mixed Conducting Ceramics* **Vol. 210** Ed. by T. A. Ramanarayanan, W. L. Worrell, H. L. Tuller, The Electrochemical Society, Inc., Pennington, 1994, p. 542; J. Maier, in: *Proc. MRS Meeting*, Boston (USA), (1990) 499.

[227]R. Andreaus and W. Sitte, *J. Electrochem. Soc.* **144** (1997) 1040.

[228]J. Jamnik, J. Maier, and S. Pejovnik, *Electrochim. Acta*, **44** (1999) 4139.

[229]J. R. Macdonald and D. R. Franceschetti, *J. Chem. Phys.* **68** (1978) 1614.

[230]J. Jamnik, X. Guo, and J. Maier, *Appl. Phys. Lett.* **82** (17) (2003) 2820.

[231]L. Heyne in: *Solid Electrolytes*, Ed. by S. Geller, Springer, Berlin Heidelberg, New York, 1977, p. 137.

[232]L.D. Burke, H. Rickert, and R. Steiner, *Z. Phys. Chem* N. F. **Z4** (1971) 146.

[233]X. Guo, J. Fleig, and J. Maier, *J. Electrochem. Soc.* **148** (2001) J50.

[234]J. Mizusaki and K. Fueki, *Rev. Chim. Min.* **17** (1980) 356.

[235]I. Riess, *Solid State Ionics* **66** (1993) 331.

[236]X. Guo, J. Fleig, and J. Maier, *Solid State Ionics* **154–155** (2002) 563.

[237]S. Crouch-Baker, *Solid State Ionics*, **45** (1991) 101; ibid. *Solid State Ionics*, **46** (1991) 309.

[238]J. Maier, P. Murugaraj, G. Pfundtner, and W. Sitte, *Ber. Bunsenges. Phys. Chem.* **93** (1989) 1350.

[239]I. Denk, W. Münch, and J. Maier, *J. Am. Ceram. Soc.* **78(12)** (1995) 3265.

[240]J. Maier, *Ber. Bunsenges. Phys. Chem.* **93** (1989) 1468.

[241]I. Riess, *Solid State Ionics* **91** (1996) 221.
[242]I. Riess and D. Cahen, *J. Appl. Phys.*, **82** (1992) 3147.
[243]I. Riess, *Z. Phys. Chem.* **219** (2005) 1.
[244]J. Maier, *J. Phys. Chem. Solids* **46** (1985) 197.
[245]J. Maier and G. Schwitzgebel, H.-J. Hagemann, *J. Solid State Chem.* **58** (1985) 1.
[246]M. Quilitz and J. Maier, *J. Superconductivity* **9(1)** (1996) 121.
[247]Usually the surface reaction is complex. If, e.g., the dissociation of O_2^- is rate determining, it has to be assumed for the described boundary conditions that the rate of the subsequent surface steps are (specifically) "faster" than the diffusion.
[248]T. Bieger, J. Maier, and R. Waser, in: *Defects in Insulating Materials* **Vol. 2** Ed. by O. Kanert, J.-M. Spaeth, World Scientific, Singapore, (1993) 964; T. Bieger, J. Maier, and R. Waser, *Ber. Bunsenges. Phys. Chem.* **97**(9) (1993) 1098.
[249]J. Maier, P. Murugaraj, G. Pfundtner, and A. Rabenau, *Solid State Ionics* **40/41** (1990) 802.
[250]J. Crank, *The Mathematics of Diffusion*, Oxford University Press, Glasgow, 1956; H. S. Carslaw and J. C. Jaeger, *Conduction of Heat in Solids*, Clarendon Press, Oxford, 1959.
[251]H. Schmalzried, *Z. Phys. Chem. N. F.* **38** (1963) 87.
[252]J. Maier and G. Schwitzgebel, *Mater. Res. Bull.* **17** (1982) 1061.
[253]I. Riess, *Solid State Ionics* **44** (1990) 207.
[254]R. Metselaar and P. K. Larsen, *J. Phys. Chem. Solids* **37** (1976) 599.
[255]G. Pfundtner, Ph.D. Thesis, University of Tübingen, 1993.
[256]J. Maier, *Solid State Ionics* **135(1–4)** (2000) 575.
[257]M. Leonhardt, Ph D thesis, Stuttgart, 1999.
[258]J. Maier, J. Jamnik, and M. Leonhardt, *Solid State Ionics* **129**(1–4) (2000) 25.
[259]M. Leonhardt, J. Jamnik, and J. Maier, *Electrochem. Solid-State Lett.* **2(7)** (1999) 333.
[260]W. Weppner and R. A. Huggins, *Annu. Rev. Mat. Sci.* **8** (1978) 269.
[261]J. Mizusaki, *J. Solid State Chem.* **131** (1997) 150.
[262]U. von Oehsen and H. Schmalzried, *Ber. Bunsenges. Phys. Chem.* **85** (1981) 7.
[263]J. Novotny, M. Rekas, and W. Weppner, *J. Am. Ceram. Soc.* **73** (1990) 1040.
[264]S. Yamaguchi, K. Terabe, A. Saito, and Y. Iguchi, *Jpn. J. Appl. Phys.* **27** (1988) L179.
[265]J. Maier and G. Pfundtner, *Adv. Mater.* **3** (1991) 292.
[266]J. Maier and M. Quilitz, in: *Proceedings of the Electroceramics IV*, **Vol. II**, Ed. by R. Waser, S. Hoffmann, D. Bonnenberg, Ch. Hoffmann **Vol. II**, Augustinus Buchhandlung (1994) 901.
[267]W. Preis and W. Sitte, *Solid State Ionics* **86–88** (1996) 779.
[268]N. Valverde, *Z. Phys. Chem. NF* **74** (1971) 146; W. Piekarcyak, W. Weppner, A. Rabenau, *Z. Naturforsch* **34a** (1979) 430; W. Weppner, and R. A. Huggins, *J. Electrochem. Soc.* **124** (1977) 10.
[269]M. Holzinger, A. Benisek, W. Schnelle, E. Gmelin, J. Maier, and W. Sitte. *J. Chem. Thermodyn.* **35** (2003) 1469.
[270]H. Schmalzried and A. Navrotsky, *Festkörperthermodynamik*, VCH, Weinheim, 1975.
[271]J. J. Egan, *J. Phys. Chem.* **68** (1964) 978; R. J. Heus, J. J. Egan, *Z. Phys. Chem.* **49** (1966) 38.
[272]U. Warhus, J. Maier, and A. Rabenau, *J. Solid St. Chem.* **72** (1988) 113.
[273]A. Schroeder, J. Fleig, H. Drings, R. Wuerschum, J. Maier and W. Sitte, *Solid State Ionics* **173** (1–4) (2004) 95.
[274]P. Knauth and G. Schwitzgebel, A. Tschöpe, and S. Villain, *J. Solid State Chem.* **140** (1998) 295.
[275]W. H. Mulder and J. H. Sluyters, *J. Electroanal. Chem.* **468** (1999) 127.

[276]T. Bieger, H. Yugami, N. Nicoloso, J. Maier, and R. Waser, *Solid State Ionics* **72** (1994) 41.

[277]J. Maier and W. Münch, *J. Chem. Soc., Faraday Trans.* **92** (12) (1996) 2143; J. Maier in: *Ionic and Mixed Conducting Ceramics*, **Vol. 94–12** Ed. by T. A. Ramanarayanan, W. L. Worrell, and H. L. Tuller, The Electrochemical Society., Pennington, 1994, p. 542.

[278]K. Funke, *Progr. Solid State Chem.* **22** (1993) 111. K. Funke et al., *Phys. Chem. Chem. Phys.* **4** (2002) 2155. Please note that in Part I^2 in text on page 116 above Eq. (128) –W and –g have to be replaced by $-\dot{W}$ and $-\dot{g}$.

[279]K. Funke, B. Roling, and M. Lange, *Solid State Ionics* **105** (1998) 195.

[280]K. Funke and I. Riess, *Z. Phys. Chem. NF* **140** (1984) 217.

[281]A. K. Jonscher, *Nature* **267** (1977) 673; K. L. Ngai, *Comments Solid State Phys.* **9** (1979) 127; W. K. Lee, J. F. Liu, A. S. Nowick, *Phys. Rev. Lett.* **67** (1991) 1559.

[282]P. Maass, J. Petersen, A. Bunde, W. Dieterich, and H. E. Roman, *Phys. Rev. Lett.* **66** (1991) 52; J. Petersen, W. Dieterich, *Philos. Mag. B* **65** (1992) 231; B. Rinn, W. Dieterich, P. Maass, *Philos. Mag. B* **77** (1998) 1283; P. Maass, M. Meyer, and A. Bunde, *Phys. Rev. B* **51** (1995) 8164.

[283]W. Dieterich and P. Maass, *Chem. Phys.* **284** (2002) 439.

[284]A. Bunde and W. Dieterich, *J. Electroceram.* **5** (2000) 81.

[285]J. Maier, *Ber. Bunsenges. Phys. Chem.* **90** (1986) 26.

[286]J. Fleig and J. Maier, *Electrochim. Acta* **41** (7/8) (1996) 1003.

[287]M. Kleitz, H. Bernard, E. Fernandez, and E. Schouler, *Adv. Ceram. Sci. Tech. Zirconia* **3** (1981) 310; M. Meyer, H. Rickert, and U. Schwaitzer, *Solid State Ionics* **9, 10** (1983) 689; J. E. Bauerle, *J. Phys. Chem. Solids* **30** (1969) 2657; H. Rickert, H.-D. Wiemhöfer, *Ber. Bunsenges. Phys. Chem.* **87** (1983) 236.

[288]R. Holm, *Electrical Contacts Handbook*, Springer, Berlin, Heidelberg, New York, 1958; J. Newman, *J. Electrochem. Soc.* **113** (1966) 501.

[289]W. G. Amey and F. Hamburger, *Proc. Am. Soc. Test. Mater.* **49** (1949) 1079.

[290]J. Fleig, and J. Maier, *Solid State Ionics* **86–88** (1996) 1351.

[291]J. Jamnik, J. Fleig, and J. Maier in: *Mater. Res. Soc. Symp. Proc.* **411** (1996) 25.

[292]J. Fleig and J. Maier, *Solid State Ionics* **85** (1996) 9.

[293]J. Fleig, S. Rodewald, and J. Maier, *J. Appl. Phys.* **87** (5) (2000) 2372.

[294]W. Zipprich, and H.-D. Wiemhöfer, *Solid State Ionics* **135** (2000) 699; H.-D. Wiemhöfer, H.-G. Bremes, U. Nigge, W. Zipprich, *Solid State Ionics* **150** (2002) 63.

[295]A. Uhlir, *Bell. Syst. Tech. J.* **34** (1955) 105.

[296]L. J. van der Pauw, *Philips Res. Repts.* **13** (1958) 1; I. Riess, D. S. Tannhauser, *Solid State Ionics* **7** (1982) 307.

[297]L. B. Valdes, in: *Proc. I.R.E.* **42** (1954) 420.

[298]J. Jamnik, H.-U. Habermeier, and J. Maier, *Physica B* **204** (1995) 57.

[299]S. V. Kalinin and D. A. Bonnell, *J. Appl. Phys.* **91** (2) (2002) 832.

[300]J. Fleig and J. Maier (eds.), *J. Electroceram.* **5** (2) (2000) 79.

[301]J. Janek and C. Korte, *Solid State Ionics* **92** (1996) 193; 15.

[302]C. H. J. Stuhrmann, H. Kreiterling, and K. Funke, *Solid State Ionics* **154** (2002) 104.

[303]H. S. Carslaw and J. C. Jäger, *Conduction of Heat in Solids*, Clarendon Press, Oxford, 1959.

Synthesis and Characterization of Nanoporous Carbon and Its Electrochemical Application to Electrode Material for Supercapacitors

Gyoung-Ja Lee and Su-Il Pyun*

Department of Materials Science and Engineering, Korea Advanced Institute of Science and Technology, Guseong-dong, Yuseong-gu, Daejeon, Republic of Korea
**Corresponding author*

I. INTRODUCTION

Nanoporous carbons have played a key role in many areas of modern science and technology such as purification of liquids and gases,[1,2] separation of mixture,[3,4] catalyst for reaction,[5-7] and storage of energy[8-17]. Especially, the highly porous carbons are widely used as electrode materials in electric double-layer capacitors (EDLCs), due to such advantageous features as high surface area, good electrical conductivity, and adequate corrosion resistance.[8-17] The pores are classified into three groups according to their size:

- pores with width (distance between the walls of a slit-shaped pore) smaller than 2 nm are called micropores,
- pores with width between 2 nm and 50 nm are called mesopores, and
- pores with width larger than 50 nm are called macropores.

Modern Aspects of Electrochemistry, Number 41, edited by C, Vayenas et al., Springer, New York, 2007.

The physical and chemical activation processes have been generally employed to prepare the porous carbons.[18-35] However, the pore structures are not easily controlled by the activation processes and the size of the pores generated by the activation processes is limited to the micropore range only. Recently, much attention has been paid to the synthesis of meso/macroporous carbons with various pore structures and pore size distributions (PSD) by using various types of such inorganic templates as silica materials and zeolites.[17,36-55]

The electro-active surface of the porous carbon electrode for EDLCs is accessible only through the cumulative resistance of the electrolyte inside the pore. Therefore, the porous structure of the porous carbon becomes one of the most important factors influencing the energy/power densities. Fractal analysis has proven to be useful to describe the geometric and structural properties of rough surfaces and pore surfaces.[56-66]

There are two conventional definitions in describing the fractality of porous material - the pore fractal dimension which represents the pore distribution irregularity[56,59,62] and the surface fractal dimension which characterizes the pore surface irregularity.[56,58,65] Since the geometry and structure of the pore surfaces are closely related to the electro-active surface area which plays a key role in the increases of capacity and rate capability in practical viewpoint, the microstructures of the pores have been quantitatively characterized by many researchers based upon the fractal theory.

In this respect, this review provides a comprehensive survey of synthetic methods and physicochemical properties of the porous carbon materials. Furthermore, as electrochemical applications of the porous carbons to electrode materials for supercapacitor, the effects of geometric heterogeneity and surface inhomogeneity on ion penetration into the pores during double-layer charging/discharging are discussed in detail by using ac-impedance spectroscopy, current transient technique, and cyclic voltammetry.

The structure of this review is as follows: in Section II, the activation and templating methods for preparing the porous carbons are briefly summarized. Section III surveys the structural characteristics of the porous carbons by using gas adsorption method. In Section IV, the molecular probe method and the image analysis method for quantitative characterization of the pore surface irregularity and the size distribution irregularity based upon the fractal theory are discussed in detail. Section V is devoted to the investigation of the electrochemical performance of the porous carbon electrodes for EDLCs considering the effects of geometric heterogeneity and surface inhomogeneity on kinetics of double-layer charging/discharging.

II. PREPARATION OF POROUS CARBONS

1. Activation method

The activated carbons have been prepared by carbonization of precursors, i.e., raw materials with subsequent activation by physical treatment and/or chemical treatment. The activation procedure is conducted in order to enlarge the volume and size of the pores which were already created during the carbonization process and to create some new porosity. Most commonly employed activation methods are divided into physical and chemical activations.[14-16,18-35]

In physical activation, the gasification of the carbonized carbon usually occurs at higher temperature above 800 °C in the presence of suitable oxidizing gases such as steam, carbon dioxide CO_2, or a mixture of these.[14-16,18,19,21-28,30,32,34,35] Activation of the carbon in atmospheres of steam and CO_2 occurs by the following endothermic reactions.

$$C\ (s) + H_2O\ (g) \rightarrow CO\ (g) + H_2\ (g) \tag{1}$$

$$C\ (s) + CO_2\ (g) \rightarrow 2CO\ (g) \tag{2}$$

Walker et al.[18] found that the reaction (2) took place uniformly throughout the entire volume of the carbons during gasification, resulting in the attainment of high internal surface area. Since the reactions of the carbon with steam and CO_2 are both endothermic, the external heating is required to drive the above reactions and to maintain the reaction temperature. The activation with CO_2 involves a less energetic reaction than that with steam and thus requires a higher temperature.

The surface area, the pore size, and the carbon yield are greatly influenced by oxidizing gas, heat-treatment temperature, and activation time. In addition, the oxidative activation introduced the surface acidic functional groups (SAFGs) on the pore surfaces of the carbon. The amounts of the SAFGs can be also controlled by changing the activation conditions. Extensive burn-off at higher temperatures increases the surface area and pore size with less development of the SAFGs.

In chemical activation, the carbonaceous precursor is impregnated with a chemical agent, and then the impregnated product is pyrolyzed between 400 °C and 800 °C in an inert atmosphere. After that, the pyrolyzed product is cooled and washed exhaustively in order to remove the chemical agent, followed by filtration of the carbon materials.[19-23,26,29,31-35] The chemical agents used are normally alkali, alkali earth metal, and some acids such as KOH,[22,26,29,32,34,35] K_2CO_3,[32] NaOH,[32,34] Na_2CO_3,[32] $ZnCl_2$,[20-23] and H_3PO_4,[23,31,33] etc.

The main advantages of the chemical activation compared to the physical activation are lower activation temperature, shorter activation time, and higher development of the porosity. Furthermore, since the chemical agents used are just substances with dehydrogenation properties which inhibit formation of tar and reduce the production of other volatile products, the carbon yields of the chemically-activated carbons are usually higher than those of the physically-activated carbons. However, the general mechanism of the chemical activation is not well understood, as compared to that of the physical activation. Other disadvantages of the chemical activation are the need of additional washing procedure in order to remove the chemical agents and the corrosiveness of the chemical activation process.

Since the heat-treatment temperature used in the chemical activation is lower than that used in the physical activation, the porous structure is more developed in the case of the chemical activation. The higher the degree of impregnation, the larger is the pore size of the activated carbon. Generally, kind of chemical agent, impregnation ratio of chemical agent to the carbonaceous material, and physical mixing method for distribution of chemical agent into the carbonaceous material largely affect the resulting pore structures of the chemically-activated carbons.

The physical and chemical activation methods are effective in preparing the microporous carbons with high surface area. However, the pore structures of the carbons are not easily controlled by the activation processes and the size of the pores generated by the activation processes is limited to the micropore range only. Under these circumstances, the templating method which will be considered in the following Section II.2 has recently attracted considerable attention in preparing porous carbons with various pore structures.

2. Templating Method

The meso/macroporous carbons have attracted much attention in their application as electrode materials in EDLCs, since the meso/macropores promote the formation of an effective double-layer or the transfer of ions into the pores, resulting in the increases in the electrolyte wettability and the rate capability.[67,68] In this regard, there has been considerable research targeted towards developing the synthetic methods of novel meso/macroporous carbons.[17,36-55,69-72] Various types of such inorganic templates as silica materials and zeolites are widely used for the synthesis of the meso/macroporous carbons, since it was revealed[17,36-55] that these inorganic templates contribute to the formation of the meso/macropores with various pore structures and broad PSD.

In templating method, many polymeric precursors including resorcinol-formaldehyde gel,[38,40,42,69,70,73-76] furfuryl alcohol,[41,49,53,55,76,77] phenolic resin,[49,72,74] melamine-formaldehyde resin,[69,72] and mesophase pitch,[17,45,46,54] etc. are generally used as carbonaceous precursors. The porous carbons are prepared by templating method according to the following procedures:

1. mixing of inorganic templates and carbonaceous precursors,
2. polymerization of the initial mixture,
3. carbonization of the polymerized composite, and
4. removal of the inorganic template.

Among the inorganic templates, zeolite produces more regulated pores as compared to the silica template. If nano-channels in zeolite are completely filled with carbonaceous precursor and then the carbon materials are extracted from the zeolite framework, one can obtain the porous carbon of which structure reflects the porosity of the original zeolite template. The ordered mesoporous silica templates, e.g., MCM-48[38,39,47] and SBA-15[47] have been employed to prepare the ordered porous carbons by the procedures involving the pore filling of the silica template with carbonaceous precursor followed by carbonization and silica dissolution. The resulting pore sizes of the ordered mesoporous carbons are smaller than about 10 nm.

In order to prepare the meso/macropores with the size of several decades of nanometers, the silica colloids have been frequently used as a template since the colloidal silica templates are reproducible and effective for the synthesis of the porous carbons with larger pore size, broad PSD, high pore volume, and high surface area by sol-gel process.[17,38,40,42,45,46,54,55] The silica gels are produced from the polycondensation of tetramethoxy silane (TMOS) or tetraethoxy silane (TEOS) in the presence of an acid or base. The shape and size of the silica are significantly influenced by the degree of silica particle agglomeration which is dependent upon the sol-gel reaction conditions, and hence the pore structure which reflects the structure of the silica template also depends upon the sol-gel reaction conditions, e.g., pH, amount of the silica template, and gelation temperature.

Recently, Pyun et al.[17,54,55] prepared various meso/macroporous carbon specimens with various pore structures using the templating method with the changes of carbonaceous precursor, colloidal silica sol solution, amount of the silica template, and pH. Figure 1 plots the PSD curves for three kinds of mesoporous carbon specimens I, II, and III synthesized by using furfuryl alcohol, $C_5H_6O_2$ and commercial colloidal silica sol solution of Ludox SM-30 as a carbonaceous precursor and silica template, respectively, with the change of pH value of initial solution mixture.[55] The mixture of furfuryl alcohol and silica sol solution was heated in a vacuum oven at 100 °C for 5 days to polymerize the furfuryl alcohol. After that, the composite was carbonized at 900 °C for 3 h under Ar gas atmosphere, followed by dissolution of silica with 48 wt. % aqueous HF solution at room temperature.

It is noticeable from Figure 1 and Table 1 that the values of average pore diameter D_{ave} and standard deviation of PSD σ for specimens I, II, and III increased as well with increasing pH. Since the silica particles in colloidal silica sol solution of Ludox SM-30 have a uniform particle size of about 7 nm, it is reasonable to say that the formation of the pores larger than silica particle size is caused by the agglomeration of the silica sol particles. Besides, as the pH increased, the values of micro- and mesoporosity decreased, whereas that value of macroporosity increased. Accordingly, it is suggested that as the pH increases, the stability of silica particles decreases and hence causes higher agglomeration of the silica particles, which results in the increase of the relative volume fraction of the pores with a larger size.

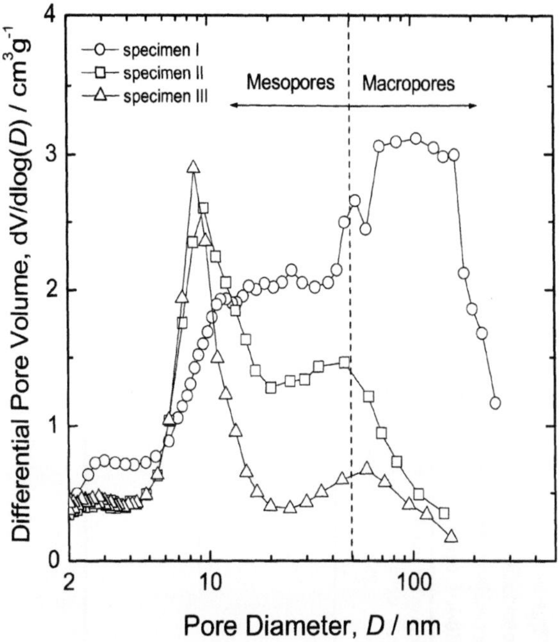

Figure 1. Plots of differential pore volume against pore diameter calculated from the N_2 gas adsorption isotherms obtained from meso/macroporous carbon specimens I (-○-), II (-□-), and III (-Δ-) using Barrett-Joyner-Halenda (BJH) method. Reprinted with permission from G. -J. Lee and S. -I. Pyun, *Carbon*, **43** (2005) 1804. Copyright © 2005, with permission from Elsevier.

III. STRUCTURAL CHARACTERISTICS OF POROUS CARBONS

1. Types of Adsorption Isotherms and Hysteresis Loops

The first systematic attempt to interpret adsorption isotherms for gas/solid equilibria was by Braunauer, Deming, Deming, and Teller (BDDT) in 1940.[78] They classified the adsorption isotherms into five types, and the BDDT classification became the core of the modern International Union of Pure and Applied Chemistry

Table 1.
The Pore Structures and Surface Fractal Characteristics of Meso/Macroporous Carbon Specimens I, II, and III Calculated from the Analyses of the N_2 Gas Adsorption Isotherms. Reprinted with permission from G. -J. Lee and S. -I. Pyun, *Carbon*, 43 (2005) 1804. Copyright © 2005, with permission from Elsevier.

Specimen	BET surface area, A_{BET} ($m^2 g^{-1}$)	Micro-porosity[a]	Meso-porosity[a]	Macro-porosity[a]	Average pore diameter, D_{ave} (nm)	Standard deviation of PSD, σ (nm)	Pore radius of the maximum differential pore volume, r_{max} (nm)	Surface fractal dimension, $d_{surf,SP}$[a]	Low-end thickness, $t_{ad,min}$ (nm)	High-end thickness, $t_{ad,max}$ (nm)
I ($SiO_2/C_5H_6O_2 = 0.7$, pH = 10.2)	849	0.01	0.49	0.50	17.0	74.2	53.4	2.51 ± 0.03	0.35	36.74
II ($SiO_2/C_5H_6O_2 = 0.7$, pH = 8.7)	826	0.04	0.75	0.21	8.8	23.2	4.8	2.62 ± 0.02	0.35	3.68
III ($SiO_2/C_5H_6O_2 = 0.7$, pH = 7.2)	790	0.06	0.80	0.14	6.4	20.9	4.3	2.66 ± 0.02	0.35	3.00

[a]Dimensionless

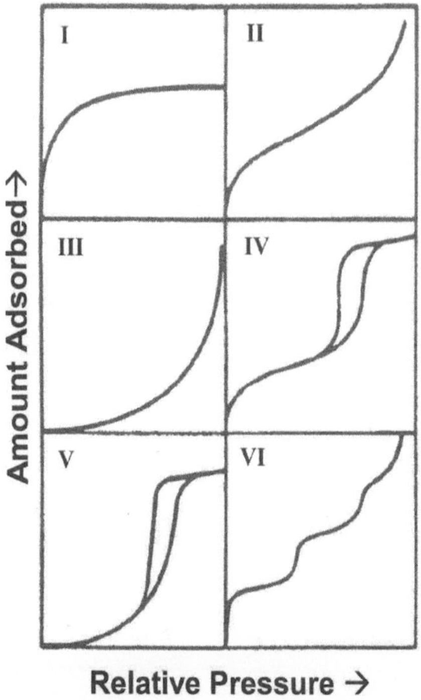

Figure 2. Classification of adsorption/desorption isotherms.

(IUPAC) classification of the adsorption isotherms[79] in Figure 2. The sixth isotherm in Figure 2 is recently added.

Type I isotherms are originally known as Langmuir isotherms since they represent Langmuir mechanism of monolayer coverage. In fact, most Type I isotherms can be attributed to the physisorption on the microporous adsorbents with relatively small external surfaces.Type II isotherms (S-shaped or sigmoid) are monolayer-multilayer isotherms, which are normally obtained from non-porous or macroporous adsorbents. Type IV isotherms are given by adsorbents which possess mesopore structure, such as many silica gels, mesoporous carbons, and other porous oxides. The remaining isotherms in the IUPAC classification exhibit particular features, which indicate weak adsorbate-adsorbent interactions in the case of Type III and V isotherms. The stepwise Type VI isotherms, indicative of a nonporous solid with an almost completely uniform surface, are quite rare.

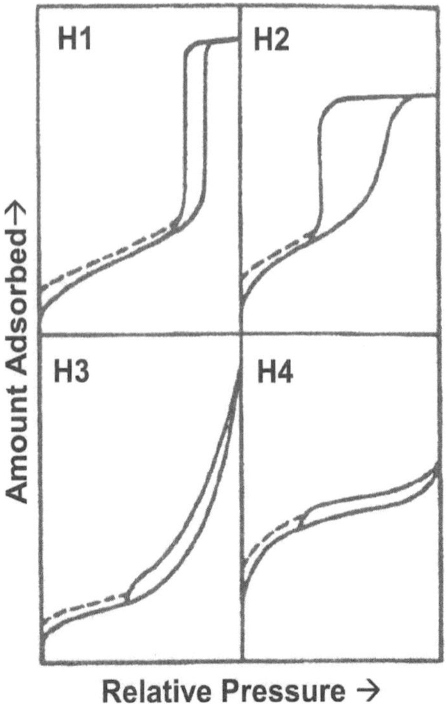

Figure 3. Classification of adsorption/desorption hysteresis loops.

For many of activated microporous carbons,[80-84] the isotherms exhibit prominent adsorption at low relative pressures and then level off, i.e., the isotherms exhibit Type I behavior. Type I isotherms may be also observed in the mesoporous materials with pore sizes close to the micropore range. In particular, in the case of gas adsorption on highly uniform cylindrical pores, the adsorption isotherms exhibit discernible steps at relative pressures down to 0.1 or perhaps even lower.[85-87] Such Type I behavior can be indicative of some degree of broadening of the mesopore size distribution.

Adsorption on macroporous carbons proceeds via multilayer formation in such a manner that the amount adsorbed increases gradually as the relative pressure increases, and then the multilayer build-up near the saturation vapor pressure may be abrupt. This unrestricted monolayer-multilayer adsorption gives rise to Type II and III isotherms. In this case, the adsorption and desorption branches of the isotherms coincide fairly well with each other, i.e., there is no hysteresis loop.

On the other hand, for the mesoporous carbons, the gas adsorption process changes with increasing relative pressure as follows: first monolayer-multilayer adsorption on the pore surfaces, then capillary condensation in the mesopores, and finally multilayer adsorption on the outer surfaces in sequence (Types IV and V isotherms). Therefore, the adsorption process for the mesoporous carbons is initially similar to that for the macroporous carbons, but at higher relative pressures the amount adsorbed rises very steeply due to the capillary condensation in the mesopores.

Capillary condensation and evaporation do not reversibly take place at the same pressure, which leads to the appearance of hysteresis loops. The hysteresis loop, which is due to the different processes between adsorption into and desorption from the mesopores, is closely related to the pore structure of the mesoporous material.[54,86,88] Thus, the pore structures of the mesoporous carbons can be estimated from the shape of the hysteresis loop.

According to the IUPAC,[79] the hysteresis loops are classified into four types from Type H1 to Type H4. The Type H1 loop in Figure 3 is characteristic of the mesoporous materials consisting of the pores with cylindrical pore geometry or the pores with high degree of pore size uniformity.[88,89] Hence, the appearance of the H1 loop on the adsorption isotherms for the porous solids generally indicates facile pore connectivity and relative narrow PSD.

Many porous adsorbents give Type H2 hysteresis loop, but in such systems PSD or pore shape is not well-defined. Indeed, the H2 loop is especially difficult to interpret. In the past it was considered to be a result of the presence of the pores with narrow necks and wide bodies (ink-bottle pores), but it is now recognized that this provides an over-simplified picture and the pore connectivity effects must be taken into account.[79]

Type H3 hysteresis loop, which does not level off near the saturation vapor pressure, is characteristic of the mesoporous materials being comprised of agglomerates of plate-like particles with slit-shaped pores.[79,86] Type H4 loop, which features parallel and almost horizontal branches, is attributable to the adsorption/ desorption in narrow slit-like pores. However, Type H4 loop was recently reported for MCM-41 being comprised of particles with internal voids of irregular shape and broad PSD,[90] and also observed for hollow spheres with walls composed of ordered mesoporous silica.[91]

2. Determinations of Surface Area and Pore Size Distribution

The Brunauer, Emmett, and Teller (BET) theory,[92] which generalized the Langmuir theory[93] by incorporating the concept of multilayer adsorption, has been generally used for the determination of the surface area of porous materials, in spite of oversimplication of the model upon which the theory is based. The derivation of the BET equation involves the following assumptions: The surface is flat; All adsorption sites exhibit the same adsorption energy; There are no lateral interactions between adsorbed molecules; The adsorption energy for all the molecules except for the first layer is equal to the liquefaction energy, and an infinite number of layers can be formed on the surface.

Assuming that the condensation rate of the gas molecules onto an already adsorbed layer is equal to the evaporation rate of the molecules from the adsorbed layer, the BET equation can be obtained in a linear form as follows

$$\frac{p}{V(p^{\circ}-p)}=\frac{1}{V_{\text{momo}}C}+\frac{C-1}{V_{\text{mono}}C}\left(\frac{p}{p^{\circ}}\right) \tag{3}$$

where V is the volume of adsorbed gas molecules at the relative pressure p/p^o; V_{mono}, the volume of monolayer coverage; p^o, the saturation pressure of the gas, and p represents the adsorption equilibrium pressure of the gas. The values of V_{mono} and constant C can be estimated from the linearity of the plot of $p/V(p^o - p)$ *vs.* p/p^o, i.e., the BET plot. According to the BET theory, the value of C is related to the enthalpy (heat) of adsorption in the first adsorbed layer Q_l

$$C \propto \exp\left(\frac{Q_1 - Q_L}{RT}\right) \tag{4}$$

where Q_L is the heat of liquefaction of the adsorbate, R and T are the gas constant and the absolute temperature, respectively. It is now recognized that the value of C gives an indication of the magnitude of the adsorbent-adsorbate interaction energy.

The total BET surface area A_{BET} is calculated from the volume of monolayer coverage V_{mono} to

$$A_{\mathrm{BET}} = \frac{V_{\mathrm{mono}} A_{\mathrm{c}} N_{\mathrm{A}}}{V_{\mathrm{m}}} \tag{5}$$

where A_{c} is the molecular cross-sectional area; N_A, the Avogadro constant (6.02×10^{23} molecules mol^{-1}), and V_{m} is the molar volume of the ideal gas (22414 cm^3). The specific surface area can be obtained by dividing A_{BET} by the mass of the adsorbent m. The values of A_{c} are given as 0.162 nm^2, 0.142 nm^2, and 0.215 nm^2 for N_2, Ar, and Kr, respectively.[94]

Now let us overview the theoretical adsorption models for characterization of the pore structures according to the pore size range. For physical adsorption of the gas molecules on such microporous solids as activated carbons and zeolites, Dubinin and Radushkevich[95] developed an empirical equation, which describes the volume filling process in the micropores. Their theory incorporates earlier work by Polanyi[96] in regard to the adsorption potential A_{ad} defined as

$$A_{\mathrm{ad}} = -\Delta G = RT \ln(p^{\circ} / p) \tag{6}$$

Here, ΔG is Gibbs' free energy. For carbon materials being comprised of slit-shaped pores, the Dubinin-Radushkevich (D-R) equation is given as

$$V = V_{\mathrm{o}} \exp\left[-\left(\frac{A_{\mathrm{ad}}}{\beta E_{\mathrm{o}}} \right)^2 \right] \tag{7}$$

where V_{o} is the total micropore volume; β, the affinity coefficient, and E_o is the characteristic adsorption potential for the reference vapor, benzene. Dubinin and Stoeckli[97] suggested the relationship between E_o and average half-width x of the slit-shaped pores, $E_o = k/x$. Here, the empirical constant k is assumed to be 12 kJ nm mol^{-1} for nitrogen gas.[94]

A more general expression, called the Dubinin-Astakhov (D-A) equation[98] is written as

$$V = V_o \exp\left[-\left(\frac{A_{ad}}{\beta E_o}\right)^n\right] \tag{8}$$

Here, n is the equation parameter. The D-R equation is a special case for $n = 2$.

In order to determine the PSD of the micropores, Horvath-Kawazoe (H-K) method has been generally used. In 1983, Horvath and Kawazoe[99] developed a model for calculating the effective PSD of slit-shaped pores in molecular-sieve carbon from the adsorption isotherms. It is assumed that the micropores are either full or empty according to whether the adsorption pressure of the gas is greater or less than the characteristic value for particular micropore size. In H-K model, it is also assumed that the adsorbed phase thermodynamically behaves as a two-dimensional ideal gas.

The H-K method is based upon the model suggested by Everett and Powl[100] which describes the interaction potential of a single adsorbate molecule between two parallel planes of the atoms of graphitized carbon. In the H-K expansion of the Everett and Powl's work, the space between the parallel carbon planes, i.e., the pore is assumed to be filled with adsorbed gas molecules. Thus, the contribution of adsorbate-adsorbate-adsorbent interaction to the total interaction potential is considered along with that of adsorbate-adsorbent interaction.

The H-K method has been modified by many researchers in consideration of the pore geometry. Saito and Foley[101] extended the H-K approach to the case of cylindrical pores. Based upon the Everett-Powl model, they assumed that the pore is perfectly cylindrical in shape and infinite in length, and the inside wall of the cylindrical pore is a single layer of atoms with a continuum of interaction potential. Cheng and Yang[102] extended the H-K equation to the case of spherical pores, notably cavities in zeolite. In Cheng and Yang model, the only interactions are considered between the adsorbate and oxygen atoms of the spherical cavity wall within the zeolite, and the cavity wall is supposed to be comprised of a single lattice plane of atoms.

In the case of the meso/macropores, the physical adsorption related to the capillary condensation is generally described with the aid of following Kelvin equation[94,103]

$$\frac{1}{r_1}+\frac{1}{r_2}=-\frac{RT}{\gamma_L V_{cond}\cos\theta}\ln\left(\frac{p^*}{p^o}\right) \tag{9}$$

where r_1 and r_2 are two primary radius of curvature of the liquid meniscus in the pores; γ_L, the liquid surface tension; V_{cond}, the molar volume of the condensed adsorbate; θ, the contact angle between the solid and condensed phase (taken to be zero for N_2, hence $\cos\theta = 1$), and p^* represents the critical condensation pressure of the gas.

The pore shape is generally assumed to be either cylindrical or slit-shaped: in the former case, the meniscus is hemispherical, and hence r_1 is equal to r_2; in the latter case, the meniscus is hemicylindrical, and thus r_1 is equal to the width of the slit and r_2 is infinite. From the relation $\frac{1}{r_{L,m}}=\frac{1}{2}\left(\frac{1}{r_1}+\frac{1}{r_2}\right)$, the mean radius of curvature of the liquid meniscus, $r_{L,m}$ is expressed as

$$r_{L,m}=\frac{2\gamma_L V_{cond}\cos\theta}{RT\ln(p^*/p^o)} \tag{10}$$

It should be recognized that the Kelvin equation provides the core radius instead of the actual pore radius. The core radius represents the radius of the inner free space in the pore, which is not yet filled with the adsorbate. Therefore, in order to obtain the actual pore radius, the Kelvin equation needs to be corrected for the actual thickness of adsorbed molecule layers t_{ad}. The radius of a cylindrical pore, r_p is given by

$$r_p = r_{L,m} + t_{ad} \tag{11}$$

and the width of a parallel-sided slit, w_p is written as

$$w_p = r_{L,m} + 2t_{ad} \tag{12}$$

Based upon the Kelvin equation, the PSD of the meso/macropores has been generally determined by Barrett, Joyner, and Halenda (BJH) method.[104] Furthermore, the density functional theory[94] which is based upon a molecular-based statistical thermodynamic theory was recently introduced in order to analyze the PSD of the porous materials from micropores to macropores in size in considerations of the adsorbate-adsorbate and adsorbate-adsorbent interaction energy parameters, the pore size, the pore geometry, and the temperature.

IV. FRACTAL CHARACTERISTICS OF POROUS CARBONS

The real objects found in nature have complex structures which Euclidean geometry can not characterize. After Mandelbrot[105] developed a new geometry, i.e., fractal geometry, which provides a new paradigm for understanding many physical phenomena in nature, fractal geometry has been widely used in a number of fields, e.g., science, art,[106-108] economics,[109-112] etc. Especially, in science, fractal analysis has proven to be useful to describe the geometric and structural properties of rough surfaces and pore surfaces.[56-66] Since the geometry and structure of the pore surfaces are closely related to the electro-active surface area which plays a key role in the increases of capacity and rate capability in practical viewpoint, the microstructures of the pores have been quantitatively characterized by many researchers based upon the fractal theory.

There are two conventional definitions in describing the fractality of porous material—the surface fractal dimension[56,58,65] and the pore fractal dimension.[56,59,62] The former fractal dimension characterizes the pore surface irregularity: the larger the value of the surface fractal dimension is, the more irregular and the rougher is the pore surface. The surface fractal dimensions of 2 and 3 mean a perfectly flat pore surface and a very rough pore surface, respectively. On the other hand, the latter fractal dimension represents the pore distribution irregularity: the larger the value of the pore fractal dimension is, the narrower is the PSD that exhibits a power-law behavior. The pore fractal dimensions of 2 and 3 indicate the porous electrode with homogeneous PSD and that electrode composed of the almost same-sized pores, respectively.

Several simple relations have been proposed for the determination of the fractal dimensions from the results of such experiments as gas molecular probe method, transmission electron microscopy (TEM), small-angle X-ray scattering, neutron scattering, and laser light scattering.[63,66,113-116] Among those techniques, gas molecular probe method and image analysis method have been widely used for the calculation of the surface and pore fractal dimensions of the porous materials, since theoretical bases are well established and their experimental procedures are relatively simple. The detailed descriptions of gas molecular probe method and image analysis method are given in the following Sections VI.1 and VI.2.

1. Molecular Probe Method Using Gas Adsorption

Since Avnir and Pfeifer's pioneer works[56,117-119] regarding the characterization of the surface irregularity at the molecular level by applying the fractal theory of surface science, molecular probe method using gas adsorption has played an important role in the determination of the surface fractal dimensions of the porous materials.

There are two molecular probe methods available for the determination of the surface fractal dimension. One is the multiprobe method (MP method),[56,117,120-133] which uses several kinds of multiprobe molecules with different molecular sizes. If the probe molecule is varied through a series of spheres with radius r_m, the surface fractal dimension is given by

$$N_{\text{mono}} \propto r_{\text{m}}^{-d_{\text{surf,MP}}} \tag{13}$$

Here, N_{mono} is the number of adsorbed molecules to form a monolayer for each probe molecule and $d_{\text{surf,MP}}$ is the surface fractal dimension determined by using the MP method. The probe molecules need not to be spherical, provided they belong to a homologous series for which the ratio [linear extent $r_m]^2$ to molecular cross-sectional area A_c is the same for all members, i.e., an isotropic series. In this case, Eq. (13) turns into

$$N_{\text{mono}} \propto A_{\text{c}}^{-\frac{d_{\text{surf,MP}}}{2}} \tag{14}$$

The MP method is widely applicable to various solid surfaces in whose system N_{mono} can be accurately determined. It was recently reported[64,134-136] that even for the micropore whose pore width is greater than the bilayer thickness of N_2 (about 0.7 nm), $d_{\text{surf,MP}}$ can be determined using Eq. (14) if the monolayer evaluation fit for the micropore is introduced.

The other molecular probe method is the single-probe method (SP method), which is separately proposed by Pfeifer et al.[56,58,60,137,138] In the SP method, a single adsorption isotherm is analyzed using a modified FHH theory. The FHH model first developed by Frenkel,[139] Halsey,[140] and Hill[141] describes a multilayer gas adsorption on the solid surface. Since the SP method is used only for one probe molecule, this method is more convenient than the MP method. However, there are many theoretical limitations in applying the SP method to determination of the surface fractal dimension. Therefore, it is really necessary to discuss about whether the SP method is an adequate tool or not before applying the SP method to certain system, in order to investigate the surface fractal dimension.

The theoretical assumptions used in the SP method are as follows:

1. The detailed structure of the layer of the adsorbed molecules is not important and is regarded as a liquid film.
2. The thickness of the layer of the adsorbed molecules is the characteristic distance scale for fractal surface.
3. Van der Waals attraction forces between solid/gas interactions and the liquid/gas surface tension forces are contributed to the grand potential of the system.

On the basis of the theory of Pfeifer et al.,[56,58,60,137,138] the surface fractal dimension can be determined from the gas adsorption isotherm according to the following equation

$$\ln\left(\frac{V}{V_{\text{mono}}}\right) = const. + C_{\text{SP}}\left[\ln\ln\left(\frac{p^{\circ}}{p}\right)\right] \tag{15}$$

where C_{SP} is the power-law exponent which is dependent upon the surface fractal dimension determined by using the SP method $d_{\text{surf,SP}}$.

At the early stage of multilayer build-up, the interface is dominantly controlled by van der Waals forces between solid/gas interactions and the liquid/gas surface tension forces are neglected. Then, the relationship between C_{SP} and $d_{\text{surf,SP}}$ is written as

$$C_{\text{SP}} = \left(\frac{d_{\text{surf,SP}} - 3}{3}\right) \tag{16}$$

For higher coverage, the interface is controlled by the liquid-gas surface tension forces (capillary condensation) and van der

Waals forces between solid/gas interactions are negligibly small. Then, the relationship between C_{SP} and $d_{surf,SP}$ changes to the following expression.

$$C_{SP} = \left(d_{surf,SP} - 3\right) \quad (17)$$

Van der Waals forces between solid/gas interactions and the liquid/gas surface tension forces represent the limiting cases, but in general both the forces competitively affect the adsorption process. Therefore, in determining the surface fractal dimension by using the SP method, it is very important to use appropriate relation between C_{SP} and $d_{surf,SP}$. According to Ismail and Pfeifer,[60] the threshold for the dominant forces between van der Waals forces and the liquid/gas surface tension forces is given as

$$\varepsilon = 3\left(1 + C_{SP}\right) - 2 \quad (18)$$

If $\varepsilon < 0$, the liquid/gas surface tension forces are dominant, while the van der Waals forces are dominant if $\varepsilon \geq 0$.

Pyun et al.[54,55,142] have investigated the surface fractal characteristics of the carbon specimens using the SP method. Figure 4 depicts the plots of ln V/V_{mono} against lnln (p^o/p) reconstructed from the nitrogen gas adsorption isotherms obtained from meso/macroporous carbon specimens (a) I, (b) II, and (c) III.[55] The values of ε for specimens I, II, and III were determined to be –0.47, –0.14, and –0.02, respectively in the first linear segment, and thus the surface fractal dimensions $d_{surf,SP}$ were estimated from combined Eqs. (15) and (17). It is noticeable from Table 1 that the carbon specimen prepared at the higher pH value gave the lower value of $d_{surf,SP}$ than that specimen prepared at the lower pH value, which is due to more agglomeration of the silica sol particles. In their work, it is confirmed that as the silica-templated pores being comprised of the carbon specimen agglomerate, the irregularity of that pore surface decreases.

The number of adsorbed molecule layers n_{ad} is calculated from the following relationship:[60]

$$n_{ad} = \frac{t_{ad}}{a_o} = \left(\frac{V}{V_{mono}}\right)^{1/\left(3 - d_{surf,SP}\right)} \quad (19)$$

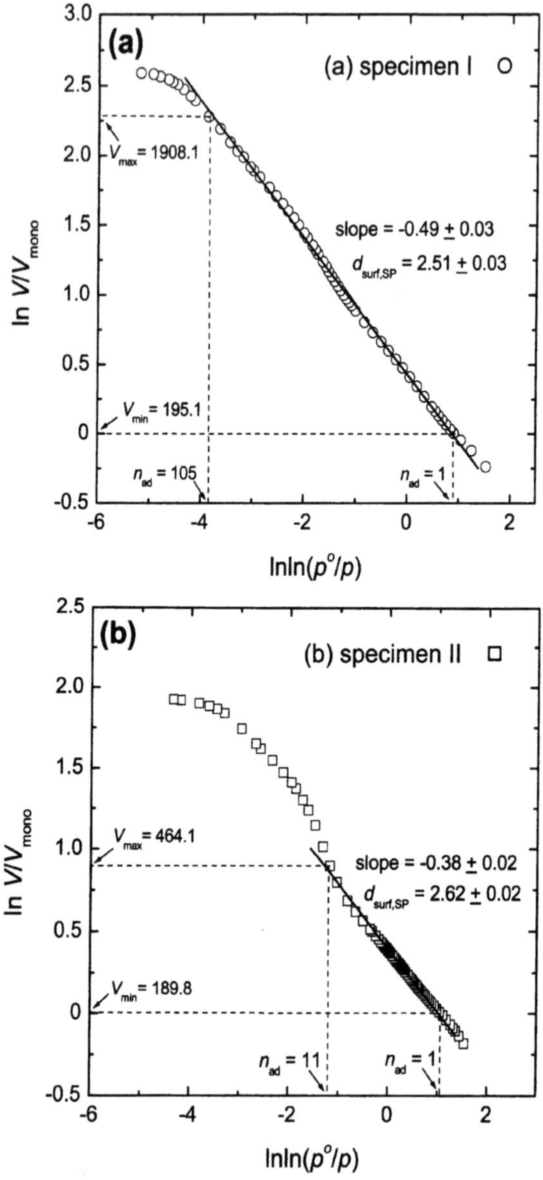

Figure 4. Plots of ln V/V_{mono} against lnln (p^o/p) reconstructed from the N_2 gas adsorption isotherms obtained from the specimens (a) I (○), (b) II (□), and (c) III (Δ).Reprinted with permission from G. -J. Lee and S. -I. Pyun, *Carbon*, **43** (2005) 1804. Copyright © 2005, with permission from Elsevier.

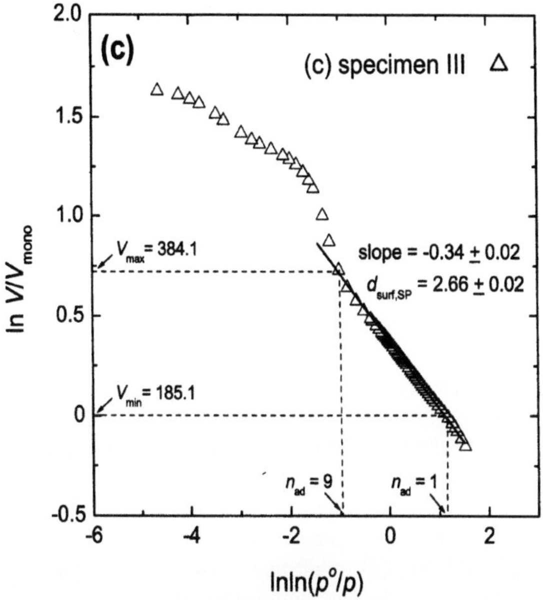

Figure 4. Continuation.

Here, the actual thickness of adsorbed molecule layers t_{ad} is obtained by multiplying n_{ad} by adsorbed molecule diameter a_o. The actual thickness range in which the fractal geometry is satisfied represents the length-scale cutoff range of fractality.[56,58] The value of the high-end thickness of the fractal regime $t_{ad,max}$ can be evaluated by using following equation

$$t_{ad,max} = t_{ad,min}\left(\frac{V_{max}}{V_{min}}\right)^{1/(3-d_{surf,SP})} \tag{20}$$

where $t_{ad,max}$ is the value of the low-end thickness of the fractal regime, V_{min} and V_{max} are the lower and upper limits of the power-law regimes in the plot of ln V against ln ln (p^o/p), respectively.

In Lee and Pyun's work,[55] it is interesting to note that $t_{ad,max}$ is slightly smaller in value than the pore radius of the maximum differential pore volume r_{max} for specimens I, II, and III, as listed in

Table 1. As the relative pressure increases, the thickness of adsorbed molecule layers on the pore surfaces increases as well, and then the pore filling process caused by capillary condensation occurs first in the small pores simultaneously with the multilayer adsorption on the larger pores. For specimen III, the value of $t_{ad,max}$ is significantly larger than those values for the other specimens, which is ascribed to the fact that as a result of the pore filling process in the larger macropores the adsorbed volume starts to increase abruptly only near the saturation vapor pressure in the gas adsorption isotherm. From the above results, they suggested that $t_{ad,max} / t_{ad,min}$ is closely related to r_{max}, that is, the larger r_{max}, the wider ranges the length-scale of the fractal regime in value.

It is reasonable to compare the length-scale range of the fractal regime each other for the carbon specimens based upon the value of $t_{ad,max}$. However, it is ambiguous that the value of $t_{ad,max}$ represents the actual outer cutoff length satisfying the fractal geometry or not. Since the gas molecule layers could more adsorb on the pore surface if the capillary condensation does not occur in the pores with maximum differential pore volume, the actual outer cutoff length could be larger in value than that of $t_{ad,max}$. Under these circumstances, for a better understanding of the outer cutoff of fractality, additional exact estimation of the outer cutoff length is still needed by using other complementary methods.

On the other hand, it is impossible to apply the SP method to the correct description of gas adsorption in the micropores, since the adsorption in the micropores does not occur by multilayer adsorption but by micropore volume filling process. In this case, the pore fractal dimension gives a physical importance for the description of structural heterogeneity of the microporous solids. Terzyk et al.[143-149] have intensively investigated the pore fractal characteristics of the microporous materials using gas adsorption isotherms theoretically simulated.

Based upon D-A equation[98] in consideration of PSD, the overall adsorption isotherm Θ of the microporous materials is written as

$$\Theta = \int_{x_{min}}^{x_{max}} \theta_L f_P(x) dx \tag{21}$$

The local adsorption isotherm θ_L is represented by the original D-A equation and $f_P(x)$ is the micropore size distribution ranging from x_{min} to x_{max} (the lower and upper limits of the slit-like micropore half-width, respectively). If $f_P(x)$ is assumed to be PSD with pore fractality derived by Pfeifer and Avnir[56], $f_P(x)$ is given by

$$f_P(x) = \rho x^{2-d_{\text{pore,SP}}} \quad (x_{\min} < x < x_{\max}) \tag{22}$$

with

$$\rho = \frac{3-d_{\text{pore,SP}}}{x_{\max}^{3-d_{\text{pore,SP}}} - x_{\min}^{3-d_{\text{pore,SP}}}} \tag{23}$$

where $d_{\text{pore,SP}}$ is the pore fractal dimension determined by using the SP method, $x_{\min}$ and $x_{\max}$ are the lower and upper limits of the pore fractality, respectively.

From the above, the overall adsorption isotherm Θ can be represented by[146]

$$\Theta = \int_{x_{\min}}^{x_{\max}} \exp(-\mu' A_{\text{ad}}^n x^n) \times \rho x^{2-d_{\text{pore,SP}}} dx \tag{24}$$

where $\mu' = (k\beta)^{-n}$ depends upon the adsorbate and the microporous structure (β is the affinity coefficient and k is an empirical constant assumed to be 12 kJ nm mol^{-1} for nitrogen gas.[94])

Terzyk et al.[143] gave first the analytical solution of Eq. (24) as follows

$$\Theta = \frac{V}{V_o} = \frac{\rho}{n}(\mu' A_{\text{ad}}^{\text{n}})^{(d_{\text{pore,SP}}-3)/n} \times \left[\gamma\left(\frac{3-d_{\text{pore,SP}}}{n}, x_{\max}^n \mu' A_{\text{ad}}^n\right) - \gamma\left(\frac{3-d_{\text{pore,SP}}}{n}, x_{\min}^n \mu' A_{\text{ad}}^n\right)\right] \tag{25}$$

where γ is an incomplete gamma function. If $n = 2$, D-R isotherm is obtained in view of PSD with pore fractality.[150]

Recently, Lee and Pyun have focused on the characterization of pore fractality of the microporous carbon powder specimens by using nitrogen gas adsorption method based upon the D-A adsorption theory in consideration of PSD with pore fractality. Figure 5 envisages the nitrogen gas adsorption isotherm obtained from the as-reactivated carbon powder specimen prepared by reactivation of the commercially as-activated carbon powder at 1000 °C in an atmosphere of CO_2/CO gas mixture for 2 h. The solid

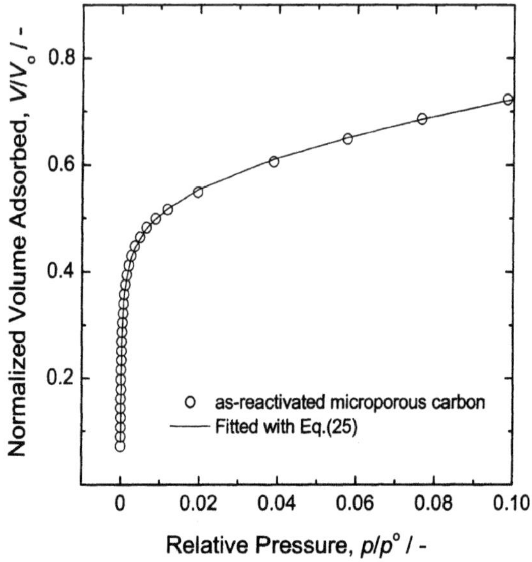

Figure 5. Nitrogen gas adsorption isotherm experimentally obtained from the as-reactivated carbon powder specimen prepared by reactivation of the commercially as-activated carbon powder at 1000 °C in an atmosphere of CO_2/CO gas mixture for 2 h. The solid lines were determined from the non-linear fitting of the experimental adsorption isotherms with Eq. (25).

line in Figure 5 was determined from the non-linear fitting of the experimental adsorption isotherm with Eq. (25). From the quantitative comparison of the adsorption isotherm experimentally obtained with that theoretically calculated, the values of $d_{pore,SP}$, x_{min}, and x_{max} were estimated to be 2.523 $\pm$ 0.035, 0.26 nm, and 0.50 nm, respectively. This will be reported in greater detail in a future publication.

2. Image Analysis Method

The solid surfaces and interfaces are investigated using surface profiler or imaging equipments such as scanning tunneling microscopy (STM), atomic force microscopy (AFM), scanning electron microscopy (SEM) and TEM in order to quantify the surface roughness. However, such analysis method as STM and AFM can characterize only the surface roughness of the nonporous materials or macroporous materials without tortuous, overlaid or deep pores. Recently, high resolution transmission electron microscopy (HRTEM) has been applied to evaluate the surface fractal dimensions d_{surf} of the microporous carbon surfaces.[54,151-153] The most attractive point in this method is that the pores ranging in different sizes can be extracted from the TEM images which include contributions from many different pore sizes by the inverse fast Fourier transform (IFFT) operation by selecting the specific frequency range.

The surface fractal dimension d_{surf} of the porous materials can be determined from the TEM image by using perimeter-area method[54,154-159]. If the scaling property of the porous materials is undoubtedly isotropic, the 3-D pore surface is simply related to the projection of the 3-D pore surface onto the 2-D surface. It is well known[154,155] that the area A and the perimeter P of the self-similar lakes are related to their self-similar fractal dimension $d^{\mathrm{L}}_{\mathrm{F,ss}}$ by

$$P = \beta_{\mathrm{P}} d^{\mathrm{L}}_{\mathrm{F,ss}} A^{d^{\mathrm{L}}_{\mathrm{F,ss}}/2} \tag{26}$$

where β_{P} is a proportionality constant. The surface fractal dimension of the original surface d_{surf} is related to $d^{\mathrm{L}}_{\mathrm{F,ss}}$ by

$$d_{\mathrm{surf}} = d^{\mathrm{L}}_{\mathrm{F,ss}} + 1 \tag{27}$$

In order to determine the value of d_{surf}, the original TEM image is digitized with the image scanner, and is then transformed to the binary images. For this, a low frequency cut off filter operation is performed along with fast Fourier transformation (FFT) to make the brightness uniform before the binary transformation, followed by the IFFT. The image is then transformed to its brightness on a scale from 0 to 255. Finally, any brightness larger than 128 is set at 255, and any brightness smaller than 127 is set at 0.

For instance, Figures 6a and 6b show the digitized TEM images of the mesoporous carbon specimen CAS30 prepared using silica imprinting method and the binary TEM images transformed from Figure 6a, respectively.[54] The TEM picture of the carbon specimen was digitized with the image scanner and then the data were transferred to 384×384 pixels. The digitized image size in

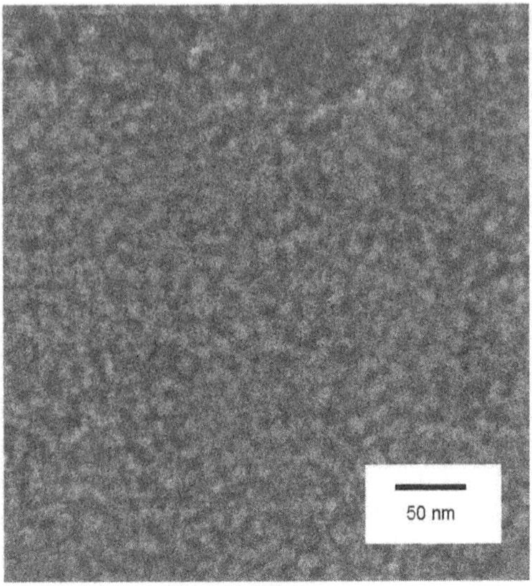

(a)

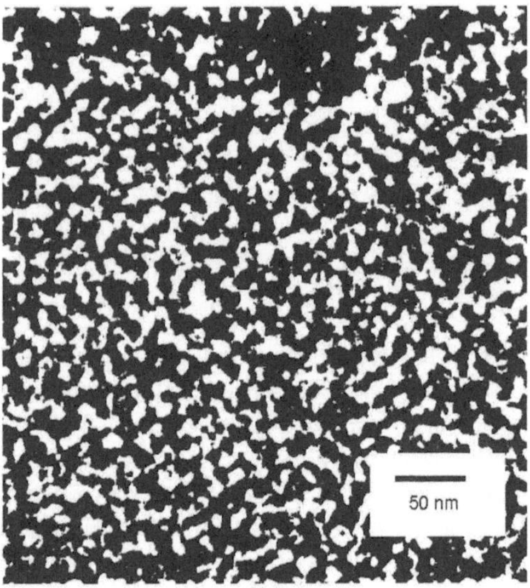

(b)

Figure 6. (a) The digitized TEM images of the mesoporous carbon specimen CAS30 prepared using silica imprinting method and (b) the binary TEM images transformed from (a). Reprinted with permission from S. -I. Pyun and C. -K. Rhee, *Electrochim. Acta,* **49** (2004) 4171. Copyright © 2004, With permission from Elsevier.

Figure 6a corresponds to an area of 400 × 400 nm^2, so that the image contains information about many pores.

Figure 7 demonstrates on a logarithmic scale the dependence of perimeter P on area A of the pores obtained from the binary TEM image of CAS30 in Figure 6b. The (log P – log A) plots obtained from the carbon specimen displayed two straight lines with different slopes that can be divided into region I and II, indicating multifractal geometry of the carbon specimen. The individual surface fractal dimensions in regions I and II were determined from Eqs. (26) and (27) to be 2.08 ± 0.018 and 2.72 ± 0.046, respectively. The transition area A_{tr} from region I to II were determined to be 108 nm^2, which corresponds to the pore diameter of 12 nm based upon spherical pore shape.

In Pyun and Rhee's work,[54] it should be emphasized that the pore diameter calculated from A_{tr} is similar to the transition pore diameters between two groups of the carbonization-induced pores and the silica-imprinted pores in PSD. It is thus indicated that the

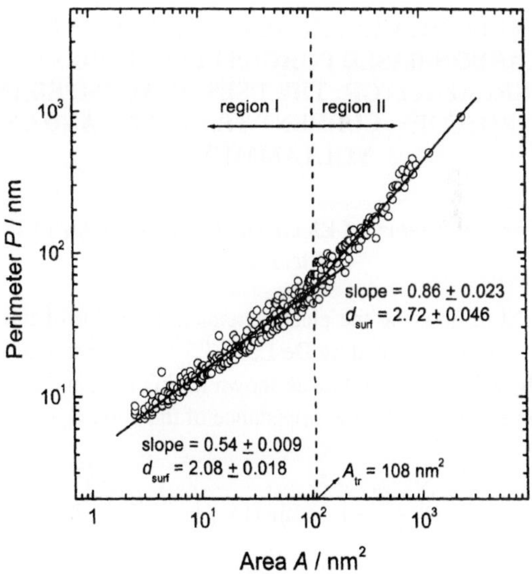

Figure 7. Dependence of perimeter P on area A of the lakes on a logarithmic scale obtained from the binary TEM images of CAS30 in Figure 6b. Reprinted with permission from Reprinted with permission from S. -I. Pyun and C. -K. Rhee, *Electrochim. Acta*, **49** (2004) 4171. Copyright © 2004, With permission from Elsevier.

values of d_{surf} in regions I and II represent the individual surface fractal dimensions of the carbonization-induced pore surface and the silica-imprinted pore surface, respectively. From the comparison of the overall surface fractal dimension evaluated by the molecular probe method using gas adsorption with the individual fractal dimension, they concluded that the overall surface fractal dimension is crucially influenced by the individual dimension of the silica-imprinted pore surface.

As mentioned above, the advantage of the image analysis method is that the information of the pores in different size ranges can be extracted from the TEM images using IFFT operation by selecting the specific frequency range. By comparing the individual surface fractal dimension determined from the TEM images with the overall surface fractal dimension estimated by using other complementary methods, we can more specify the cutoff length scale of the surface fractality and examine the individual surface fractal dimension which dominantly influences the overall surface fractal dimension.

V. ELECTROCHEMICAL CHARACTERISTICS OF CARBON-BASED POROUS ELECTRODES FOR SUPERCAPACITOR: THE USES OF AC-IMPEDANCE SPECTROSCOPY, CURRENT TRANSIENT AND CYCLIC VOLTAMMETRY

1. General Theory of Electrochemical Behavior of Porous Electrodes

The model to describe the electrochemical behavior of the porous electrode was first treated by De Levie.[160-162] He represented a pore surface by a transmission line as shown in Figure 8, and derived the following expression for the impedance of the pore, Z_o

$$Z_o = (1-j)\left(\frac{R'}{2\omega C'}\right)^{1/2} \coth\left[(1+j)\left(\frac{\omega R'C'}{2}\right)^{1/2} l_p\right] \tag{28}$$

with

$$R' = \frac{1}{k_e \pi r^2} \text{ and } C' = C_d 2\pi r \tag{29}$$

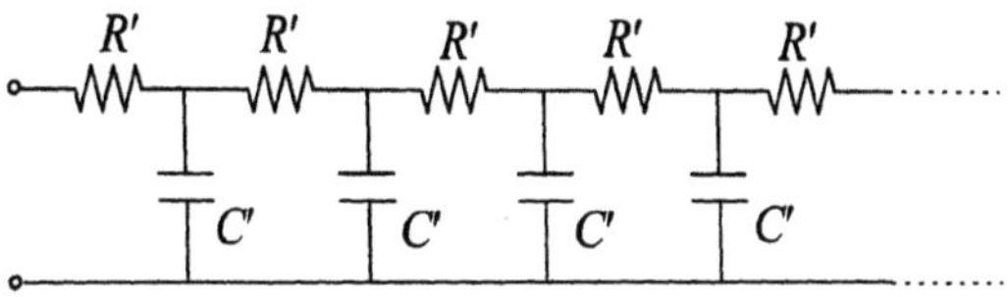

Figure 8. Equivalent circuit of a transmission line network representing the ion migration into the pores. R' and C' denote the resistance of the electrolyte inside a pore and the double-layer capacitance of the electrode/electrolyte interface, respectively, both of which are taken per unit length.

where R' is the resistance of the electrolyte inside a pore per unit pore length; C', the double-layer capacitance of the electrode/electrolyte interface per unit pore length; ω, angular frequency; l_p, the pore length; r, the pore radius; k_e, the conductivity of the electrolyte, and C_d represents the specific double-layer capacitance per unit area.

The assumptions of this model include the cylindrical pore shape, uniformly distributed R' and C', the lack of tortuosity of the pore, the lack of distributions of pore size and length, and no curvature of the equipotential surface in a pore. The resulting impedance spectra theoretically calculated based upon Eq. (28) is plotted in Figure 9. The straight inclined line at high frequencies is attributable to the semi-infinite ion migration through the pores. The capacitive line at low frequencies is due to the accumulation of ions at the bottom of the pores. At high frequencies, Eq. (28) reduces to

$$Z_o(\text{high freq.}) = (1-j)\left(\frac{R'}{2\omega C'}\right)^{1/2} \tag{30}$$

and the interfacial impedance has a phase angle of 45°, whereas at low frequencies Eq. (28) can be written as

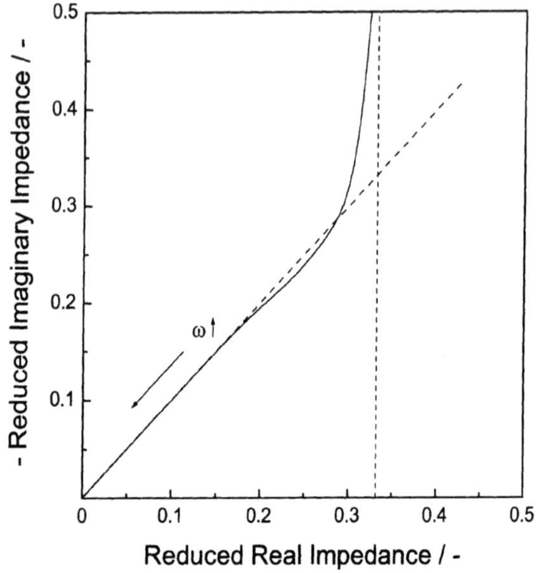

Figure 9. Nyquist plot of the impedance spectra of a cylindrical pore theoretically calculated with Eq. (28).

$$Z_o(\text{low freq.}) = \frac{1}{j\omega C' l_p} + \frac{R' l_p}{3} \tag{31}$$

and the phase angle tends towards 90°.

Although De Levie's work succeeds in explaining qualitatively the observed impedance behavior of the porous electrode, many assumptions and approximations limited the model's ability to represent accurately the measured impedance spectra. In particular, the impedance spectra experimentally measured on the porous carbon electrodes for EDLC do not show an ideal impedance behavior of a cylindrical pore. Over the years, such non-ideal impedance behavior of the porous electrode, i.e., frequency dispersion has been investigated by many researchers.[9,14-17,163-178] Generally, it has been suggested that the geometric heterogeneity such as pore shape and pore size/length distributions and surface inhomogeneity such as different crystallographic orientations and the SAFGs can cause time constant distributions, thus causing frequency dispersion. The details are described in the following Sections V.2 and V.3.

2. Effect of Geometric Heterogeneity on Ion Penetration into the Pores during Double-Layer Charging/Discharging

The electrochemical impedance of a real electrode is frequently represented by an equivalent circuit containing constant phase element (CPE) showing power-law frequency dependence as follows

$$Z(\omega) = (1/C_{\mathrm{CPE}})(j\,\omega)^{-\alpha} \tag{32}$$

where C_{CPE} and α mean the CPE coefficient and the CPE exponent, respectively, and ω represents the angular frequency. Since the CPE or capacitance dispersion of the impedance spectra causes time constant distributions, and thus significantly influences the kinetics of double-layer charging/discharging of the porous electrode, it is necessary to understand the physical origins of the capacitance dispersion. It has been considered that one of the most possible causes of the CPE is known to be of geometric origin: An irregular and porous electrode geometry causes current density inhomogeneities and thus yields deviations from ideal capacitive behavior.

Keiser et al.[164] first showed that the more occluded the shape of the pore, the more distorted the impedance locus from the ideal capacitive behavior. However, the pore shapes in real system turn out to be much complicated and thus a straightforward analytical calculation is not usually possible of the overall impedance for those complicated pores. In connection with this problem, the fractal geometry has given a powerful tool for the analysis of the CPE behavior of the porous electrode. A number of theoretical papers[166,179-191] have devoted to investigate the relationship between the fractal geometry of the electrode and the CPE impedance on the basis of the electrolytic resistive distribution due to the surface irregularity.

Among various models proposed, a model of self-similar fractal electrode from Nyikos and Pajkossy[166] gives the following relation of the CPE exponent α to the self-similar fractal dimension $d_{\mathrm{F,ss}}$ as

$$\alpha = \frac{1}{d_{\mathrm{F,ss}} - 1} \tag{33}$$

For a perfectly smooth surface with $d_{\mathrm{F,ss}} = 2$ at all scales, Eq. (33) predicts $\alpha = 1$, i.e., purely capacitive behavior. In other limit as $d_{\mathrm{F,ss}} = 3$, $\alpha = 0.5$ which is de Levie's well-known result for the electrode with cylindrical pore. Eq. (33) also implies that the surfaces with different morphologies but with same surface fractal dimension are equivalent as far as the impedance is concerned. The relationship between $d_{\mathrm{F,ss}}$ and α is dependent upon the model used in the fractal characterization of the electrode.

However, Kerner and Pajkossy[167,171,176] have recently shown that for the electrode with microscopic roughness in aqueous electrolyte, the capacitance dispersion due to the surface irregularity should appear at such a high frequency range as kHz to GHz which is much higher than those frequencies at which the capacitance dispersion is usually experimentally observed. In their work, it was recognized that the capacitance dispersion is closely related to the surface inhomogeneity rather than surface irregularity. Pyun et al.[17,178] also suggested that the contribution of the surface inhomogeneity is much higher than the contribution of the surface irregularity to the capacitance dispersion on the porous carbon electrodes.

As a matter of fact, for porous carbon electrode it is still a troublesome issue to relate the determined surface fractal dimension $d_{\mathrm{F,ss}}$ with the CPE exponent α. The effect of the surface inhomogeneity on the ion penetration into the pores during double-layer charging/discharging will be discussed in detail in the following Section V.3.

In recent years it has been demonstrated by many researchers[16,172,173] that the frequency dispersion or capacitance dispersion is intimately related to PSD or pore length distribution (PLD). In this case, the frequency dispersion is not called 'CPE behavior' since the phase angle of the impedance spectra did not show a constant value over the whole frequency range. The phase angle of the impedance spectra measured on the porous electrode with broad PSD or PLD is larger than 45° in value at high frequencies and smaller than 90° in value at low frequencies.

Considering PSD or PLD, the total impedance of the pores, Z_{tot} is given by

$$\frac{1}{Z_{\text{tot}}} = \int_{-\infty}^{\infty} \frac{1}{Z_o} f(x)dx \approx \sum_i \frac{1}{Z_{o,i}} f_i \Delta x_i \equiv \sum_i \frac{1}{Z_{o,i}} n_i \tag{34}$$

where $f(x)$ and $f(x)dx$ represent a distribution density function of PSD or PLD and the number of the pores between x and $x + dx$, respectively.

Song et al.[172] theoretically calculated the impedance spectra based upon Eq. (34) with such distribution functions of PSD as normal, lognormal, Lorentzian, log Lorentzian distributions. They concluded that the impedance spectra simulated based upon the transmission line model (TLM) with different PSD functions share a common point that the wider PSD leads to the more frequency dispersion in the impedance spectra.

Recently, Lee et al.[16] investigated the effect of PLD on the kinetics of double-layer charging/discharging of the activated-carbon fiber cloth electrode (ACFCE). In order to disregard the effect of the pore shape on the electrochemical performance of the electrode, they used the ACFCE as an electrode material since the shape of the pores comprising the activated carbon fiber is cylindrical and homogeneous throughout the bulk of the fiber[152,192]. It is assumed that the pore length l_p takes a lognormal distribution $f(x')$ as follows

$$f(x') = \frac{V_{\text{tot}}}{\pi r^2 l_p} \frac{1}{\sqrt{2\pi}\sigma} \exp\left[-\frac{1}{2}\left(\frac{x' - \mu}{\sigma}\right)^2\right] \tag{35}$$

with

$$x' = \ln(l_p / l_o) \tag{36}$$

where x' denotes the natural logarithm of l_p divided by an arbitrary unit pore length l_o to obtain a dimensionless quantity; V_{tot}, the total pore volume; μ and σ are the mean value and the standard deviation of the distribution variable x', respectively.

Substituting $y = \dfrac{x' - \mu}{\sigma}$ into Eq. (35), we get

$$f(y) = \frac{V_{\text{tot}}}{\pi r^2 l_p} \frac{1}{\sqrt{2\pi}} \exp\left(-\frac{1}{2} y^2\right) \tag{37}$$

Hence, the Z_{tot} leads to

$$\frac{1}{Z_{tot}} = \int_{-\infty}^{\infty} \frac{1}{Z_o} f(y) dy = \int_{-\infty}^{\infty} \frac{1}{Z_o} \frac{V_{tot}}{\pi r^2 l_p} \frac{1}{\sqrt{2\pi}} \exp\left(-\frac{1}{2} y^2\right) dy \quad (38)$$

with

$$\int_{-\infty}^{\infty} \frac{1}{\sqrt{2\pi}} \exp\left(-\frac{1}{2} y^2\right) dy = 1 \quad (39)$$

From the quantitative coincidence of the impedance spectrum experimentally measured with that theoretically calculated with Eq. (38) as depicted in Figure 10, they suggested that such non-ideal

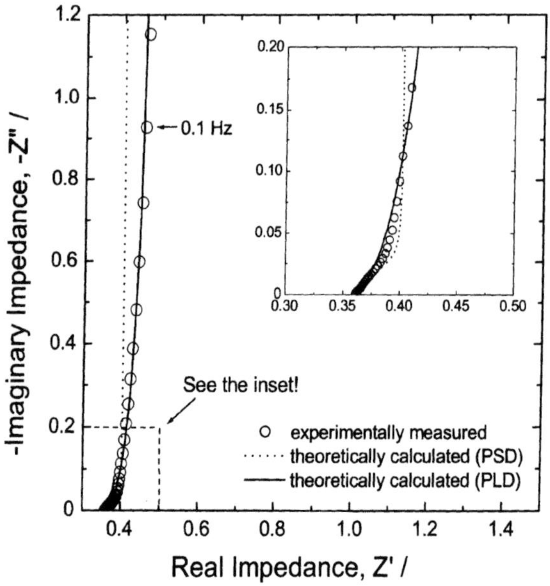

Figure 10. Nyquist plot of the impedance spectrum experimentally measured on the ACFCE at an applied potential of 0.1 V (*vs.* SCE) in a 30 wt % H_2SO_4 solution. Dotted and solid lines represent the impedance spectra theoretically calculated based upon the transmission line model (TLM) in consideration of pore size distribution (PSD) and pore length distribution (PLD), respectively. Reprinted with permission from G. -J. Lee, S. -I. Pyun, and C. -H. Kim, *J. Solid State Electrochem.*, **8** (2004) 110. Copyright © 2003, with kind permission of Springer Science and Business Media.

impedance behavior of the ACFCE is mainly due to PLD, rather than due to PSD.

When PLD is considered together with the solution resistance R_{sol}, between reference electrode (R.E.) and working electrode (W.E.), the total current $I_{tot}(t)$ is written as

$$I_{tot}(t) = \int I(t) f(y) dy \quad (40)$$

$$= \int \frac{2\Delta E}{R_{sol}} \sum_{n=1}^{\infty} \frac{\Lambda}{\Lambda^2 + \Lambda + m_n^2} \times \exp\left(-m_n^2 k_e r t \Big/ 2 C_d l_p^2 \right) \frac{V_{tot}}{\pi r^2 l_p} \frac{1}{\sqrt{2\pi}} \exp(-\frac{1}{2} y^2) dy$$

with

$$\Lambda = \frac{R_p}{R_{sol}} \text{ and } R_{sol} = \frac{l_s}{k_e \pi r^2} \quad (41)$$

where ΔE is the potential step; R_p, the resistance of the electrolyte inside a pore; l_s, the distance between R.E. and W.E.; m_n, the nth positive root of ($m \tan m - \Lambda = 0$); $f(y)$ and $f(y)dy$ represent a distribution density distribution of the pore length and the number of the pores between y and $y + dy$, respectively.

Lee et al.[16] theoretically calculated the cathodic current transients based upon the TLM as a function of σ of PLD. Figures 11a and 11b illustrate the logarithmic cathodic current transients and the derivatives of the logarithmic cathodic current transients calculated from Eq. (40) with the change of σ of PLD, respectively. It is noted that as σ increases, the current decays more rapidly with time due to the dominant contribution of the pores with smaller length in the region I, whereas the current decays more slowly with time due to the dominant contribution of the pores with larger length in the region II.

In the case of potential scanning with a scan rate ν, in consideration of PLD, the total current $I_{tot}(t)$ is given by

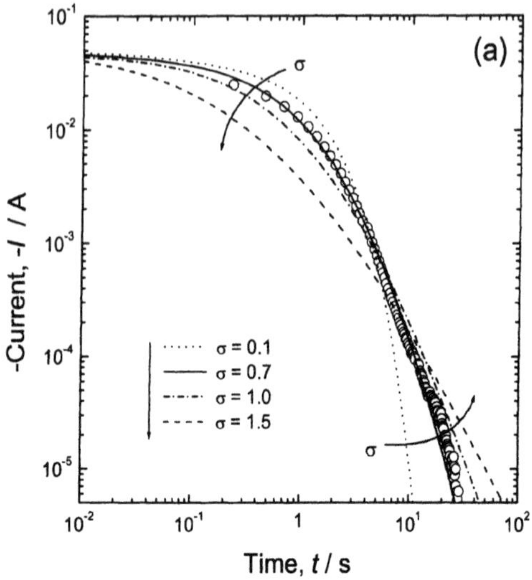

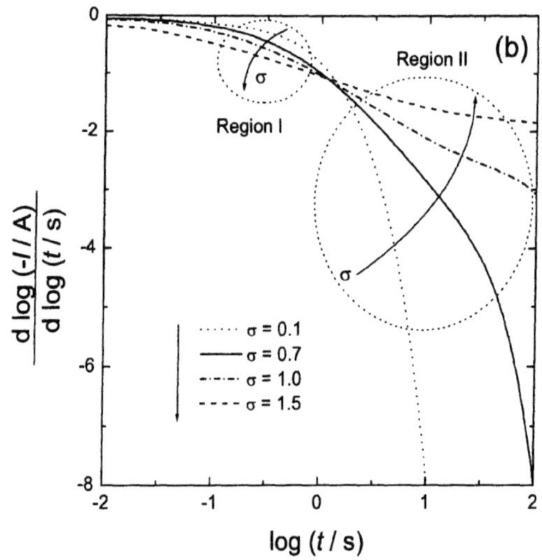

Figure 11. (a) The cathodic current transients on a logarithmic scale and (b) the derivatives of the logarithmic cathodic current transients theoretically calculated from Eq. (40) based upon the TLM as a function of standard deviation σ of PLD. Open circles in Figure 11 represent the cathodic current transient experimentally measured on the ACFCE. Reprinted with permission from G. -J. Lee, S. -I. Pyun, and C. -H. Kim, *J. Solid State Electrochem.*, **8** (2004) 110. Copyright © 2003, with kind permission of Springer Science and Business Media.

$$I_{\text{tot}}(t) = \int I(t) f(y) dy \tag{42}$$

$$= \int (\beta_1 - \beta_2) \nu C_{\text{d}} 2\pi r l_{\text{p}} \frac{V_{\text{tot}}}{\pi r^2 l_{\text{p}}} \frac{1}{\sqrt{2\pi}} \exp\left(-\frac{1}{2} y^2\right) dy$$

with

$$\beta_1 = 1 - \sum_{n=1}^{\infty} \frac{2\Lambda^2}{m_n^2[\Lambda^2 + \Lambda + m_n^2]} \exp\left({-m_n^2 k_{\text{e}} rt} \middle/ {2C_{\text{d}} l_{\text{p}}^2} \right) \tag{43}$$

$$\beta_2 = \sum_{n=1}^{\infty} \left(\frac{\Lambda^2 + m_n^2}{\Lambda^2 + \Lambda + m_n^2} \right) \exp\left({-m_n^2 k_{\text{e}} rt} \middle/ {2C_{\text{d}} l_{\text{p}}^2} \right) \left(\frac{2}{m_n^2} \right) \sin^2 m_n \tag{44}$$

The rate capability γ_{cap} defined as the quotient of the reduced charge for the ACFCE divided by the reduced charge for the ideal double layer capacitor increased with increasing σ of PLD, as shown in Figure 12. Consequently, from the above theoretical and experimental results, it is concluded that the ion penetration into the pores during double-layer charging/discharging is more impeded as σ of PLD increases.

3. Effect of Surface Inhomogeneity on Ion Penetration into the Pores during Double-Layer Charging/Discharging

It has been reportedly known[17,178,193-198] that for the carbon electrodes the basal and edge planes are randomly distributed on the electrode surface and the values of the specific double-layer capacitance per unit area C_{d} range from about 1 μF cm^{-2} for the basal planes to about 70 μF cm^{-2} for the edge planes. The conductive and capacitive properties of the basal and edge planes are different from each other. Along the direction of the edge planes, the carbon specimen can be treated as a conductor, while it shows the properties of semiconductor along the direction of the basal planes. These randomly distributed capacitive elements give rise to the distribution in the current density along the electrode surface.

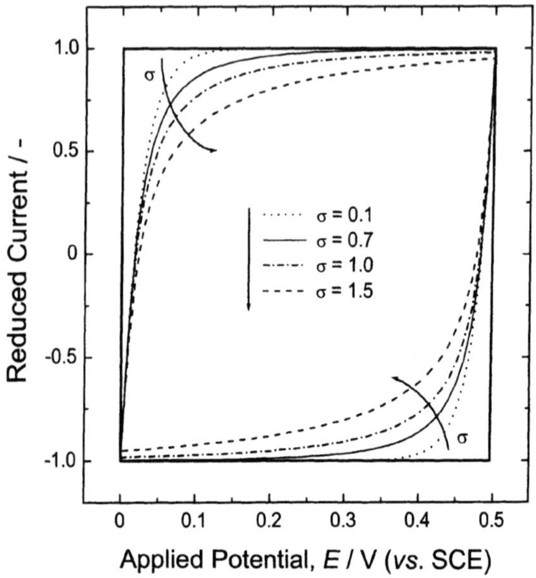

Figure 12. Plots of the reduced current against the applied potential theoretically calculated based upon the TLM as functions of σ of PLD at a scan rate of 20 mV s^{-1}. Solid bold line denotes the ideal double-layer capacitor where the time constant is zero. Reprinted with permission from G. -J. Lee, S. -I. Pyun, and C. -H. Kim, *J. Solid State Electrochem.*, **8** (2004) 110. Copyright © 2003, with kind permission of Springer Science and Business Media.

Furthermore, the basal plane is flat on the atomic scale and can be considered as a nearly perfect single crystal since there are no unsaturated chemical bonds, free electrons and/or SAFGs. On the other hand, the edge planes are associated with unpaired electrons or have residual valencies which are very active. Therefore, various defects or SAFGs are bound to the edge planes, causing more increase in the surface inhomogeneity. The surface inhomogeneities resulting from these crystallographic orientations can cause time constant distributions, and hence they may influence the capacitance dispersion.

Recently, Lee and Pyun[17] have intensively investigated the effect of microcrystallite structures on electrochemical characteristics of the mesoporous carbon electrodes for EDLC. They effectively controlled the microcrystallite sizes of the carbon specimens in *a*- and *c*-axis direction, i.e., L_a (parallel to the basal

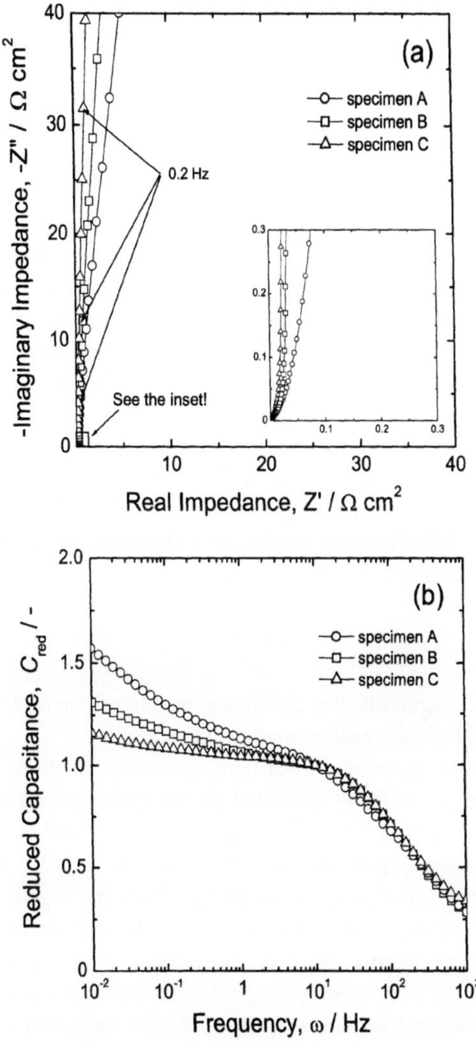

Figure 13. (a) Nyquist plots of the impedance spectra and (b) plots of reduced capacitance C_{red} *vs.* frequency ω experimentally measured on carbon specimens A (-○-), B (-□-), and C (-Δ-) at an applied potential of 0.2 V (vs. SCE) in a 30 wt.% H_2SO_4 solution. Here, the solution resistance was subtracted from the measured impedance spectra. The reduced capacitance in (b) was determined from the normalization of the capacitance with respect to the value of the capacitance calculated from the impedance spectra at 10 Hz. Reprinted from Ref. [17], Copyright (2006), with permission from Elsevier.

planes) and L_c (parallel to the edge planes) by using different carbonaceous precursors with various heat-treatment temperatures.

The impedance spectra measured on three kinds of carbon specimens exhibited the CPE behavior in low frequencies (Figure 13a), and the frequency dispersion of the impedance spectra at low frequencies increased in the order of specimens C, B, and A (Figure 13b). The CPE exponent α decreased with increasing L_c/L_a, i.e., with increasing surface inhomogeneity as listed in Table 2. From their work, it is strongly indicated that the value of α is closely related to the amounts of the edge planes, that is, the frequency dispersion of capacitance increases with increasing surface inhomogeneity due to the wider time constant distribution. Furthermore, it is also found from Table 2 that the value of C_d calculated at the lowest frequency, i.e., 10^{-2} Hz in the impedance spectra increased with increase in L_c/L_a, which is due to the higher value of C_d for the edge planes than that for the basal planes.

For the time domain responses of the CPE, the current density difference (Δi)/scan rate (ν) relation is expressed by the following power-law during the potential scanning[199]

$$\Delta i \propto \nu^{\alpha} \tag{45}$$

where Δi represents the difference in current density between anodic and cathodic peak currents of the CVs.

Figure 14 gives on a logarithmic scale the variations of Δi of the CVs experimentally measured on the carbon electrodes as a function of the scan rate ν.[17] For specimens A, B, and C, Δi of the CVs were linearly proportional to the scan rate ν to the power of 0.894, 0.916, and 0.942, respectively, which are slightly smaller than those values of α determined from the impedance spectra. From the qualitative coincidence between the values of α determined from the impedance spectra and the CVs, it is suggested that the distributive characteristics of the capacitance in the frequency domain greatly influence the capacitance dispersion in the time domain which is related to the kinetics of double-layer charging/discharging.

Figure 15 presents the plots of the rate capability γ_{cap} against the scan rate ν calculated from the CVs experimentally measured on carbon specimens A, B, and C.[17] The higher the value of L_c/L_a was, the lower exhibited the rate capability γ_{cap}, regardless of the scan rate ν. Besides, as L_c/L_a increased, the rate capability γ_{cap}

Table 2.
The Mass of the Electrode m, the BET Surface Area A_{BET}, the Total Pore Volume V_{tot}, the Average Pore Diameter D_{ave}, the Size of the Microcrystallite L_a (Parallel to the Basal Plane), L_c (Parallel to the Edge Plane), the Ratio of L_c/L_a, the CPE Exponent α and the Specific Double-Layer Capacitance C_d for Carbon Specimens A, B and C. Reprinted from Ref. [17], Copyright (2006), with permission from Elsevier.

Specimen	Mass of the electrode, m (mg)	BET surface area, A_{BET} ($m^2 g^{-1}$)	Total pore volume, V_{tot} ($cm^3 g^{-1}$)	Average pore diameter, D_{ave} (nm)	L_a (nm)	L_c (nm)	L_c/L_a /-	CPE exponent α /-	Specific double-layer capacitance, C_d ($F m^{-2}$)
A	1.33	750	1.80	12.4	1.6	1.3	0.81	0.919	0.20
B	2.53	394	1.39	12.0	1.9	1.2	0.63	0.945	0.11
C	2.10	477	1.46	11.2	3.9	2.0	0.51	0.964	0.04

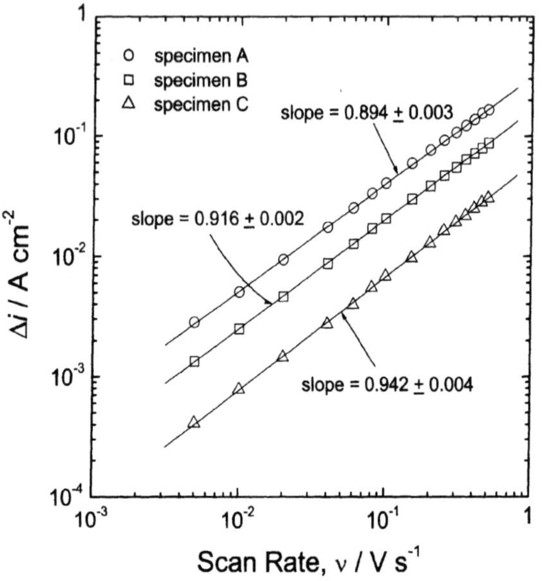

Figure 14. Dependence of the current density difference Δi between anodic and cathodic peak currents of the cyclic voltammograms (CVs) on the scan rate ν for carbon specimens A (○), B (□), and C (Δ). Reprinted from Ref. [17], © (2006), with permission from Elsevier.

decayed more rapidly with the scan rate ν. From the above results, it is concluded that the ion penetration into the pores is more impeded with increasing surface inhomogeneity, which is attributable to both the higher value and wider distribution of the time constant.

It has been generally reported[14,196,200-203] that the chemical heterogeneities (SAFGs) significantly influence the electrochemical performance of the porous carbon electrodes. The SAFGs are mostly bound to lateral planes (edges of the aromatic sheet) of the carbon electrode, because these sites are associated with unpaired electrons or have residual valencies which are very reactive. The SAFGs contribute to the increase in the total capacitance of EDLC,[203] and the electrolyte wettability can be improved by increasing amounts of the oxygen content associated with hydrophilic surface groups.[204] However, the presence of the SAFG

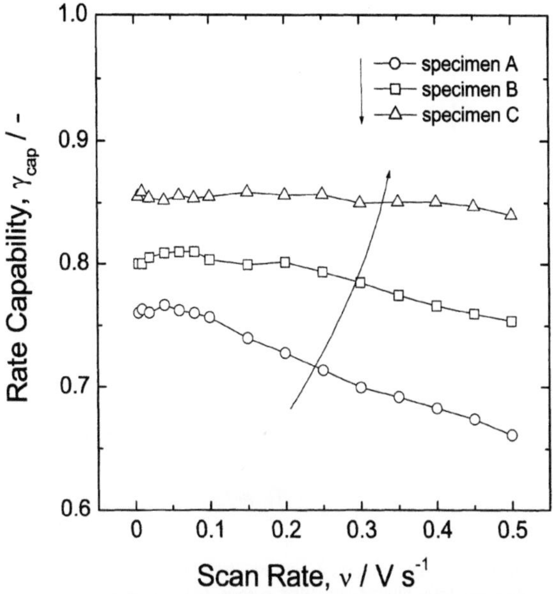

Figure 15. Plots of the rate capability γ_{cap} against the scan rate ν calculated from the CVs experimentally measured on carbon specimens A (○), B (□), and C (Δ). Reprinted from Ref. [17], Copyright (2006), with permission from Elsevier.

on the carbon surface is not always beneficial since they serve to increase the leakage current.[205]

The SAFGs formed on the carbon electrode are the intermediate product of such gaseous oxides as CO and CO_2 appearing as final carbon oxidation products. The concentration of the SAFG increases with an increase in the oxidation temperature and reaches a maximum at 400 °C to 500 °C. At higher temperatures, the SAFG is thermally unstable and hence decomposes to CO and CO_2 gases.[196,197] The SAFGs can be also formed on the carbon electrode by electrochemical activation. Sullivan et al.[202] electrochemically activated the glassy carbon electrodes at 1.85 to 2.05 V (*vs*. SCE) and then reduced at –0.2 to –0.4 V (vs. SCE) in sulfuric acid solution. During repetitive activation/reduction processes, the large amounts of the SAFGs were formed on the glassy carbon electrodes and those SAFGs

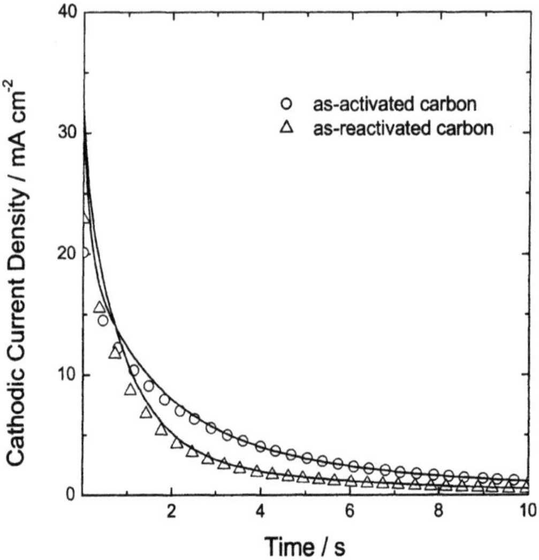

Figure 16. The cathodic current transients experimentally obtained from the as-activated carbon (○) and as-reactivated carbon (Δ) electrode specimens in a 30 wt.% H_2SO_4 solution by dropping the applied potential from 0.1 to 0.08 V (*vs*. SCE). Solid lines represent the simulated cathodic current transients. Reprinted with permission from C.-H. Kim, S.-I. Pyun, and H.-C. Shin, *J. Electrochem. Soc.* **149** (2002) A93. Copyright © 2001, with permission from The Electrochemical Society.

significantly affected the impedance spectra and the CVs in shape and value, depending upon the potential. It has been reported[200-202] that the double-layer capacitance which is related to the state of protonation of the SAFGs is strongly dependent upon the potential and exhibits the maximum value in potential ranges between 0.3 and 0.5 V (*vs*. SCE).

Kim et al.[14] have attempted to investigate the role of the SAFG in the kinetics of double-layer charging/discharging of activated carbon powder electrode specimens. In order to leave the effect of PSD out of consideration, they prepared two kinds of as-activated and as-reactivated carbon powder specimens, which are characterized by almost the same PSD, but by different concentration of the oxygen functional groups. The concentration of the oxygen functional groups of the as-activated carbon powder specimen was measured to be 1.35 mmol g^{-1} which is much higher in value than that of the as-reactivated carbon powder specimen (0.34 mmol g^{-1}).

The cathodic current transients (designated as open symbols) experimentally determined from the as-activated and as-reactivated carbon electrode specimens are given in Figure 16, together with those simulated (designated as solid lines). The simulation was performed from the circuit analysis using the SPICE based upon the six-RC-element ladder network at a potential step by taking the values of circuit elements obtained from the complex nonlinear least-squares (CNLS) fitting method of experimental impedance spectra. For the as-activated carbon electrode specimen, the current density decayed more slowly with time than that for the as-reactivated carbon electrode specimen due to higher time constant for the former carbon electrode specimen than that for the latter carbon electrode specimen.

In Figure 17, the rate capability γ_{cap} was determined to be 0.60 for the as-activated carbon electrode specimen. This value is lower than that for the as-reactivated carbon electrode specimen 0.76. From these results, it is confirmed that the SAFG reduces the ion penetration depth, and hence impedes the ion penetration into the pores during double layer charging of the carbon electrodes.

VI. CONCLUDING REMARK

The present article first provided the brief overview of the synthetic methods of the porous carbons. In order to prepare the microporous carbons with high surface area, the physical/chemical activation methods have been widely used for a long time.[18-35] Recently, the meso/macroporous carbons with various pore structures are prepared by templating methods by using various templates and changing sol-gel reaction conditions, e.g., pH, amount of template, and gelation temperature.[17,36-55]

Subsequently, the characterization of the pore structures of the porous materials using gas adsorption method was discussed in detail. The types and characteristics of the adsorption isotherms and the hysteresis loops were introduced. In addition, the BET (Braunauer, Emmett, and Teller) theory[92] for the determination of the surface area and various theoretical models for characterization of the pore structures according to the pore size range were summarized based upon the adsorption theory.

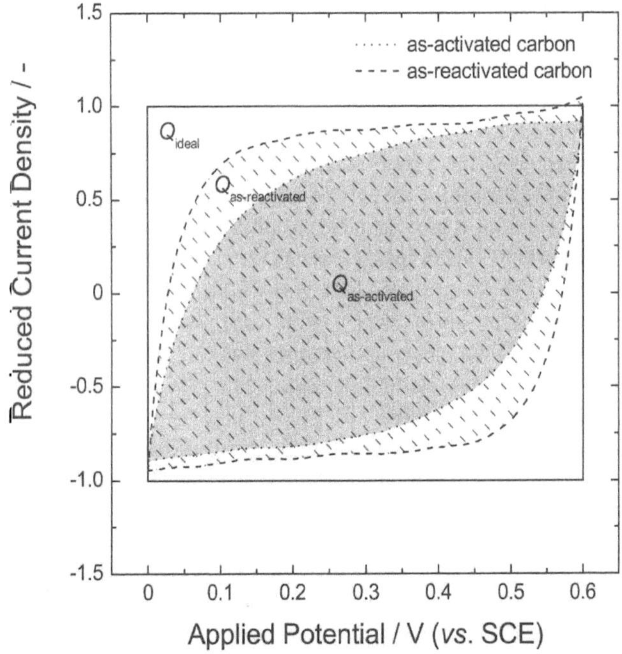

Figure 17. Plots of the reduced current density against the applied potential obtained from the as-activated carbon (dotted line) and as-reactivated carbon (dashed line) electrode specimens. Solid line represents the ideal double layer capacitor where the time constant is zero. Reprinted from C.-H. Kim, S.-I. Pyun, and H.-C. Shin, *J. Electrochem. Soc.* **149** (2002) A93. Copyright © 2001, with permission from The Electrochemical Society.

And then, the molecular probe method and the image analysis method for quantitative characterization of pore surface irregularity and size distribution irregularity were discussed based upon the fractal theory. The surface fractal dimensions[56,58,65] which characterize the pore surface irregularity were successfully determined by the multiprobe (MP) method[56,117,120-133] that uses several kinds of multiprobe molecules with different molecular sizes and the single-probe (SP) method[56,58,60,137,138] based upon the modified FHH (Frenkel,[139] Halsey,[140] and Hill[141]) theory describing multilayer gas adsorption.

On the other hand, for the microporous carbons with pore size distribution (PSD) with pore fractality, the pore fractal dimensions[56,59,62] which represent the size distribution irregularity can be theoretically calculated by non-linear fitting of experimental adsorption isotherm with Dubinin-Astakhov (D-A) equation in consideration of PSD with pore fractality.[143-149] The image analysis method[54,151-153] has proven to be also effective for the estimation of the surface fractal dimension of the porous materials using perimeter-area method.[154-159]

Finally, as electrochemical applications of the porous carbons to electrode materials for supercapacitor, this review covered the effects of geometric heterogeneity and surface inhomogeneity on ion penetration into the pores during double-layer charging/discharging. It is noted that for the reliable fractal characterization of the pore surfaces in terms of the constant phase element (CPE) exponent α, it must be rigorously checked whether the surface state of the electrode is homogeneous or not and whether the effect of the surface roughness on the CPE is more dominant rather than that of the surface inhomogeneity or not.

From the theoretical and experimental findings, it is obvious that the wider PSD or pore length distribution (PLD) leads to the more time constant distribution, thus causing higher retardation of ion penetration into the pores during double-layer charging/discharging. Furthermore, it is recognized that the surface inhomogeneities such as microcrystallite sizes and the surface acidic functional groups (SAFGs) impede the ion penetration into the pores during double-layer charging/discharging of the carbon electrodes, and hence decrease the rate capability γ_{cap}.

ACKNOWLEDGEMENTS

This work was supported by a grant from the Center for Advanced Materials Processing (CAMP) of the 21st Century Frontier R&D Program funded by the Ministry of Commerce, Industry and Energy (MOCIE), Republic of Korea. Furthermore, this work was partly supported by the Brain Korea 21 project.

NOTATION

a_o	Diameter of adsorbed molecule
A	Area
A_{ad}	Adsorption potential
A_{BET}	Total Braunauer, Emmett, and Teller (BET) surface area
A_c	Molecular cross-sectional area
A_{tr}	Transition area at which the linear relationship between the logarithm of perimeter and the logarithm of area changes
α	Constant phase element (CPE) exponent
ACFCE	Activated carbon fiber cloth electrode
AFM	Atomic force microscopy
β	Affinity coefficient
β_P	Proportionality constant in the relationship between the area and the perimeter
BDDT	Braunauer, Deming, Deming, and Teller
BET	Braunauer, Emmett, and Teller
BJH	Barrett, Joyner, and Halenda
C	Constant in BET equation
C'	Double-layer capacitance of the electrode/electrolyte interface per unit pore length
C_{CPE}	CPE coefficient
C_d	Specific double-layer capacitance per unit area
C_{red}	Reduced double-layer capacitance
C_{SP}	Power-law exponent which is dependent upon the surface fractal dimension determined by using the single-probe method using gas adsorption
CNLS	Complex nonlinear least-squares
CPE	Constant phase element
CV	Cyclic voltammogram
$D_{F,ss}$	Self-similar fractal dimension
$d^{L}_{F,ss}$	Self-similar fractal dimension of the two-dimensional self-similar lakes
$d_{pore,SP}$	Pore fractal dimension determined by using the single-probe method using gas adsorption
d_{surf}	Surface fractal dimension
$d_{surf,MP}$	Surface fractal dimension determined by using the multiprobe method using gas adsorption
$d_{surf,SP}$	Surface fractal dimension determined by using the single-probe method using gas adsorption

D	Pore diameter
D_{ave}	Average pore diameter
D-A	Dubinin-Astakhov
D-R	Dubinin- Radushkevich
E	Potential
E_o	Characteristic adsorption potential for the reference vapor, benzene
ΔE	Potential step
ε	Threshold for the dominant forces between van der Waals forces and the liquid/gas surface tension forces in the multilayer gas adsorption
EDLC	Electric double-layer capacitor
$f(x)$	Distribution density function of pore size distribution or pore length distribution
$f(x)dx$	Number of the pores between x and $x + dx$
$f_P(x)$	Fractal pore size distribution
FFT	Fast Fourier transform
FHH	Frenkel, Halsey, and Hill
ΔG	Gibbs' free energy
γ	Incomplete gamma function
γ_{cap}	Rate capability
γ_L	Liquid surface tension
HRTEM	High resolution transmission electron microscopy
H-K	Horvath-Kawazoe
I	Current
I_{tot}	Total current
Δi	Current density difference between anodic and cathodic peak currents of the cyclic voltammogram
IFFT	Inverse fast Fourier transform
IUPAC	International Union of Pure and Applied Chemistry
j	$\sqrt{-1}$
k	Empirical constant
k_e	Conductivity of the electrolyte
l_o	Arbitrary unit pore length
l_p	Pore length
l_s	Distance between reference electrode and working electrode
L_a	Microcrystallite sizes in a- axis direction (parallel to the basal planes)
L_c	Microcrystallite sizes in c- axis direction (parallel to the edge planes
m	Mass

m_n	nth positive root of $(m \tan m - \Lambda) = 0$
μ	Mean value of the distribution variables
MP	Multiprobe
n	Equation parameter in Dubinin-Astakhov equation
n_{ad}	Number of adsorbed molecule layers
N_A	Avogadro constant
N_{mono}	Number of adsorbed molecules to form a monolayer
ν	Potential scan rate
p	Adsorption equilibrium pressure of the gas
p^o	Saturation pressure of the gas
p^*	Critical condensation pressure of the gas
P	Perimeter
pH	Potential of hydrogen
PLD	Pore length distribution
PSD	Pore size distribution
Q	Reduced charge
Q_1	Enthalpy (heat) of adsorption in the first adsorbed layer
Q_L	Enthalpy (heat) of liquefaction of adsorbate
r	Pore radius
r_1	First radius of curvature of the liquid meniscus in the pores
r_2	Second radius of curvature of the liquid meniscus in the pores
r_{ave}	Average pore radius
$r_{L,m}$	Mean radius of curvature of the liquid meniscus
r_m	Radius of the sphere probe molecule
r_{max}	Pore radius of the maximum differential pore volume
r_p	Radius of a cylindrical pore
R	Gas constant
R'	Resistance of the electrolyte inside a pore per unit pore length
R_p	Resistance of the electrolyte inside a pore
R_{sol}	Solution resistance between reference electrode and working electrode
R.E.	Reference electrode
σ	Standard deviation of the distribution variables
SAFG	Surface acidic functional group
SCE	Saturated calomel electrode
SEM	Scanning electron microscopy
SP	Single-probe
SPICE	Simulation program with integrated circuit emphasis

STM	Scanning tunneling microscopy
t	Time
t_{ad}	Actual thickness of adsorbed molecule layers
$t_{ad,max}$	High-end thickness of the fractal regime
$t_{ad,min}$	Low-end thickness of the fractal regime
T	Absolute temperature
θ	Contact angle between the solid and condensed phase
θ_L	Local adsorption isotherm
Θ	Overall adsorption isotherm
TEM	Transmission electron microscope
TEOS	Tetraethoxy silane
TLM	Transmission line model
TMOS	Tetramethoxy silane
V	Volume of adsorbed gas molecule
V_{cond}	Molar volume of the condensed adsorbate
V_m	Molar volume of the gas
V_{max}	Upper limits of the power-law regimes in the plot of $\ln V$ against $\ln \ln(p^o/p)$
V_{min}	Lower limits of the power-law regimes in the plot of $\ln V$ against $\ln \ln(p^o/p)$
V_{mono}	Volume of monolayer coverage
V_o	Total micropore volume
V_{tot}	Total pore volume
w_p	Width of a parallel-sided slit
ω	Angular frequency
W.E.	Working electrode
x	Average half-width of the slit-shaped pores
x'	Natural logarithm of pore length l_p divided by an arbitrary unit pore length l_o to obtain a dimensionless quantity
x_{max}	Upper limits of the pore fractality
x_{min}	Lower limits of the pore fractality
Z	Electrochemical impedance
Z'	Real impedance
Z''	Imaginary impedance
Z_o	Impedance of the pore
Z_{tot}	Total impedance of the pores

REFERENCES

[1]A. Adachi and T. Kobayashi, *Bull. Environ. Contam. Toxicol.* **54** (1995) 440.
[2]K. Kaneko, *Stud. Surf. Sci. Catal.* **120** (1998) 635.
[3]J. Sun, M. J. Rood, and A.A. Lizzio, *Gas Sep. Purif.* **10** (1996) 91.
[4]S. Sircar, T. C. Golden, and M. B. Rao, *Carbon* **34** (1996) 1.
[5]J.M. Martín-Martínez, L. Singoredjo, M. Mittelmeijer-Hazeleger, F. Kapteijn, and J. A. Moulijn, *Carbon* **32** (1994) 897.
[6]D. Mehandjiev, E. Bekyarova, and M. Khristova, *J. Colloid Interface Sci.* **192** (1997) 440.
[7]S. V. Mikhalovsky and Yu. P. Zaitsev, *Carbon* **35** (1997) 1367.
[8]M. Nakamura, M. Nakanishi, and K. Yamamoto, *J. Power Sources* **60** (1996) 225.
[9]D. Qu and H. Shi, *J. Power Sources* **74** (1998) 99.
[10]B.E. Conway, *Electrochemical Supercapacitors: Scientific Fundamentals and Technological Applications*, Kluwer Academic/Plenum Publishers, New York, 1999.
[11]G. Salitra, A. Soffer, L. Eliad, Y. Cohen, and D. Aurbach, *J. Electrochem. Soc.* **147** (2000) 2486.
[12]K. Honda, T. N. Rao, D. A. Tryk, A. Fujishima, M. Watanabe, K. Yasui, and H. Masuda, *J. Electrochem. Soc.* **148** (2001) A668.
[13]M. Endo, T. Maeda, T. Takeda, Y. J. Kim, K. Koshiba, H. Hara, and M. S. Dresselhaus, *J. Electrochem. Soc.* **148** (2001) A910.
[14]C. -H. Kim, S. -I. Pyun, and H. -C. Shin, *J. Electrochem. Soc.* **149** (2002) A93.
[15]S. -I. Pyun, C. -H. Kim, S. -W. Kim, and J. -H. Kim, *J. New Mat. Electrochem. Systems* **5** (2002) 289.
[16]G. -J. Lee, S. -I. Pyun, and C. -H. Kim, *J. Solid State Electrochem.* **8** (2004) 110.
[17]G. -J. Lee and S. -I. Pyun, *Electrochim. Acta* **51** (2006) 3029.
[18]P. L. Walker, Jr., F. Rusinko, Jr., and E. Raats, *J. Phys. Chem.* **59** (1955) 245.
[19]J. S. Mattson and H. B. Mark, Jr., *Activated Carbon: Surface Chemistry and Adsorption from Solution*, Marcel Dekker, New York, 1971.
[20]F. Caturla, M. Molina-Sabio, and F. Rodriguez-Reinso, *Carbon* **29** (1991) 999.
[21]F. Rodríguez-Reinoso and M. Molina-Sabio, *Carbon* **30** (1992) 1111.
[22]A. Ahmadpour and D.D. Do, *Carbon* **34** (1996) 471.
[23]C. Namasivayam and K. Kadirvelu, *Bioresource Technol.* **62** (1997) 123.
[24]S. R. Tennison, *Applied Catalysis A: General* **173** (1998) 289.
[25]C. -F. Chang, C.-Y. Chang, and W. -T. Tsai, *J. Colloid Interface Sci.* **232** (2000) 45.
[26]H. Teng and S. -C. Wang, *Carbon* **38** (2000) 817.
[27]Z. Ryu, J. Zheng, M. wang, and B. Zhang, *J. Colloid Interface Sci.* **230** (2000) 312.
[28]A. R. Sánchez, A. A. Elguézabal, and L. de La Torre-Saenz, *Carbon* **39** (2001) 1367.
[29]D. Lozanó-Castelló, M. A. Lillo-Ródenas, D. Cazorla-Amorós, and A. Linares-Solano, *Carbon* **39** (2001) 741.
[30]F. Suárez-García, J. I. Paredes, A. Martínez-Alonso, and Juan M.D. Tascón, *J. Mater. Chem.* **12** (2002) 3213.
[31]A. M. Puziy, O. I. Poddubnaya, A. Martínez-Alonso, and F. Suárez-García, J. M. D. Tascón, *Carbon* **41** (2003) 1181.
[32]K. Okada, N. Yamamoto, Y. Kameshima, and A. yasumori, *J. Colloid Interface Sci.* **262** (2003) 179.
[33]M. C. Baquero, L. Giraldo, J. C. Moreno, F. Suárez-García, A. Martínez-Alonso,

and J. M. D. Tascón, *J. Anal. Appl. Pyrolysis* **70** (2003) 779.
[34]J. A. Maciá-Agulló, B. C. Moore, D. Cazorla-Amorós and A. Linares-Solano, *Carbon* **42** (2004) 1367.
[35]F. -C. Wu, R. -L. Tseng, and R. -S. Juang, *J. Colloid Interface Sci.* **283** (2005) 49.
[36]E. Ruckenstein and Y. H. Hu, *Carbon* **36** (1998) 269.
[37]H. Tamon, H. Ishizaka, T. Araki, and M. Okazaki, *Carbon* **36** (1998) 1257.
[38]J.W. Lee, S.H. Yoon, T.H. Hyeon, S.M. Oh, and K.B. Kim, *Chem. Commun.* (1999) 2177.
[39]R. Ryoo, S. -H. Joo, and S. Jun, *J. Phys. Chem. B*, **103** (1999) 7743.
[40]S. J. Han and T. H. Hyeon, *Carbon* **37** (1999) 1645.
[41]D. Kawashima, T. Aihara, Y. Kobayashi, T. Kyotani, and A. Tomita, *Chem. Mater.* **12** (2000) 3397.
[42]S. J. Han, K. N. Sohn, and T. H. Hyeon, *Chem. Mater.* **12** (2000) 3337.
[43]J. W. Lee, K. N. Sohn, and T. H. Hyeon, *J. Am. Chem. Soc.* **123** (2001) 5146.
[44]S. H. Joo, S. J. Choi, I. Oh, J. Kwak, Z. Liu, O. Terasaki, and R. Ryoo, *Nature* **412** (2001) 169.
[45]Z. Li and M. Jaroniec, *Carbon* **39** (2001) 2077.
[46]Z. Li and M. Jaroniec, *J. Am. Chem. Soc.* **123** (2001) 9208.
[47]M. Kang, S. H. Yi, H. I. Lee, J. E. Yie, and J. M. Kim, *Chem. Commun.* (2002) 1944.
[48]W. W. Lukens and G. D. Stucky, *Chem. Mater.* **14** (2002) 1665.
[49]A. B. Fuertes and D. M. Nevskaia. *Micropor. Mesopor. Mater.* **62** (2003) 177.
[50]J. W. Lee, J. Y. Kim, and T. H. Hyeon, *Chem. Commun.* (2003) 1138.
[51]J. Yamashita, T. Ojima, M. Shioya, H. Hatori, and Y. Yamada, *Carbon* **41** (2003) 285.
[52]P. M. Barata-Rodrigues, T. J. Mays, and G. D. Moggridge, *Carbon* **41** (2003) 2231.
[53]Y. X. Wang, S. H. Tan, D. L. Jiang, and X. Y. Zhang, *Carbon* **41** (2003) 2065.
[54]S. -I. Pyun and C. -K. Rhee, *Electrochim. Acta* **49** (2004) 4171.
[55]G. -J. Lee and S. -I. Pyun, *Carbon* **43** (2005) 1804.
[56]P. Pfeifer and D. Avnir, *J. Chem. Phys.* **79** (1983) 3558.
[57]C. W. Lung and S. Z. Zhang, *Physica D* **38** (1989) 242.
[58]P. Pfeifer, Y. J. Wu, M. W. Cole, and J. Krim, *Phys. Rev. Lett.* **62** (1989) 1997.
[59]M. Jaroniec, *Fuel* **69** (1990) 1573.
[60]I. M. K. Ismail and P. Pfeifer, *Langmuir* **10** (1994) 1532.
[61]M. K. Wu, *Aerosol Sci. Technol.* **25** (1996) 392.
[62]R. Wojsz and A. P. Terzyk, *Comput. Chem.* **20(4)** (1996) 427.
[63]B. Sahouli, S. Blacher, and F. Brouers, *Langmuir* **13** (1997) 4391.
[64]M. Sato, T. Sukegawa, T. Suzuki, and K. Kaneco, *J. Phys. Chem. B* **101** (1997) 1845.
[65]P. Pfeifer and K. Y. Liu, *in Equilibria and Dynamics of Gas Adsorption on Heterogeneous Solid Surfaces,* Ed. by W. Rudziński, W.A. Steele, and G. Zgrablich, Elsevier, New York, 1997, p. 625.
[66]N. R. Khalili, M. Pan, and G. Sandí, *Carbon* **38** (2000) 573.
[67]C. Lin, J. A. Ritter, and B. N. Popov, *J. Electrochem. Soc.* **146** (1999) 3639.
[68]S. Shiraishi, H. Kurihara, L. Shi, T. Nakayama, and A. Oya, *J. Electrochem. Soc.* **149** (2002) A855.
[69]R. W. Pekala, C. T. Alviso, F. M. Kong, and S. S. Hulsey, *J. Non-Cryst. Solids* **145** (1992) 90.
[70]R. W. Pekala, J. C. Farmer, C. T. Alviso, T. D. Tran, S. T. Mayer, J. M. Miller, and B. Dunn, *J. Non-Cryst. Solids* **225** (1998) 74.

[71]R. Zhang, Y. Lu, L. Zhan, X. Liang, G. Wu, and L. Ling, *Carbon* **41** (2002) 1660.

[72]R. Zhang, W. Li, X. Liang, G. Wu, Y. Lü, L. Zhan, C. Lu, and L. Ling, *Micropor. Mesopor. Mater.* **62** (2003) 17.

[73]R. W. Pekala, *J. Mater. Sci.* **24** (1989) 3221.

[74]R. W. Pekala, C. T. Alviso, X. Lu, J. Gross, and J. Fricke, *J. Non-Cryst. Solids* **188** (1995) 34.

[75]X. Lu, R. Caps, J. Fricke, C.T. Alviso, and R. W. Pekala, *J. Non-Cryst. Solids* **188** (1995) 226.

[76]S. M. Manocha, D. Y. Vashistha, and L. M. Manocha, *J. Mater. Sci. Lett.* **16** (1997) 705.

[77]H. Müller, P. Rehak, C. Jäger, J. Hartmann, N. Meyer, and S. Spange, *Adv. Mat.* **12** (2000) 1671.

[78]S. Braunauer, L. S. Deming, W. E. Deming, and E. Teller, *J. Am. Chem. Soc.* **62** (1940) 1723.

[79]K. S. W. Sing, D.H. Everett, R. A. W. Haul, L. Moscou, R.A. Pierotti, J. Rouquérol, and T. Siemieniewska, *Pure Appl. Chem.* **57** (1985) 603.

[80]F. Stoeckli, T. A. Centeno, A. B. Fuertes, and J. Muňiz, *Carbon* **34** (1996) 1201.

[81]Z. Li, M. Kruk, M. Jaroniec, and S. -K. Ryu, *J Colloid Interface Sci.* **204** (1998) 151.

[82]Z. Ryu, J. Zheng, and M. Wang, *Carbon* **36** (1998) 427.

[83]M. El-Merraoui, M. Aoshima, and K. Kaneko, *Langmuir* **16** (2000) 4300.

[84]A. H. Lu and J. T. Zheng, *J Colloid Interface Sci.* **236** (2001) 369.

[85]M. Kruk and M. Jaroniec, *Chem. Mater.* **12** (2000) 222.

[86]M. Kruk and M. Jaroniec, *Chem. Mater.* **13** (2001) 3169.

[87]R. Ryoo, I. -S. Park, S. Jun, C. W. Lee, M. Kruk, and M. Jaroniec, *J. Am. Chem. Soc.* **123** (2001) 1650.

[88]Q. Huo, D. I. Margolese, and G. D. Stucky, *Chem. Mater.* **8** (1996) 1147.

[89]M. Kruk and M. Jaroniec, *Langmuir* **13** (1997) 6267.

[90]H. -P. Lin, S. -T. Wong, C. -Y. Mou, and C. -Y. Tang, *J. Phys. Chem. B* **104** (2000) 8967.

[91]P. J. Kooyman, M. J. Verhoef, and E. Prouzet, *Stud. Surf. Sci. Catal.* **129** (2000) 535.

[92]S. Brunauer, P. H. Emmett, and E. Teller, *J. Am. Chem. Soc.* **60** (1938) 309.

[93]I. Langmuir, *J. Am. Chem. Soc.* **40** (1918) 1631.

[94]P. A. Webb and C. Orr, *Analytical Methods in Fine Particle Technology*, Micrometritics Instrument Corporation, Norcross, 1997.

[95]M.M. Dubinin and L.V. Radushkevich, *Proc. Acad. Sci. USSR* **55** (1947) 331.

[96]M. Polanyi, *Verb. Deutsch. Phyrik. Ges.* **16** (1914) 1012.

[97]M. M. Dubinin and H. F. Stoeckli, *J. Colloid Interface Sci.* **75** (1980) 34.

[98]M. M. Dubinin and B. A. Astakhov, *Izv. AN. SSSR. Ser. Khim.* **5** (1971) 11.

[99]G. Horvath and K. Kawazoe, *J. Chem. Eng. Jpn.* **16** (1983) 470.

[100]D. H. Everett and J. C. Powl, *J. Chem. Soc., Faraday Trans.* **72** (1976) 619.

[101]A. Saito and H. C. Foley, *AIChE Journal* **37** (1991) 429.

[102]L. S. Cheng and R. T. Yang, *Chem. Eng. Sci.* **49** (1994) 2599.

[103]S. J. Gregg and K. S. W. Sing, *Adsorption, surface area and porosity*, Academic Press, New York, 1982.

[104]E. P. Barrett, L. G. Joyner, and P. P. Halenda, *J. Am. Chem. Soc.* **73** (1951) 373.

[105]B. B. Mandelbrot, *The Fractal Geometry of Nature*, W.H. Freeman and Company, New York, 1983.

[106]H. O. Peitgen and D. Saupe, Ed., *The Science of Fractal Images*, Springer Verlag, New York, 1988.

[107]C. A. Pickover, *Computers, Pattern, Chaos and Beauty*, St. Martin's Press, New York, 1990.
[108]H. O. Peitgen, H. Jürgens, and D. Saupe, *Fractals for the Classroom: Introduction to Fractals and Chaos*, Springer Verlag, New York, 1992.
[109]R. N. Mantegna and H. E. Stanley, *Nature* **376** (1995) 46.
[110]M. H. R. Stanley, L. A. N. Amaral, S.V. Buldyrev, S. Havlin, H. Leschhorn, P. Maass, M. A. Salinger, and H. E. Stanley, *Nature* **379** (1996) 804.
[111]H. Takayasu and K. Okuyama, *Fractals* **6** (1998) 67.
[112]H. Takayasu, M. Takayasu, M.P. Okazaki, K. Marumo, and T. Shimizu, in *Paradigms of Complexity*, Ed. by M.M. Novak, World Scientific, 2000, p. 243.
[113]T. Lours, J. Zarzycki, A. Craievich, D. I. dos Santos, and M. Aegerter, *J. Non-Cryst. Solids* **95** (1987) 1151.
[114]W. I. Friesen and R. J. Mikula, *J Colloid Interface Sci.* **120(1)** (1987) 263.
[115]J. J. Friel and C. S. Pande, *J. Mater. Res.* **8(1)** (1993) 100.
[116]P. Tang, N. Y. K. Chew, H. K. Chan, and J.A. Raper, *Langmuir* **19** (2003) 2632.
[117]D. Avnir and D. Farin, *J. Chem. Phys.* **79** (1983) 3566.
[118]P. Pfeifer, D. Avnir, and D. Farin, *Surf. Sci.* **126** (1983) 569.
[119]D. Avnir, D. Farin, and P. Pfeifer, *Nature* **308** (1984) 261.
[120]D. Avnir, D. Farin, and P. Pfeifer, *J. Colloid Interface Sci.* **103** (1985) 112.
[121]D. Farin, S. Pelog, D. Yavin, and D. Avnir, *Langmuir* **1** (1985) 399.
[122]A. Y. Myer, D. Farin, and D. Avnir, *J. Am. Chem. Soc.* **108** (1986) 7897.
[123]D. Avnir, *J. Am. Chem. Soc.* **109** (1987) 2931.
[124]D. Farin and D. Avnir, *J. Phys. Chem.* **91** (1987) 5517.
[125]D. Farin and D. Avnir, *J. Am. Chem. Soc.* **110** (1988) 2039.
[126]A. V. Neimark, *Adsorpt. Sci. Technol.* **7** (1990) 210.
[127]D. K. Ludlow and T. P. Moberg, *Anal. Instrum.* **19** (1990) 113.
[128]K. Kaneko, *Langmuir* **7** (1991) 109.
[129]M. Sato, T. Sukegawa, T. Suzuki, S. Hagiwara, and K. Kaneko, *Chem. Phys. Lett.* **181** (1991) 526.
[130]T. Suzuki and T. Yano, *Biol. Chem.* **55** (1991) 967.
[131]D. Avnir, D. Farin, and P. Pfeifer, *New J. Chem.* **16** (1992) 439.
[132]Y. Lefebvre, S. Lacelle, and C. Jolicoeur, *J. Mater. Res.* **7** (1992) 1888.
[133]W. C. Conner and C. O. Bennett, *J. Chem. Soc., Faraday Trans.* **89** (1993) 4109.
[134]R. F. Cracknell, P. Gordon, and K. E. Gubbins, *J. Phys. Chem.* **97** (1993) 494.
[135]K. Kaneko, R. F. Cracknell, and D. Nichlson, *Langmuir* **10** (1994) 4606.
[136]T. Suzuki, K. Kaneko, N. Setoyama, M. Maddox, and K. E. Gubbins, *Carbon* **34** (1996) 909.
[137]P. Pfeifer and M. W. Cole, *New J. Chem.* **14** (1990) 221.
[138]P. Pfeifer, J. Kenntner, and M. W. Cole, in *Fundamentals of Adsorption*, Ed. by A. B. Mersmann and S. E. Scholl, American Institute of Chemical Engineers, New York, 1991, p. 689.
[139]J. Frenkel, *Kinetic Theory of Liquids*, Clarendon, Oxford, 1946.
[140]G. D. Halsey, *J. Chem. Phys.* **16** (1948) 931.
[141]T. L. Hill, *Adv. Catal.* **4** (1952) 211.
[142]S. -B. Lee, S. -I. Pyun, and C. -K. Rhee, *Carbon* **41** (2003) 2446.
[143]A. P. Terzyk, R. Wojsz, G. Rychlicki, and P. A. Gauden, *Colloids and Surfaces A: Physicochemical and Engineering Aspects* **119** (1996) 175.
[144]R. Wojsz and A. P. Terzyk, *Comput. Chem.* **21(2)** (1997) 83.
[145]A. P. Terzyk, P. A. Gauden, G. Rychlicki, and R. Wojsz, *Colloids Surf. A: Physicochemical and Engineering Aspects* **136** (1998) 245.
[146]A. P. Terzyk, P. A. Gauden, G. Rychlicki, and R. Wojsz, *Colloids Surf.* **152** (1999) 293.

[147]P. A. Gauden, A. P. Terzyk, and G. Rychlicki, *Carbon* **39** (2001) 267.
[148]P. A. Gauden, A. P. Terzyk, G. Rychlicki, and P. Kowalczyk, *J. Colloid Interface Sci.* **244** (2001) 439.
[149]A. P. Terzyk and P.A. Gauden, *The Arabian Journal for Science and Engineering* **28** (2003) 133.
[150]M. M. Dubinin, *J. Colloid Interface Sci.* **23** (1967) 487.
[151]M. Endo, K. Takeuchi, Y. Sasuda, K. Matsubayashi, K. Oshida, and M. S. Dresselhaus, *Electron. Commun. Jpn.* **77** (1994) 139.
[152]K. Oshida, K. Kogiso, K. Matsubayashi, K. Takeuchi, S. Kobayashi, M. Endo, M.S. Dresselhaus, and G. Dresselhaus, *J. Mater. Res.* **10** (1995) 2507.
[153]M. Endo, T. Furuta, F. Minoura, and C. Kim, K. Oshida, G. Dresselhaus, and M.S. Dresselhaus, *Supermolecular Science* **5** (1998) 261.
[154]B. B. Mandelbrot, *Phys. Scr.* **32** (1985) 257.
[155]J. Feder, *Fractals*, Plenum Press, New York, 1988.
[156]J. M. Gómez-Rodríguez, A. M. Baró, and R. C. Salvarezza, *J. Vac. Sci. Technol. B* **9** (1991) 495.
[157]P. Ocon, P. Herrasti, L. Vázquez, R. C. Salvarezza, J. M. Vara, and A. J. Arvia, *J. Electroanal. Chem.* **319** (1991) 101.
[158]J. M. Gómez-Rodríguez, A. M. Baró, L. Vázquez, R. C. Salvarezza, J. M. Vara, and A. J. Arvia, *J. Phys. Chem.* **96** (1992) 347.
[159]P. Herrasti, P. Ocón, R.C. Salvarezza, J. M. Vara, L. Viázquez, and A. J. Arvia, *Electrochim. Acta* **37** (1992) 2209.
[160]R. de Levie, *Electrochim. Acta* **9** (1964) 1231.
[161]R. de Levie, *Electrochim. Acta* **10** (1965) 113.
[162]R. de Levie, *Advances in Electrochemistry and Electrochemical Engineering*, Ed. by P. Delahay, Vol. VI, John Wiley & Sons, New York, 1967, p. 329.
[163]W. Scheider, *J. Phys. Chem.* **79** (1975) 127.
[164]H. Keiser, K. D. Beccu, and M. A. Gutjahr, *Electrochim. Acta* 21 (1976) 539.
[165]M. Kramer and M. Tomkiewicz, *J. Electrochem. Soc.* **131** (1984) 1283.
[166]Nyikos and T. Pajkossy, *Electrochim. Acta* **30** (1985) 1533.
[167]T. Pajkossy, *J. Electroanal. Chem.* **364** (1994) 111.
[168]K. Eloot, F. Debuyck, M. Moors, and A. P. Van Peteghem, *J. Appl. Electrochem.* **25** (1995) 326.
[169]K. Eloot, F. Debuyck, M. Moors, and A. P. Van Peteghem, *J. Appl. Electrochem.* **25** (1995) 334.
[170]T. Momma, X. Liu, T. Osaka, Y. Ushio, and Y. Sawada, *J. Power Sources* **60** (1996) 249.
[171]Z. Kerner and T. Pajkossy, *J. Electroanal. Chem.* **448** (1998) 139.
[172]H. -K. Song, Y. -H. Lee, and Le H. Dao, *Electrochim. Acta* **44** (1999) 3513.
[173]H. -K. Song, H. -Y. Hwang, K. -H. Lee, and Le H. Dao, *Electrochim. Acta* **45** (2000) 2241.
[174]J. Bisquert, *Phys. Chem. Chem. Phys.* **2** (2000) 4185.
[175]J. Bisquert, G. Garcia-Belmonte, F. Fabregat-Santiago, N. S. Ferriols, P. Bogdanoff, and E. C. Pereira *J. Phys. Chem. B* **104** (2000) 2287.
[176]Z. Kerner and T. Pajkossy, *Electrochim. Acta* **46** (2000) 207.
[177]D. Qu, *J. Power Sources* **109** (2002) 403.
[178]C. -H. Kim, S. -I. Pyun, and J. -H. Kim, *Electrochim. Acta* **48** (2003) 3455.
[179]A. Le Mehaute and G. Crepy, *Solid State Ionics* **9&10** (1983) 17.
[180]A. Le Mehaute, *J. Stat. Phys.* **36** (1984) 665.
[181]S. H. Liu, *Phys. Rev. Lett.*, **55** (1985) 529.
[182]B. Sapoval, *Solid State Ionics* **23** (1987) 253.
[183]T. C. Halsey, *Phys. Rev. A* **35** (1987) 3512.

[184]T. Kaplan, L. J. Gray, and S.H. Liu, *Phys. Rev. B* **35** (1987) 5379.
[185]J. C. Wang, *Electrochim. Acta* **33** (1988) 707.
[186]R. M. Hill and L. A. Dissado, *Solid State Ionics* **26** (1988) 295.
[187]W. H. Mulder and J. H. Sluyters, *Electrochim. Acta* **33** (1988) 303.
[188]Y. T. Chu, *Solid State Ionics* **26** (1988) 299.
[189]M. Blunt, *J. Phys. A: Math. Gen.* **22** (1989) 1179.
[190]A. Maritan and F. Toigo, *Electrochim. Acta* **35** (1990) 141.
[191]T. Pajkossy and L. Nyikos, *Phys. Rev. B* **42** (1990) 709.
[192]M. A. Daley, D. Tandon, J. Economy, and E. J. Hippo, *Carbon* **34** (1996) 1191.
[193]J. P. Randin and E. Yeager, *J. Electroanal. Chem. and Interfacial Electrochem.* **36** (1972) 257.
[194]J. P. Randin and E. Yeager, *J. Electroanal. Chem. and Interfacial Electrochem.* **58** (1975) 313.
[195]A. J. Groszek, *Carbon* **25** (1987) 717.
[196]K. Kinoshita, *Carbon: Electrochemical and Physicochemical Properties*, John Wiley & Sons, New York, 1988, p. 293.
[197]H. Marsh, E. A. Heintz, and F. R. Reinoso, *Introduction to Carbon Technologies*, Universidad de Alicante, Secretariado de Publicaciones, 1997, p. 36
[198]H. Marsh and F. R. Reinoso, *Sciences of Carbon Materials*, Universidad de Alicante, Secretariado de Publicaciones, 2000, p. 8.
[199]A. Sadkowski, *Electrochim. Acta* **38** (1993) 2051.
[200]C.P. Smith and H.S. White, *Langmuir* **9** (1993) 1.
[201]M.G. Sullivan, B. Schnyder, M. Bärtsch, D. Alliata, C. Barbero, R. Imhof, and R. Kötz, *J. Electrochem. Soc.* **147** (2000) 2636.
[202]M.G. Sullivan, R. Kötz, and O. Haas, *J. Electrochem. Soc.* **147** (2000) 308.
[203]C. -T. Hsieh and H. Teng, *Carbon*, **40** (2002) 667.
[204]K. Kinoshita and J. A. S. Bett, *Carbon*, **13** (1975) 405.
[205]K. Kinoshita and X. Chu, "Carbon for Supercapacitors"; in *Proceedings of the Symposium on Electrochemical Capacitors*, F. M. Denlick and M. Tomkiewicz, Editors, PV 95-29, The Electrochemical Society Proceeding Series, Pennington, NJ, 1995, p. 171-180.

The Use of Graphs in the Study of Electrochemical Reaction Networks

Joseph D. Fehribach

Fuel Cell Center, Departments of Chemical Engineering and Mathematical Sciences, Worcester Polytechnic Institute

I. INTRODUCTION

The purpose of this chapter is to review and discuss the uses of graphs in the study of chemical and particularly electrochemical reaction networks. Such reaction networks are defined by (often elementary) reaction steps, and in turn the total chemical process associated with a given set of reaction steps is its reaction network. Such a network should normally determine at least one overall reaction. Certain steps in a given reaction network may occur at specified locations, and the overall process may involve transport between these locations. Graphs have long been used in various ways to clarify all of these concepts.

In what follows, any graph used to study a reaction network will be termed a *reaction graph*. Although there have been many such uses over the years, there are three general categories which largely cover all uses of graphs:

1. reaction species graphs,
2. reaction mechanism graphs, and
3. reaction route graphs.

Modern Aspects of Electrochemistry, Number 41, edited by C. Vayenas et al., Springer, New York, 2007.

The names of these categories are roughly those given by previous authors, however, previous names have not necessarily been consistent or correctly reflected similarities between graphs. In particular, the terms "reaction graph" and "reaction diagram" have been used by several authors in the past to identify particular graphs described below. These uses are sometimes inconsistent, and the term seems too general to associate with any specific graph representing a reaction. This chapter will focus on representative examples from each of the above categories, with a particular emphasis on comparing and contrasting their distinctions as graphs. The ordering of the presentation is basically chronological.

Before defining the three categories, it is perhaps useful to give some basic graph-theoretic definitions. For our purposes, a *graph* is a mathematical structure consisting of points (usually called nodes or vertices) and line segments (usually called edges or branches)—each line segment connecting (identifying) a pair of these points. The key issue in dividing the various uses into the three categories mentioned above is what the nodes and edges of a particular graph represent. The standard definition of a graph is here extended in two major ways: First, most of our graphs will be *digraphs*, meaning that a positive direction will be associated with each edge. Second, some of our graphs will have *free* or *half* edges, meaning that these edges will be associated with a single node and thus have a free end.

Reaction species graphs are generally the earliest reaction graphs in use. As its name suggests, graphs in this category tend to emphasize the species in the process. In species graphs, individual species are represented by nodes, while edges represent reaction steps, the connections between species and reactions, or in some cases, more loosely the flow of the reaction process. Because of the sometimes-vague role of the edges, reaction species graphs perhaps are the most amorphous of the three categories. It is not clear when the earliest reaction species graph was published, but such graphs go back at least to the Krebs cycle in biochemistry in the 1950s and probably much earlier.

The second category to be considered here is the reaction mechanism graphs. In these graphs, nodes represent (elementary) reaction steps, and edges represent individual species. Reaction mechanism graphs are useful in depicting the locations of reaction steps and the separations between steps. In the third category, that of reaction route graphs, nodes represent combinations of species (sometimes termed *complexes*) and their component potentials (defined below), and edges represent reaction steps. Graphs in this third category are most helpful in computing the overall reaction rate for a reaction network using the rates of the individual reaction steps, or for finding an equivalent reaction circuit in much the same way one might find an equivalent electrical circuit. While these latter two categories are very distinct, graphs in both can be seen as extensions of those in the first category.

There are several issues that tend to run across all categories of graphs and should probably be addressed in general. For one, the use of directed versus undirected edges is not consistent among the various authors. In some cases, directed edges are used only for reaction steps that run essentially to completion; in others, directed edges indicate the forward directions for reversible steps. Which of these two approaches is better will not be decided here. Also some authors concentrate on whether or not certain reaction steps are or are not "elementary." While this is often an important distinction, it is not particularly relevant here. In addition it should be pointed out that individual molecules, ions or intermediates can never be tracked around any reaction graphs; all reaction graphs represent only the net reaction process, not the paths of individual molecules, ions or intermediates. This is a particularly important point when there are cycles in species and mechanism graphs. Finally a number of the graphs discussed below are used in various ways to help in making quantitative calculations. Such calculations can be described most generally in terms of reaction route graphs, and therefore quantitative and/or mathematical issues will be covered in the context of those graphs.

There are other uses of graphs in chemistry—the most obvious being those depicting molecules using atoms as nodes and bonds as edges. Also graphs have been used to describe that structure of layers or regions. These are of course important uses, but since they are very different from graphs used to study reactions, they are not considered here.

In some cases, graphs discussed in this work were presented as diagrams in their original publication, rather than explicitly as graphs. Also in some cases, essentially the same graph can be drawn in different ways. Part of the goal of this chapter is to reconcile such differences without distinctions, and the three categories presented here attempt to do this. Graphs in the same category share essential similarities, even if the style of their respective drawing makes them look different, or they are described with differing wording. There are, however, several types of reaction graphs appearing in recent publications that do not fit into any of the three categories. Some of these graphs are mentioned briefly in the Section V.

Finally two previous reviews of reaction graphs should be cited in general: Bonchev and Mekenyan (1994)[1] and Temkin, Zeigarnik and Bonchev (1996).[2]

II. REACTION SPECIES GRAPHS

As mentioned above, reaction species graphs are characterized by having some or all of the individual species involved in the reaction network represented by nodes in the graph. Various authors have developed a number of graphs that fall into this category. Perhaps the most basic of these depicts the general flow (both chemically and physically) of an overall process without being too exact about the rules for drawing the graph. These are often drawn as diagrams rather than precisely as graphs, and they have been in use for many years. Again many graphical depictions of the Krebs cycle are of this type. A more-recent example was used by Liu and Macdonald (2001)[3] to describe the formation of passive films on iron in a borate buffer in the presence of EDTA

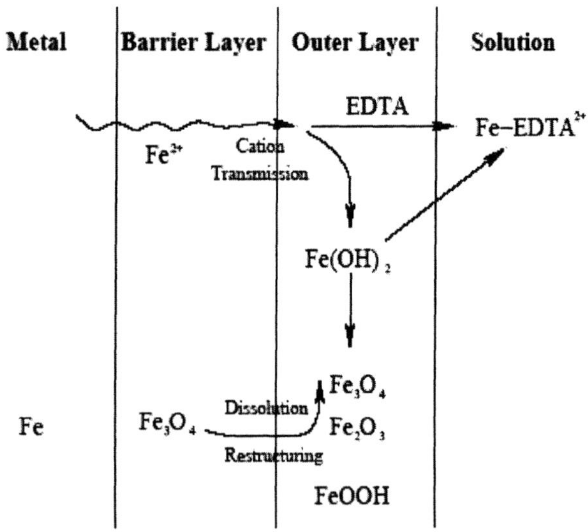

Figure 1. A species graph (or diagram) for the formation of passive films on iron: species are at nodes (not explicitly shown), reaction steps and flow are represented by directed edges. Reproduced from Ref. [3], Copyright (2001), by permission of The Electrochemical Society.

(cf Figure 1). Other recent examples of this type in electrochemistry include work by Pavlov and Petkova (2002),[4] Uchida, et al. (2002),[5] (2003),[6] Chaplin and Wragg (2003),[7] Voncken et al. (2004)[8] and Lei, Ju and Ikeda (2005).[9] While these sorts of graphs may be very helpful in understanding the processes they describe, their definitions are too vague to be dealt with in any general way, and so they will not be discussed in detail here.

1. Kinetic Graphs

A more precisely defined grouping in this category consists of graphs which concentrate on the intermediate species. Temkin (1965)[10] defined *kinetic* graphs where each node represents a specific intermediate except for a single node which represents the so-called *0-species*. Two nodes are connected by an edge if a reaction step involves one of the two associated intermediates as a reactant and the other as a product. If a reaction step has no intermediates among its reactants, then a node representing an intermediate among the products is connected by an edge to the 0-species node. Similarly, if a reaction step has no intermediates among its products, then a node representing an intermediate among the reactants is connected by an edge to the 0-species node.

As an example of the construction of kinetic graphs, consider the following set of reaction steps that yield the electrochemical oxidation of hydrogen:

$$\begin{aligned} &s_1: H_2 + S \rightleftarrows H_2{\cdot}S \\ &s_2: H_2{\cdot}S + S \rightleftarrows 2(H{\cdot}S) \\ &s_3: H_2 + 2S \rightleftarrows 2(H{\cdot}S) \\ &s_4: H_2O + H{\cdot}S \rightleftarrows S + H_3O^+ + e^- \\ &s_5: H_2O + H_2{\cdot}S \rightleftarrows H{\cdot}S + H_3O^+ + e^- \\ &s_6: H_2O + H_2 + S \rightleftarrows H{\cdot}S + H_3O^+ + e^- \end{aligned} \tag{1}$$

The overall reaction is $H_2 + 2\ H_2O \rightleftarrows 2\ H_3O^+ + 2\ e^-$. Fishtik, Callaghan and Datta (2005)[11] discuss this reaction in another context, taking $H{\cdot}S$ and $H_2{\cdot}S$ as the intermediates. For this choice of intermediates, a kinetic graph can be constructed as in Figure 2. The tradition for kinetic graphs is to use directed edges only when

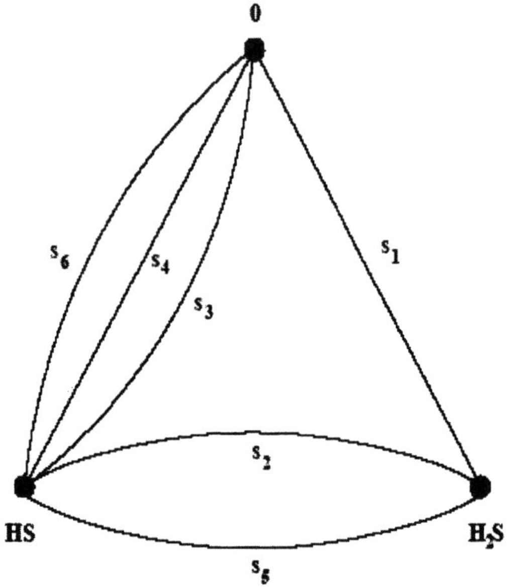

Figure 2. Kinetic graph for the electrochemical oxidation of hydrogen: nodes represent individual intermediates, an edge connects two nodes exactly when the two corresponding intermediates are the only intermediates on opposite sides of the corresponding reaction step. If no intermediate is on one side of a given step, an edge corresponding to that step connects to the 0-species node.

a reaction step proceeds effectively to completion, and undirected edges for steps when reaction steps are reversible. Since all steps in this example are of the latter type, the graph in Figure 2 contains no directed edges.

The above example illustrates a significant limitation on kinetic graphs: the mechanisms they depict must have reaction steps that are linear in their intermediates to guarantee that a kinetic graph can be drawn. *Linear in their intermediates* means that at most one intermediate can be present among the reactants or products of any single reaction step, and the stoichiometric coefficient of such an intermediate must be one. Thus if one views the reaction site S as an intermediate as was done in an example given by Temkin, Zeigarnik and Bonchev (1996, p. 61),[2] one would not be able to construct the kinetic graph for the above reaction network. Indeed because of the 2(H·S) in steps s_2 and s_3, this reaction network is not linear in its intermediates, and one must be careful in interpreting the kinetic graph in Figure 2.

Kinetic graphs have several uses: As is the case with all reaction species graphs, they show the topological relationship between the intermediate species. In addition they can be helpful in computing overall reaction rates. In both these respects, kinetic graphs can be regarded as preliminary versions of reaction route graphs (discussed in Section IV). A more extensive discussion of these and other uses of kinetic graphs is given in Temkin, Zeigarnik and Bonchev (1996).[2] Graphs similar to kinetic graphs were used by, among others, Christiansen (1953),[12] Balaban, Farcasiu and Banica (1966)[13] and more recently Helfferich (1989).[14] See also Balaban (1994)[15] for an overview of similar graphs.

2. Bipartite Graphs

The final grouping in the category of species graphs is that of *bipartite graphs*, meaning that the nodes are divided into two sets and edges are only permitted to connect nodes from one set with those of the other set. Nodes in one set (often drawn as disks) represent reacting species while those in the other set (often drawn as squares) represent reaction steps. Each species and each step is represented by a single node. A directed edge leads from a disk-node to a square-node exactly when the molecule or ion represented by the disk-node is a reactant in the step represented by the square-node. Similarly a directed edge leads from a square-node to a disk-node exactly when the molecule or ion represented by the disk-node is a product in the step represented by the square-node. So for graphs in this grouping, edges do not represent an element in the reaction process, but rather just indicate the connectivity between species and reaction steps. Since the second set of nodes represent reaction steps, bipartite graphs could be thought of as intermediate between the present category of species graphs and the second category of mechanism graphs. But since reaction locations are not represented in bipartite graphs (this is a major feature of mechanism graphs), and since bipartite graphs represent individual species as nodes, it seems appropriate to include bipartite graphs in the category of species graphs.

As an example of this sort of bipartite graph, let us again consider the electrochemical oxidation of hydrogen; the elementary steps are given in (1). A bipartite graph for this reaction is given in Figure 3.

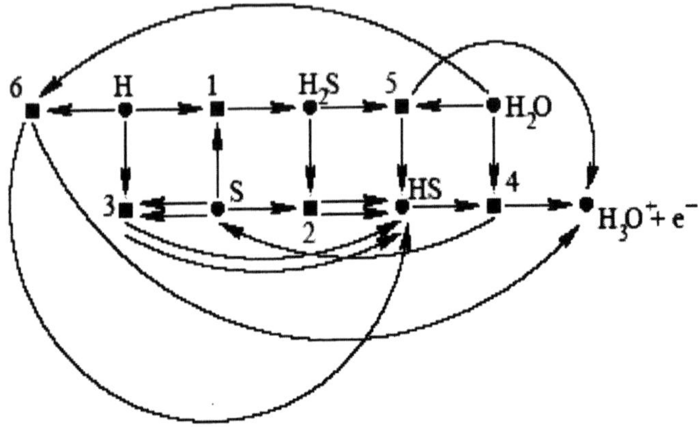

Figure 3. Bipartite graph for the electrochemical oxidation of hydrogen: disk-nodes represent individual species, square-nodes represent reaction steps. A directed edge connects a disk-node to a square-node when species represented by the disk-node is a reactant in the reaction step represented by the square-node. On the other hand, a directed edge connects a square-node to a disk-node when species represented by the disk-node is a product in the reaction step represented by the square-node.

Bipartite graphs have advantages over kinetic graphs in that they can be drawn both for reaction networks that are linear in their intermediates and those that are not. In addition, they depict all species involved in a reaction, not just intermediates. On the other hand, there are also disadvantages, for example that they are not helpful in computing reaction rates (they can not generally be associated with potentials as can reaction route graphs). And as was seen in Figure 3, there is still no guarantee for a given process that the corresponding bipartite graph will be *planar*, meaning that it is possible to draw the graph so that edges do not intersect except at nodes. (Kinetic graphs may also not be planar.) Even if a graph is planar, it may be difficult to draw the graph without unwanted intersections that make the graph appear more complicated. Still bipartite graphs do give roughly equal emphasis to both the reacting species and the reaction steps. In this regard they can be thought of as intermediate between other species graphs and the mechanism graphs of the following Section.

Balandin (1970)[16] appears to have used bipartite graphs to study reactions in the 1930s (though his work was not published until much later). Clarke (1980)[17] used a type of bipartite graph which he termed a *current diagram* to study the stability of reaction networks. He found that for a reaction process to be unstable (i.e., for the process to oscillate or behave chaotically), the corresponding current diagram must containing a *current cycle*—an irreversible cycle among the species. Sinanoglu (1981)[18] purposed a small variation on the sort of bipartite graph discussed above where reaction steps are represented by wavy edges connecting square nodes which represent the beginning and ending of the reaction step. Fan, Bertok and Friedler (2002)[19] and Friedler et al. (1992)[20] used *p-graphs* (bipartite graphs with bars rather than squares representing steps) in their algorithm for synthesizing a reaction network from elementary steps. Other applications of bipartite graphs in the study of reactions are discussed in Temkin, Zeigarnik and Bonchev (1996).[2]

III. REACTION MECHANISM GRAPHS

Reaction mechanism graphs are defined by the following characteristic properties: Reaction steps are represented by nodes, while the species involved in these steps are represented by directed edges. The direction of the edge indicates movement of the species from reaction step where it is a product to a presumably separate step where it is a reactant. A given species may appear on any number of distinct edges on a graph, and either multiple edges or edges weighted by the appropriate stoichiometric numbers may represent multiples of a given species moving between reaction steps. Each step, however, is represented by a unique node. Terminal species (those that make up the overall reaction) are represented by free edges (defined above) which connect to only one node. The overall products are represented by edges directed outward, the species consumed by the overall reaction are represented by edges directed inward, and species which cycle through the reaction are represented by edges directed inward and equally-weighted edges directed outward. Nodes can be positioned to indicate reaction sites where a particular reaction takes place, i.e., on a specific surface or in a particular bulk phase.

Reaction mechanism graphs are particularly useful in comparing several different reaction mechanisms that yield the same overall reaction, and also in considering the relative locations of various steps. This is illustrated through several examples from molten carbonate and proton-exchange membrane (PEM) fuel cell.

1. MCFC Cathodic Reactions

As a first example of the use of reaction mechanism graphs, consider the electrochemistry of molten carbonate fuel cell (MCFC) cathodes. These cathodes are typically nickel-oxide porous electrodes with pores partially filled with a molten carbonate electrolyte. Oxygen and carbon dioxide are fed into the cathode through the vacant portions of the pores. The overall cathodic reaction is $O_2 + 2CO_2 + 4e^- \rightleftarrows 2CO_3^=$. This overall reaction can be achieved through a number of reaction mechanisms; two such mechanisms are the peroxide mechanism and the superoxide-peroxide mechanism, and these are considered next.

(*i*) *Peroxide Mechanism*

The peroxide mechanism achieves the overall reaction through the following sequence of five reaction steps:

$$\begin{aligned} &s_1\text{: } CO_2(g) \rightleftarrows CO_2(e)\\ &s_2\text{: } O_2 + 2CO_3^= \rightleftarrows 2O_2^= + 2CO_2(e)\\ &s_3\text{: } O^= + CO_2(e) \rightleftarrows CO_3^=\\ &s_4\text{: } O_2^= + e^- \rightleftarrows O^= + O^-\\ &s_5\text{: } O^- + e^- \rightleftarrows O^= \end{aligned} \qquad (2)$$

Notice that four of these steps (s_2 through s_5) are chemical or electrochemical steps, while step s_1 denotes the physical dissolution of carbon dioxide as it moves from the gas phase into the electrolyte phase.

The relationship between the five peroxide mechanism reaction steps can be seen in the reaction mechanism graph in Figure 4. As defined above, each step occurs at one of the five nodes, and the directed edges give the forward direction for the mechanism. Current-carriers for the overall mechanism are in boxes, while carbonate ions that continue from one cycle to the next are circled. Dashed vertical lines represent interfaces between phases. Nodes on the gas-electrolyte interface represent reaction steps occurring at that interface; nodes attached to the electrolyte-solid interface represent reaction steps occurring at sites on the surface of the solid phase. The location of each reaction on this reaction mechanism graph follows the description of the

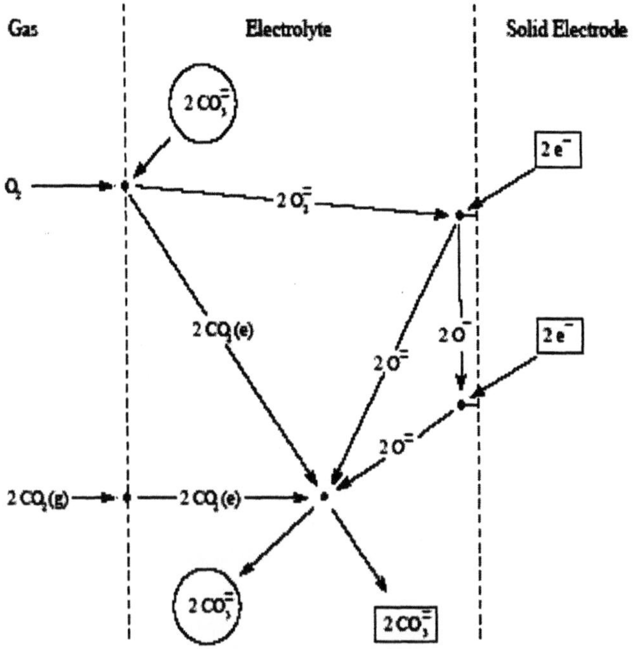

Figure 4. Reaction mechanism graph for the peroxide mechanism: reactions occur at dots, arrows show the forward direction for reactions, dashed lines separate phases, current-carriers are in boxes, and carbonate ions which continue from one cycle to the next are circled.

mechanism given by Yuh and Selman (1984),[21] except for the oxide/carbon-dioxide recombination step, s_3. Following Lee, Selman and Plomp (1993),[22] the recombination step occurs *in* the electrolyte phase, instead of at the solid-electrolyte interface.

Reaction mechanism graphs such as the one in Figure 4 can lead to a number of insights which would be difficult or impossible to make simply from studying a list of reaction steps such as (2). Hemmes, Peelen and de Wit (1999)[23] observed from the graph that since the step that produces peroxide, s_2, and the recombination step, s_3, occur at different locations in the cathode, one can separate the inlet channels for oxygen and carbon dioxide. The advantage to this separation is that since carbon dioxide is also a product of s_2, the rate of this reaction might be increased by not feeding carbon dioxide near the location of this step. This work eventually led to the design of a new type of MCFC cathode, the iMCFC cathode. In addition, the mechanism graph for the peroxide mechanism illustrates how carbonate ions (those circled in Figure 4) cycle through the overall reaction. Indeed because of the role carbonate in initiating the process, these graphs were originally called reaction cycle graphs.[24] Again, it is difficult or impossible to discover the role carbonate simply by studying the list (2).

(*ii*) *Superoxide-Peroxide Mechanism*

As reaction mechanisms become more complicated, the need for effective graphical depictions increases. To illustrate this, let us consider the superoxide-peroxide mechanism of Adanuvor, White and Appleby (1990).[25] The overall reaction for this mechanism is the same as in the previous cases: $O_2 + 2CO_2 + 4e^- \rightleftarrows 2CO_3^=$. This mechanism consists of seven steps, three of which appear to be the same as in the peroxide mechanism:

$$\begin{aligned}
&s_1\text{: } CO_2(g) \rightleftarrows CO_2(e)\\
&s_2\text{: } O_2 + 2CO_3^= \rightleftarrows 2O_2^= + 2CO_2(e)\\
&s_3\text{: } O^= + CO_2(e) \rightleftarrows CO_3^=\\
&s_4\text{: } O_2 + O_2^= \rightleftarrows 2O_2^-\\
&s_5\text{: } 3O_2 + 2O^= \rightleftarrows 4O_2^-\\
&s_6\text{: } O_2^- + e^- \rightleftarrows O_2^=\\
&s_7\text{: } O_2^= + 2e^- \rightleftarrows 2O^=
\end{aligned} \tag{3}$$

The reaction mechanism graph for this mechanism is given in Figure 5; again all of the steps except s_3 (recombination) occur at an interface.

The graphs in Figures 4 and 5 are useful in comparing the peroxide and superoxide-peroxide mechanisms. Although there are a number of similarities (essentially the same recombination reaction, for example), the graphs show the differences in the roles of peroxide in the two mechanisms. In the first case, peroxide must be transported across the electrolyte phase, while in the second, it can exists only on the gas-electrolyte interface. Indeed oxygen can not exist in the electrolyte phase. The graphs also indicate that the superoxide-peroxide mechanism requires a great deal more ion transport between the gas-electrolyte and electrolyte-solid interfaces.

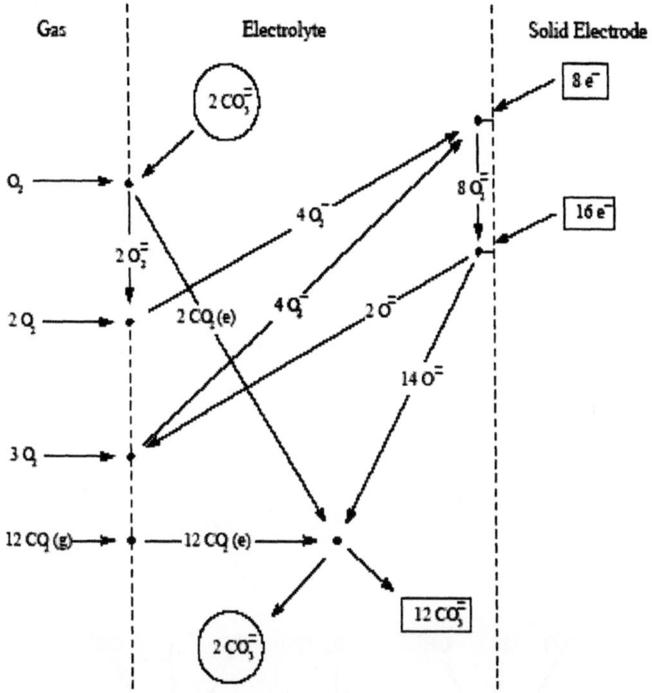

Figure 5. Superoxide-peroxide reaction mechanism graph: notation and symbols as in Figure 4.

2. HER Reactions

Reaction mechanism graphs may be useful in studying reaction mechanisms even when there is no species cycling through the reaction. To see this, consider the simple example of the electrochemical hydrogen evolution reaction (HER) mechanism. Here the overall reaction is $2H_2O + 2e^- \rightleftarrows 2OH^- + H_2$. The elementary reaction steps that combined to yield HER (cf. Fishtik et al., 2005)[26] are

$$
\begin{aligned}
&s_1: 2(H{\cdot}S) \rightleftarrows H_2 + 2S \\
&s_2: S + H_2O + e^- \rightleftarrows H{\cdot}S + OH^- \\
&s_3: H_2O + H{\cdot}S + e^- \rightleftarrows S + H_2 + OH^-
\end{aligned}
\qquad (4)
$$

The subscripts which distinguish the steps honor, respectively, Tafel, Volmer and Heyrovsky. Unlike the MCFC cathodic reaction mechanisms, however, these steps combine pairwise to yield the overall reaction. The reaction mechanism graphs for each of the three reaction mechanisms are shown in Figure 6. Notice that it is not possible to represent the entire mechanism by a single reaction mechanism graph. This is because, unlike in the MCFC case, there are now independent full reaction routes which yield the over all reaction. In both of the MCFC examples, there was only one. Still the three separate graphs do clearly convey the three HER reaction routes.

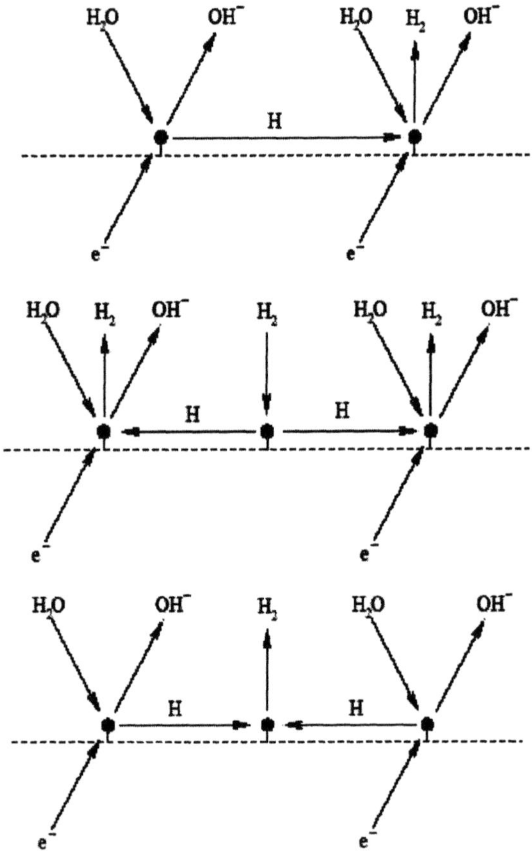

Figure 6. Reaction mechanism graphs for HER: again reaction steps occur at nodes, arrows show the forward direction for reactions, and the dashed line represents the metal-electrolyte boundary.

IV. REACTION ROUTE GRAPHS

Reaction route graphs are defined as follows: Reversible reaction steps are represented by directed edges, and nodes represent combinations of species—in particular the component potentials, i.e., sums and differences of electrochemical potentials weighted by the appropriate stoichiometric numbers. This implies that the potential difference between any two adjacent nodes is the affinity of the associated step. Indeed this potential difference must be the reaction step affinity no matter where the associated directed edge appears in the graph. The direction of each edge indicates the forward direction of the associated step. Nodes fall into two types: terminal nodes correspond to combinations of species, i. e., those that are either the products or the reactants for the overall reaction; intermediate nodes correspond to combinations that contain at least in part intermediate (non-terminal) species. Reaction route graphs can be *multi-graphs*—multiple edges are permitted to connect the same nodes. Taken as a whole, each reaction route graph must satisfy four fundamental defining conditions:

- The sum of the reaction rates for the steps incident on/from each terminal node must be some stoichiometric multiple of the overall reaction rate.
- The sum of the reaction rates for steps incident on/from each intermediate node must be zero.
- The sum of the affinities around any closed cycle must be zero.
- The sum of the affinities for any trail, path or walk between two terminal nodes must be the affinity of the overall reaction.

These four conditions are equivalent to Kirchoff's laws (the first two are equivalent to Kirchoff's current law; the last two are equivalent to Kirchoff's voltage law). This analogy implies that through the use of reaction route graphs, one can study reaction mechanisms using the tools that are traditionally used to study electrical circuits. Not all reaction networks, however, may have reaction route graphs; it is still not clear how to characterize exactly for which networks one can construct an appropriate reaction route graph. But when a reaction route graph can be drawn, it will be very helpful in understanding the reaction network.

The applications of reaction route graphs are in general among the most advance uses of graphs in the study of reaction networks

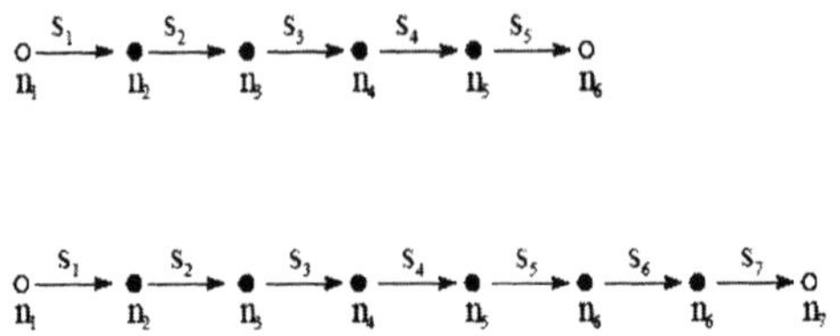

Figure 7. Reaction route graphs for the peroxide and superoxide-peroxide mechanisms: reaction steps occur on directed edges; nodes n_i represent the component potentials, the difference between these potentials for adjacent nodes is the affinity of the associated reaction step; and terminal nodes are open, intermediate nodes, closed.

to date. Reaction route graphs allow one to depict several independent full routes on a single graph. In addition, reaction route graphs are more quantitative than any of the graphs previously discussed in that they allow one to calculate or verify overall reaction rates and affinities from the reaction rates and affinities of the reaction steps. One can also define simpler equivalent reaction route graphs that yield the same overall reaction. To see this, let us return to the examples from the previous Section.

1. MCFC Cathodic Reactions

The reaction steps for both the peroxide and superoxide-peroxide mechanisms were given in the previous Section by (2) and (3). Notice that because the various intermediates exist only on or in certain phases, the full reaction route can only be achieved using the full set of reaction steps, although there is not a unique ordering of the steps. Because each of these mechanisms has only a single full reaction route, each of the reaction route graphs is simply a *chain* or *path*—an alternating node-edge sequence with no repetition of nodes or edges; these are shown in Figure 7. Despite their simplicity, these reaction route graphs are still useful in modelling the reaction electrochemistry. Consider the rates of the reaction steps for the peroxide mechanism. Typically certain steps are slow (rate limiting) while the others are relatively fast and thus can be viewed as being in equilibrium. Following Yuh and Selman (1984),[21] (1992)[27] and Kunz and Murphy (1988),[28] let us assume that steps s_3 and s_4 are both of comparable rate and rate limiting, while all of the other steps are in equilibrium. Using this assumption, one can define component potentials by identifying all of the nodes that are in equilibrium, i.e., nodes whose associated

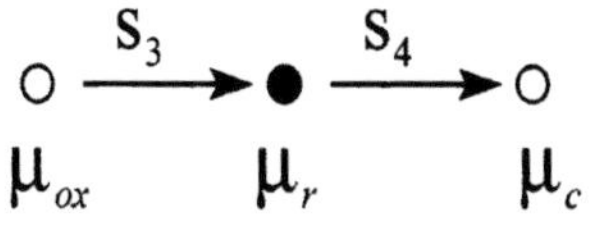

Figure 8. Equivalent component-potential reaction route graph for the peroxide and superoxide-peroxide mechanisms.

potentials are essentially equal. This allows one to draw an equivalent component-potential reaction route graph for the peroxide mechanism as in Figure 8 and simplifies ones view of the reaction network. This development (without the aid of the actual graphs) was used by the author and his colleagues in Delft to model MCFC cathodes (2000–2001).[24,29,30] They were able to develop a relatively simple model with a minimum number of variables and parameters, and to determine what electrode structures are most favorable for good performance. Depending on which of its steps are considered rate limiting, a similar derivation can be given for the superoxide-peroxide mechanism.

The reaction route graphs, however, do have certain limitations. It is not in general possible, for example, to depict the physical location of the various reactions and species. It is not easy to distinguish on reaction route graphs that the peroxide ions, which must move across the electrolyte in the peroxide mechanism, exist only on the phase interfaces (gas-electrolyte and electrolyte-solid) in the superoxide-peroxide mechanism. This depiction is one of the strong points for reaction mechanism graphs.

2. HER Reactions

A reaction route graph for HER is given in Figure 9. Since the HER mechanism has three independent reaction routes, the reaction route graph for this mechanism can not simply be a chain. Also Fishtik et al. (2005)[26] explicitly drew the overall reactions as two edges connecting the two terminal nodes; these edges are not included here so that the HER reaction route graph more closely resembles similar graphs of other authors (cf. below). Note that it is possible to display all three distinct full routes on a single reaction route graph.

Construction of reaction route graphs when there are a number (perhaps many) independent reaction routes can be difficult and is

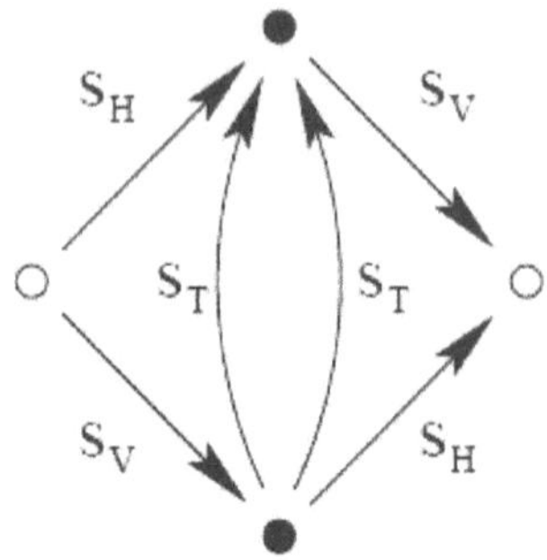

Figure 9. Reaction route graph for HER: again reaction steps occur on directed edges; nodes represent the component potentials; and terminal nodes are open, intermediate nodes, closed.

discussed by Fishtik, Callaghan and Datta (2004–2005).[11,31,32] For the present discussion, it is sufficient to show that the graph in Figure 9 satisfies the four defining properties for reaction route graphs. This can be seen directly from the reaction steps and the overall reaction: The sums of the rates for the steps which produce terminal species (here H_2, H_2O and OH^-) must be the overall reaction rate r_{OR} weighted by the corresponding stoichiometric coefficient from the overall reaction. This implies that

$$\begin{aligned} r_{H_2} &= r_{OR} = r_T + r_H \\ r_{OH^-} &= 2r_{OR} = r_V + r_H \\ r_{H_2O} &= -2r_{OR} = -r_V - r_H \end{aligned} \tag{5}$$

In addition, the sum of the rates for the steps that produce intermediate species (here HS) must be zero; thus $r_V - r_H - 2r_T = 0$. Clearly the expressions for r_{OR} corresponding to H_2O and OH^- are equivalent, and using the intermediate-species condition, one can see that the H_2 expression in (5) is also equivalent with these two. So the terminal and intermediate-species conditions yield two independent requirements; these two guarantee that the first two conditions defining a reaction route graph (which amount to Kirchoff's current law) are satisfied by the graph in Figure 9. The intermediate-species condition guarantees that the sum of the rates at both intermediate nodes in Figure 9 are zero. The terminal-species condition, on the other hand, guarantees that the sum of the rates at both terminal nodes in Figure 9 equal $2r_{\mathrm{OR}}$.

To see that the latter two defining properties are also satisfied, notice that one combination of the reaction steps yields a empty (null) route where all of the species cancel, while three other combinations yield the overall reaction:

$$\begin{aligned} \mathbf{s}_{\mathrm{V}} - \mathbf{s}_{\mathrm{H}} + \mathbf{s}_{\mathrm{T}} &= 0 \\ \mathbf{s}_{\mathrm{V}} + \mathbf{s}_{\mathrm{H}} &= \mathrm{OR} \\ 2\mathbf{s}_{\mathrm{V}} + \mathbf{s}_{\mathrm{T}} &= \mathrm{OR} \\ 2\mathbf{s}_{\mathrm{H}} - \mathbf{s}_{\mathrm{T}} &= \mathrm{OR} \end{aligned} \tag{6}$$

These relationships between the reaction steps imply corresponding relationships between the affinities:

$$\begin{aligned} \mathcal{A}_{\mathrm{V}} - \mathcal{A}_{\mathbf{H}} + \mathcal{A}_{\mathbf{T}} &= 0 \\ \mathcal{A}_{\mathrm{V}} + \mathcal{A}_{\mathbf{H}} &= \mathcal{A}_{\mathbf{OR}} \\ 2\mathcal{A}_{\mathrm{V}} + \mathcal{A}_{\mathbf{T}} &= \mathcal{A}_{\mathbf{OR}} \\ 2\mathcal{A}_{\mathbf{H}} - \mathcal{A}_{\mathbf{T}} &= \mathcal{A}_{\mathbf{OR}} \end{aligned} \tag{7}$$

All of this amounts to Kirchoff's voltage law. Clearly both (6) and (7) are consistent with Figure 9, and therefore all four defining conditions are satisfied.

Reaction route graphs and the associated systems of equations for the reaction rates and affinities can be used to compute overall rates based on rates and affinities for the individual steps. In addition, using the graph, one can determine which routes are most significant (offer the least resistance) and which can be eliminated as unnecessary because the rates are too small to be significant. For HER this was done by Fishtik et al. (2005);[26] for lower potentials it was found that the Volmer and Tafel steps are dominant, while for higher potentials, the Volmer and Heyrovsky are dominant. The process described above allows one to quantify this result—to calculate precisely how much more the reaction "current" flows through one full route rather than another. In this regard, reaction route graphs can be seen as an extension of the kinetic graphs (described above) which are only guaranteed for reactions that are linear in their intermediates to more-general reaction networks.

Reaction route graphs as developed above were defined by Fishtik, Callaghan and Datta (2004–2005);[26,31,32] similar graphs in this category have been used by other authors. Horn (1973)[33,34] proved a number of results about *reaction diagrams* and *complex graphs* which are essentially reaction route graphs, though he considered reversible, weakly reversible and irreversible reactions. Oster, Perelson and Katchalsky (1973)[35] referred to these graphs as *topological graphs*. Both of these two studies used reaction route graphs as a stepping stone to other types of graphs mentioned in the next Section. More recently Qian, Beard and Liang (2003)[36] used a simple reaction route graph in their study of a three-state kinetic cycle (cf. Figure 10). Also Palsson[37] has authored or co-authored a series of articles that propose the use of graphs similar to reaction route graphs for the study of reaction networks. Palsson's work covers both reversible and irreversible networks and defines two types of sets of routes:

1. elementary modes and
2. extreme pathways.

The extreme pathways are a basis (in the sense of vector spaces) for all routes through the reaction network, i.e., any route can be written as a linear combination of the extreme pathways. When all the reaction steps are irreversible, these two sets are equivalent, i.e., they have the same elements. When some of the steps are reversible, there are more (perhaps far more) elementary modes, though this set is still much smaller than the countably infinite set of all routes. In any event, all routes can be obtained as combinations of the elements of either set, and thus if one studies these relatively small sets of routes, one can hope to obtain results for all routes.

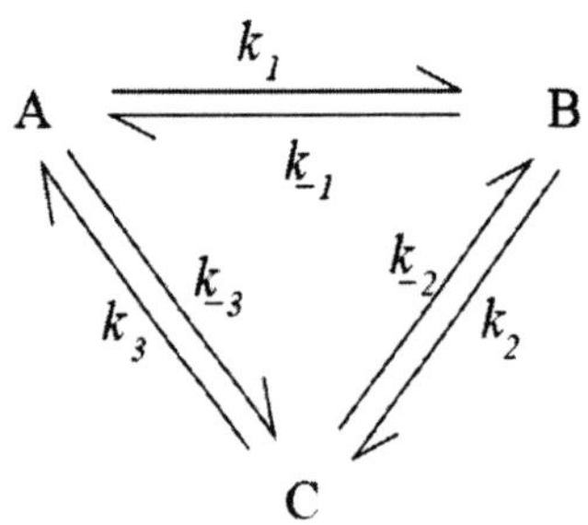

Figure 10. Three state kinetic cycle showing the relationship between three potential states A, B and C. The k_i are rate constants.[36]

V. DISCUSSION: OTHER REACTION GRAPHS

As mentioned above, the categories discussed in this chapter cover most but not all graphs used to study chemical or electrochemical reaction network. Among those *not* covered are the *bond graphs* defined by Oster, Perelson and Katchalsky (1973).[35] Bond graphs are a separate generalization of reaction route graphs (or topological graphs as Oster, Perelson and Katchalsky term them) containing two types of nodes (or junctions): one-junctions whose bonds to other elements of the graph correspond to edges on a topological graph, and zero-junctions whose bonds to one-junctions have no topological-graph counterpart. Bond graphs are discussed at length by Thoma (1975).[38] Separately Oster and Perelson[39] describe a type of directed graph (which they do not name) which is associated with a form of the stoichiometric matrix for a reaction network. In this graph, both species *and* reaction steps are represented by distinct edges. Another use of graphs that does not fit into any of the three categories defined above was given by Elkin (1987)[40] whose graphs represent linear variational systems associated with reaction processes (change in potential, change in current, change in flow rate, etc.). The implications for chemical networks, though claimed by the author, still need to be developed more fully.

Several of the works discussed above include graph-theoretic calculations of, for example, the complexity of a graph. Unfortunately it is not clear in many cases what the implications are for the reaction network for differences in the complexity of various associated graphs, particularly when the differences are small. In some cases, the results seem counterintuitive in that the more complex graph is constructed from the physically more important reaction process. More study is needed of these issues.

Finally, taken as a whole, the above discussion makes clear that no particular graph can completely represent all aspects of a reaction network. Each type of graph illustrates well certain details while leaving others less clear. To some extent, both reaction mechanism graphs and reaction route graphs can be seen as distinct generalizations of species graphs that were generally developed earlier. Reaction mechanism graphs emphasize the interconnections and locations of the species and reaction steps, while reaction route graphs emphasize the routes through the reaction network leading from the reactants to the products. It is not clear that reaction route graphs can be drawn for all reaction networks, but when they can, route graphs are a powerful tool in the study of reaction networks.

ACKNOWLEDGMENTS

This material is based upon work supported by the National Science Foundation under Grant No. DMS-0426132. Any opinions, findings, and conclusions or recommendations expressed in this material are those of the author(s) and do not necessarily reflect the views of the National Science Foundation.

The author wishes to thank Ravi Datta, Ilie Fishtik, Caitlin Callaghan, Kas Hemmes and Bill Martin for many useful discusses leading to this chapter, and Kim Ware for help in finding many of the references given here.

REFERENCES

[1]D. Bonchev and O. Mekenyan (eds.), *Graph Theoretical Approaches to Chemical Reactivity,* Kluwer, Dordrecht, 1994.
[2]O. N. Temkin, A. V. Zeigarnik, and D. Bonchev, *Chemical Reaction Networks, a Graph-Theoretical Approach,* CRC, Boca Raton, 1996.
[3]J. Liu and D. D. Macdonald, *J. Electrochem. Soc.* **148** (2001) B425.
[4]D. Pavlov and G. Petkova, *J. Electrochem. Soc.* **149** (2002) A654.
[5]H. Uchida, Y. Mizuno and M. Watanabe, *J. Electrochem. Soc.* **149** (2002) A682.
[6]H. Uchida, Y. Ueno, H. Hagihara, and M. Watanabe, *J. Electrochem. Soc.* **150** (2003) A57.
[7]R. P. S. Chaplin and A. A. Wragg, *J. Appl. Electrochem.* **52** (2003) 1107.
[8]M. M. A. J. Voncken, J. J. Schermer, A. T. J. van Niftrik, G. J. Bauhuis, P. Mulder, P. K. Larsen, T. P. J. Peters, B. de Bruin, A. Klaassen, and J. J. Kelly, *J. Electrochem. Soc.* **151** (2004) G347.
[9]J. Lei, H. Ju, and O. Ikeda, *J. Electroanal. Chem.* **567** (2005) 331.
[10]M. I. Temkin, *Dokl, Akad. Nauk. SSR* **165** (1965) 615.
[11]I. Fishtik, C. A. Callaghan, and R. Datta, *J. Phys. Chem. B.* **109** (2005), 2710.
[12]J. A. Christiansen, in *Advances in Catalysis,* Vol. 5, Ed. by W. G. Frankenburg, V. I. Komarewsky, and E. K. Rideal, Academic Press, New York, 1953, pp. 311-353.
[13]A. T. Balaban, D. Farcasiu, and D. Banica, *Rev. Roum. Chim.* **11** (1966) 1205.
[14]F. G. Helfferich, *J. Phys. Chem.* **93** (1989) 6676.
[15]A. T. Balaban, in *Graph Theoretical Approaches to Chemical Reactivity*, Ed. by D. Bonchev and O. Mekenyan, Kluwer, Dordrecht, 1994, pp. 137-180.
[16]A. A. Balandin, *Multiplet Theory of Catalysis: Theory of Hydrogenation,* Moscow State Univ., Moscow, 1970, in Russian.
[17]B. L. Clarke, in *Advances in Chemical Physics*, Vol. 43, Ed. by I. Prigogine and S. A. Rice, Wiley, Chichester, 1980, pp. 1-215.
[18]O. Sinanoglu, *J. Math. Phys.* **22** (1981) 1504.
[19]L. T. Fan, B. Bertok, and F. Friedler, *Computers and Chemistry* **26** (2002) 265.
[20]F. Friedler, K. Tarrjan, Y. W. Huang, and L. T. Fan, *Chem. Eng. Sci.* **47** (1992) 1973.

[21]C. Y. Yuh and J. R. Selman, *J. Electrochem. Soc.* **131** (1984) 2062.
[22]G. L. Lee, J. R. Selman, and L. Plomp, *J. Electrochem. Soc.* **140** (1993) 390.
[23]K. Hemmes, W. H. A. Peelen, and J. H. W. de Wit, *Electrochem. Solid-State Lett.* **2** (1999) 103.
[24]J. D. Fehribach, J. A. Prins-Jansen, K. Hemmes, J. H. W. de Wit, and F. W. Call, *J. Appl. Electrochem.* **30** (2000) 1015.
[25]P. K. Adanuvor, R. E. White, and A. J. Appleby, *J. Electrochem. Soc.* **137** (1990) 2095.
[26]I. Fishtik, C. A. Callaghan, J. D. Fehribach, and R. Datta, *J. Electoanal. Chem.* **576** (2005) 57.
[27]C. Y. Yuh and J. R. Selman, *J. Electrochem. Soc.* **139** (1992) 1373.
[28]H. R. Kunz and L. A. Murphy, *J. Electrochem. Soc.* **135** (1988) 1124.
[29]J. D. Fehribach, *European J. Appl. Math.* **12** (2001) 77.
[30]J. D. Fehribach and K. Hemmes, *J. Electrochem. Soc.* **148** (2001) A783.
[31]I. Fishtik, C. A. Callaghan, and R. Datta, *J. Phys. Chem. B.* **108** (2004) 5671.
[32]I. Fishtik, C. A. Callaghan, and R. Datta, *J. Phys. Chem. B* **108** (2004) 5683.
[33]F. Horn, *Proc. R. Soc. Lond.* A. **334** (1973) 299.
[34]F. Horn, *Proc. R. Soc. Lond.* A. **334** (1973) 313.
[35]G. F. Oster, A. S. Perelson, and A. Katchalsky, *Q. Reviews Biophys.* **6** (1973) 1.
[36]H. Qian, D. A. Beard, and S. D. Liang, *European J. Biochem.* **270** (2003) 415.
[37]J. A. Papin, J. Stelling, N. Price, S. Klant, S. Schuster, and B. O. Palsson, *TRENDS Biotech.* **22** (2004) 400. See also the references therein.
[38]J. U. Thoma, *Introduction to Bond Graphs and their Applications*, Pergamon press, Oxford, 1975.
[39]G. F. Oster and A. S. Perelson, *IEEE Trans. Circuits Sys.* **CAS-21** (1974) 709.
[40]V. V. Elkin, *Elektrokhimiya* **23** (1987) 351, in Russian.

Approximate Analytical Solutions for Models of Three-Dimensional Electrodes by Adomian's Decomposition Method

Keith Scott [a] and Yan-Ping Sun [b]

[a]*School of Chemical and Engineering and Advanced Materials, University of Newcastle upon Tyne, Newcastle, Great Britain*
[b]*Chemical Engineering Department, Taiyuan University of Technology, Taiyuan, Shanxi, China*

Three dimensional electrode structures are used in several applications, where high current densities are required at relatively low electrode and cell polarisations, e.g. water electrolysis and fuel cells. In these applications it is desirable to fully utilize all of the available electrode area in supporting high current densities at low polarisation. However conductivity limitations of three-dimensional electrodes generally cause current and overpotential to be non-uniform in the structure. In addition the reaction rate distribution may also be non-uniform due to the influence of mass transfer.[1]

In this chapter generalized mathematical models of three dimensional electrodes are developed. The models describe the coupled potential and concentration distributions in porous or packed bed electrodes. Four dimensionless variables that characterize the systems have been derived from modeling; a dimensionless conduction modulus μ, a dimensionless diffusion (or lateral dispersion) modulus s, a dimensionless transfer coefficient α and a dimensionless limiting current density γ. The first three are

Modern Aspects of Electrochemistry, Number 41, edited by C. Vayenas et al., Springer, New York, 2007.

used for porous electrodes and all four are used for packed bed electrodes. These variables are useful for three-dimensional (3-D) electrode reactor analysis.

Adomian's Decomposition Method is used to solve the model equations that are in the form of nonlinear differential equation(s) with boundary conditions.[2,3] Approximate analytical solutions of the models are obtained. The approximate solutions are in the forms of algebraic expressions of infinite power series. In terms of the nonlinearities of the models, the first three to seven terms of the series are generally sufficient to meet the accuracy required in engineering applications.

I. INTRODUCTION

Nonlinear problems frequently arise in engineering, but many texts are oriented towards linear problems due to the difficulty of non-linearity. For chemical systems as well as electrochemical systems, the mathematical models are typically nonlinear and even strongly nonlinear due to the nature of kinetics influenced by transport phenomena.

Solving the nonlinear mathematic models by a numerical method depends on the trial values that are chosen by experience and is not based upon a particular theory. The results obtained are greatly influenced by the initial values selected that is a characteristic of strong nonlinear problems. By using numerical techniques to solve nonlinear models, iterations must be implemented. With the incorrect selection of a trial value, divergence in solution can appear. The accuracy of the numerical results cannot be estimated theoretically for nonlinear problem since there is no analytical solution available. This is why an approximate analytical solution is extremely useful for a theoretical analysis of nonlinear problem! If an approximate analytical solution can be obtained, then this has a number of benefits:

1. a numerical program can be dispensed with, or a much sampler algebraic equation(s) rather than differential equations (DEs) can be used;
2. we can be free from the difficulty of iteration and diverging solution if the approximate solution is theoretically proved to be convergent; and
3. an algebraic expression is obtained that can be analyzed mathematically over the whole region of interest.

This review considers what we believe to be a suitable method to solve a range of electrochemical related problems in science and engineering, i.e., Adomian decomposition. The method is applied to several problems related to the analysis of three dimensional electrodes.[4,5] The typical structure of three dimensional electrodes is shown schematically in Figure 1, in terms of two types of electrode. Figure 1a, is appropriate for electrodes connected by an electrolyte as typically used in synthesis or in batteries, while Figure 1b is for electrodes as used in fuel cells, e.g., polymer electrolyte fuel cells (PEMFC). In general the models are concerned with determining the concentration and potential (and current) distributions in the structure.

The solution of models of these electrodes continues to be achieved using methods such as finite difference, finite elements, finite volume, shooting etc.[6,7] Here, we present an alternative approach and make comparisons with solutions of models using alternative numerical methods.

II. ADOMIAN'S DECOMPOSITION METHOD (ADM)

G. Adomian developed the decomposition method to solve the deterministic or stochastic differential equations.[3] The solutions obtained are approximate and fast to converge, as shown by Cherrault.[8] In general, satisfactory results can be obtained by using the first few terms of the approximate, series solution. According to Adomian's theory, his polynomials can approximate the

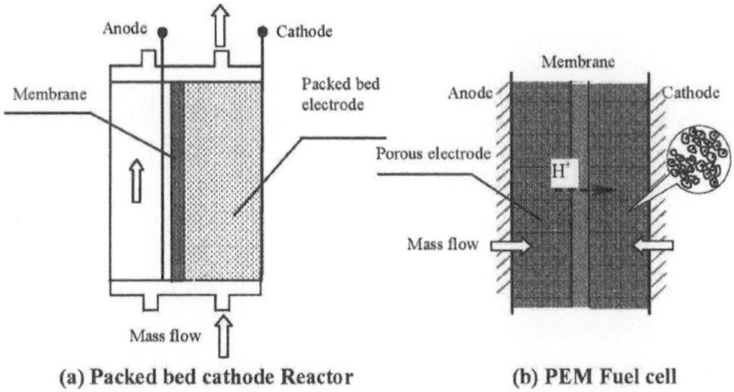

Figure 1. The typical structure of three-dimensional electrodes.

analytical solution, with a higher accuracy, by increasing the number of terms in the polynomials.

In general a differential equation can be expressed by an operator equation.[3,9] With the first decomposition, the original deterministic non-linear differential equation can be written in the Adomian's general form as:

$$\widetilde{F}u = Lu + Ru + Nu = g(x) \tag{1}$$

where, $L$is the highest order derivative which is assumed to be easily invertible, R is a linear differential operator of order less than L, N is an nonlinear operator of an analytic nonlinearity, and $g(x)$ is an inhomogeneous or a source term. Applying the inverse operator L^{-1}, to both sides of Eq. (1), and using the given initial and/or boundary conditions we obtain

$$L^{-1}Lu = L^{-1}g(x) - L^{-1}Ru - L^{-1}Nu \tag{2}$$

$$u = f_i - L^{-1}(Ru) - L^{-1}(Nu) \tag{3}$$

where, L^{-1} simply symbolizes n-fold integrations for an nth-order L. The function f_i represents the terms arising from integrating the inhomogeneous term $g(x)$ and from using the given conditions, all are assumed to be prescribed. Secondly, the solution process is assumed, i.e., the standard Adomian method defines the solution $u(x)$, as the infinite series

$$u = \sum_{n=0}^{\infty} u_n \tag{4}$$

where the components u_0, u_1, u_2, ... are usually determined recursively by

$$u_0 = f_i \tag{5}$$

$$u_{n+1} = -L^{-1}(Ru_n) - L^{-1}(Nu_n), \qquad n \geq 0 \tag{6}$$

Furthermore the decomposition method identifies the nonlinear term by the decomposition series

$$Nu = f[u(x)] = \sum_{n=0}^{\infty} A_n \tag{7}$$

where A_n are the so-called Adomian polynomials. Adomian formally introduced formulas that can generate Adomian polynomials for all forms of nonlinearity. A_n is defined as[2,3]

$$A_n = (\frac{1}{n!})(d^n/d\lambda^n) f(u(\lambda))|_{\lambda=0} \quad n = 1, 2, \ldots\ldots \tag{8}$$

where u is parameterized as $u = \sum_{i=0}^{n} u_i \lambda^i$. To apply ADM to practical nonlinear problems, the approximate solution with finite $m + 1$ terms is

$$u_{m+1} = \sum_{n=0}^{m} u_n \simeq u \tag{9}$$

that can be obtained by an finite m term approximate of Adomian Polynomials $\sum_{n=0}^{m-1} A_n$. Given a nonlinear term $f[u(x)]$, the first few polynomials are given by

$$\begin{aligned}
A_0 &= f(u_0) \\
A_1 &= u_1 f'(u_0) \\
A_2 &= u_2 f'(u_0) + \frac{1}{2!} u_1^2 f''(u_0) \\
A_3 &= u_3 f'(u_0) + u_1 u_2 f''(u_0) + \frac{1}{3!} u_1^3 f'''(u_0) \\
A_4 &= u_4 f'(u_0) + (\frac{1}{2!} u_2^2 + u_1 u_3) f''(u_0) + \frac{1}{2!} u_1^2 u_2 f'''(u_0) + \frac{1}{4!} u_1^4 f''''(u_0) \\
&\vdots
\end{aligned} \tag{10}$$

Other polynomials can be generated in a similar manner. A_0 depends only on u_0, A_1 depends only on u_0 and u_1, A_2 depends only on u_0, u_1, and u_2, and so on. It is clear that A_0 is always determined independently of the other polynomials A_n, $n \geq 1$, so that A_0 is always defined by $A_0 = f(u_0)$.

III. EXAMPLE OF APPLICATIONS TO CATALYTIC REACTIONS

An important problem in catalysis is to predict diffusion and reaction rates in porous catalysts when the reaction rate can depend on concentration in a non-linear way.[6] The heterogeneous system is modeled as a solid material with pores through which the reactants and products diffuse. We assume for diffusion that all the microscopic details of the porous medium are lumped together into the effective diffusion coefficient D_e for reactant.

Figure 2 shows a schematic of a porous catalyst slab that is supplied by reactant from the outer surface and in which reaction takes place at the internal catalytic surface. It is known in such systems that, when diffusion of species internally in the structure is slow in comparison to the rate of reaction, a variation in reactant concentration will occur in the catalyst. This variation in concentration changes the rate locally in the electrode.

In the system considered it is assumed that there is no variation in the rate constant locally in the structure. For electrochemical systems this is equivalent to assuming a constant potential in the structure and is an approach used in “plane”or “agglomerate” models of electrocatalysts.

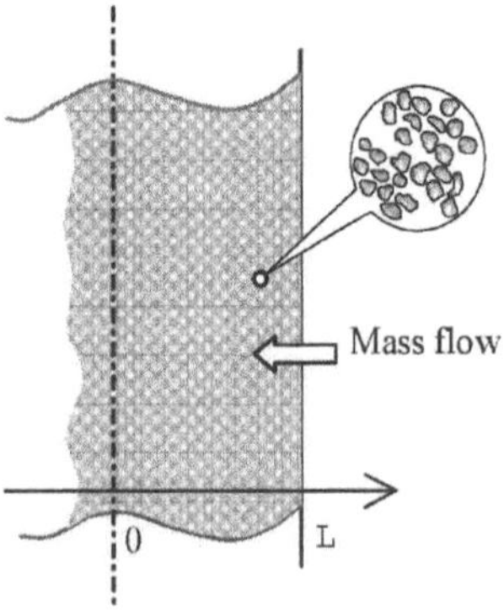

Figure 2. Schematic of a porous catalyst slab.

Assuming that diffusion, in the porous catalyst can be described by Fick's law, we have

$$\vec{N_i} = -D_{e_i} \nabla c_i \tag{11}$$

where $\vec{N_i}$ is the vector of local flux of reactant i, ∇c_i is the gradient of concentration of reactant i, and D_{ei} is the effective diffusion coefficient of i. For mass transfer only in the form of diffusion, the divergence of $\vec{N_i}$ should be

$$\nabla \cdot \vec{N_i} = R(c_i) \tag{12}$$

where $R(c_i)$ is the rate of reaction (consumption) of i per unit volume. On the assumption that D_{ei} is constant

$$\nabla \cdot \vec{N_i} = -\nabla \cdot D_{e_i} \nabla c_i = -D_{e_i} \nabla^2 c_i \tag{13}$$

With this approximation a mass balance on a volume of the porous medium gives

$$D_{e_i} \nabla^2 c_i - R(c_i) = 0 \tag{14}$$

We consider the reaction A$\rightarrow$B, with the rate depending on the nth power of concentration of A, denoted by $R(c) = kc^n$, where k is the reaction rate constant. The goal is to predict the overall reaction rate, or the mass transfer in and out of the catalyst pellet.

Let us next assume that the diffusion occurs, at a steady state, in a porous slab that is infinite in two directions, giving a large plane sheet with diffusion through the thickness of the sheet. We thereby simplify Eq. (14) to one dimension by assuming negligible variation of the concentrations in the y and z directions, to give,

$$D_e d^2 c / dx^2 - kc^n = 0 \tag{15}$$

where, x is diffusion distance. Here we consider one side (or the center) of the slab as impermeable (no flux) and the concentration is held fixed at the other side. The two boundary conditions are

$$x = 0 \qquad D_e dc / dx = 0 \tag{16}$$

$$x = L \qquad c = c^0 \tag{17}$$

where L is a half of thickness of catalyst slab, and c^0 is a bulk concentration.

Equations (15)–(17) represent an ordinary differential equation and a boundary-value problem. For a suitable solution, dimensionless equations can be derived from Eqs. (15)–(17) by letting $X = x/L$, $C = c/c^0$. For an isothermal reaction, that is nth-order and irreversible, in planar geometry we obtain

$$d^2C / dX^2 - \phi^2 C^n = 0 \tag{18}$$

$$X = 0 \qquad dC / dX = 0 \tag{19}$$

$$X = 1 \qquad C = 1 \tag{20}$$

where the C is a dimensionless concentration variable, and the Thiele modulus , $\phi^2 = k_0 L^2 {c^0}^{n-1} / D_e$. In the Thiele modulus, the group $L(k_0 {c^0}^{n-1})$ is a characteristic property for reaction, while L/D_e is a characteristic property for diffusion. The Thiele modulus thus measures the relative importance of the diffusion and reaction phenomena.

As a consequence of diffusion there is a reduction in the reaction rate as we progress inside the catalyst with a result that the overall rate is much less than would be achieved if the reactant were at a concentration as supplied at the outer surface. Thus the catalyst regions are not effectively used and the concept of effectiveness is introduced. Effectiveness is defined as the average reaction rate, i.e., with diffusion, divided by the reaction rate if the rate of reaction is evaluated at the boundary condition value at $X = 1$. The effectiveness factor can be generally given by

$$\xi = \frac{\int_0^1 C^n(X) X^{a_g - 1} dX}{\int_0^1 C^n(1) X^{a_g - 1} dX} \tag{21}$$

or

$$\xi = \frac{a_g}{\phi^2} \frac{dC}{dX}\bigg|_{X=1} \tag{22}$$

The parameter $a_g = 1, 2, 3$, for flat plate, cylindrical, or spherical geometry respectively. In the Thiele modulus, the character distance $L = L$ (a half of thickness of catalyst pellet) for flat plate, $L = r_0$ (the radius of catalyst pellet) for cylindrical, or spherical geometry.

1. Model Solution

(i) Catalyst Slab[10]

In the linear case, for reaction orders $n = 0$ or 1, analytical solutions can be obtained. For $n = 1$, the solutions of dimensionless concentration distribution $C(X)$ and effectiveness, ξ, are

$$C(X) = \frac{\cosh(\phi X)}{\cosh(\phi)} \tag{23a}$$

$$\xi = \frac{\tanh(\phi)}{\phi} \tag{23b}$$

By applying the ADM to the linear problem, we obtain[10]

$$C_n = \frac{1}{(2n)!}\phi^{2n} C_0 x^{2n} \tag{24a}$$

$$\sum_{n=1}^{\infty} C_n = C_0[1 + \sum_{n=1}^{\infty}[\frac{1}{(2n)!}\phi^{2n} C_0 x^{2n}] = C_0 \cosh(\phi X) \tag{24b}$$

where n is the number of terms, and C_0 is a constant which equals to the value of C at $X = 0$, where the boundary condition is $dC/dX = 0$. Substituting the boundary conditions, Eqs. (19) and (20), C_0 can be determined and the same analytical solution with Eq. (23) is obtained. Clearly the linear differential equation(s) can be solved by ADM. Whereas in the non-linear case, ($n \neq 0, 1$), analytical solutions do not exist but Adomian's method is available. In Adomian's method, u_0, the root of the polynomial, is C_0, the value of C at $X = 0$.

We can only obtain all of the forms of partial solutions if C_0 is known. It is obvious that, the stronger the non-linearity of the problem, the more terms will be required in the approximate solution. The non-linearity depends on both the function and the value of the parameter ϕ. The solution, the concentration distribution in the catalyst pellet, for the general nth-order reaction by using a $m + 1$ term approximation is given as

$$\begin{aligned}\Lambda_{m+1,n} &= C_0 + C_1 + C_2 + \ldots + C_m \\ &= C_0 + \frac{1}{2!}C_0{}^n\phi^2X^2 + \frac{n}{4!}C_0^{2n-1}\phi^4X^4 \\ &+ \frac{4n^2-3n}{6!}C_0^{3n-2}\phi^6X^6 + \frac{19n^3-18n^2}{8!}C_0^{4n-3}\phi^8X^8 \\ &+ \frac{496n^4-1554n^3+1689n^2-630n}{10!}C_0^{5n-4}\phi^{10}X^{10} + \ldots + C_m\end{aligned} \tag{25}$$

and the effectiveness with a m term approximation will be

$$\begin{aligned}\xi_{m,n} &= \frac{d\Lambda_{m+1,n}/dX|_{X=1}}{\phi^2} \\ &= C_0^n + \frac{n}{3!}\phi^2C_0^{2n-1} + \frac{4n^2-3n}{5!}\phi^4C_0^{3n-2} \\ &+ \frac{19n^3-18n^2}{7!}\phi^6C_0^{4n-3} \\ &+ \frac{496n^4-1554n^3+1689n^2-630n}{9!}\phi^8C_0^{5n-4} \\ &+ \ldots + \frac{1}{\phi^2}dC_m/dX|_{X=1}\end{aligned} \tag{26}$$

where n is the reaction order, and C_0 is the root of the polynomial. For instance, consider the second order reaction $R(c) = kc^2$, the nonlinear term should be $NC = \phi^2C^2$. The particular nonlinearity is easily given by using Adomian polynomials according to Eq. (10):

$$\begin{aligned}
A_0 &= \phi^2 C_0^2 \\
A_1 &= 2\phi^2 C_0 C_1 \\
A_2 &= \phi^2 (2C_0 C_2 + C_1^2) \\
A_3 &= 2\phi^2 (C_0 C_3 + C_1 C_2) \\
A_4 &= \phi^2 (2C_0 C_4 + 2C_1 C_3 + C_2^2) \\
A_5 &= 2\phi^2 (C_0 C_5 + C_1 C_4 + C_2 C_3) \\
&\;\;\vdots
\end{aligned} \tag{27}$$

According to Eqs. (18)–(20), a generalized isothermal model of reaction and diffusion in a catalyst slab with the second order reaction can be solved to obtain the solution series:

$$C \simeq \Lambda_{m+1,2} = \sum_{n=0}^{m} C_n \tag{28}$$

The analytical approximations of the dimensionless concentration distribution, $C(X)$ and effectiveness, ξ, by using the front three terms, namely $m = 2$, are listed below:

$$C(X) \simeq \Lambda_{3,2} = C_0 + \frac{1}{2!}\phi^2 C_0^2 X^2 + \frac{2}{4!}\phi^4 C_0^3 X^4 \tag{29}$$

where $C_0 = \sqrt[3]{\frac{8}{\phi^6} + \frac{12}{\phi^4}} - \frac{2}{\phi^2}$, ϕ is the Thiele modulus and X is the dimensionless distance in the catalyst slab,

$$\xi \simeq \xi_{2,2} = \left(\sqrt[3]{\frac{8}{\phi^6} + \frac{12}{\phi^4}} - \frac{2}{\phi^2}\right)^2 + \frac{\phi^2}{3}\left(\sqrt[3]{\frac{8}{\phi^6} + \frac{12}{\phi^4}} - \frac{2}{\phi^2}\right)^3 \tag{30}$$

Equations (29) and (30) can be used when $\phi < 2$, whereas more terms are needed for the approximation to meet the requirement of accuracy when $\phi > 2$.

(ii) Spherical Catalyst Pellet[11]

Equation (18) can be written in a general form for the parameter a_g = 1, 2, 3 for flat plate, cylindrical, or spherical geometry, respectively:

$$\left(\frac{d^2C}{dX^2}+\frac{a_g-1}{X}\frac{dC}{dX}\right)-\phi^2C^n=0 \tag{31}$$

The boundary conditions are:

$$X=0 \quad \frac{dC}{dX}=0 \quad \text{(center of catalyst)} \tag{32}$$

$$X=1 \quad C=1 \quad \text{(surface of catalyst)} \tag{33}$$

For a spherical geometric catalyst pellet, $a_g = 3$, $X = r/r_0$, where r_0 is the radius of catalyst pellet. In the linear case (n = 1), the solutions of the dimensionless concentration distribution $C(X)$ and effectiveness ξ are

$$C=\frac{\sinh(\phi X)}{X\sinh(\phi)} \tag{34}$$

$$\xi=\frac{3}{\phi}\left(\frac{1}{\tanh(\phi)}-\frac{1}{\phi}\right) \tag{35}$$

In a nonlinear case ($n \neq 0$ or 1), to simplify the solution procedure and for a rapid convergence of the Adomian decomposition solution, let $\mathrm{Y} = CX$ and rewrite Eq. (31) as

$$\frac{d^2\mathrm{Y}}{dX^2}=\phi^2\mathrm{Y}^nX^{1-n} \tag{36}$$

$$X=0, \qquad \mathrm{Y}=0 \tag{37}$$

$$X=1, \qquad \mathrm{Y}=1 \tag{38}$$

The approximate expressions for the concentration and effectiveness are obtained, by using *m* terms for an *n*th-order reaction, as

$$
\begin{aligned}
\Lambda_{m+1,n} &= C_0 + C_1 + C_2 + \ldots + C_m \\
&= C_0 + \frac{1}{3!} C_0 \phi^2 X^2 + \frac{n}{5!} C_0{}^{2n-1} \phi^4 X^4 \\
&+ \frac{8n^2 - 5n}{3 \times 7!} C_0{}^{3n-2} \phi^6 X^6 \\
&+ \frac{122n^3 - 183n^2 + 70n}{9 \times 9!} C_0{}^{4n-3} \phi^8 X^8 \\
&+ \frac{5032n^4 - 1264n^3 + 10805n^2 - 3150n}{45 \times 11!} C_0{}^{5n-4} \phi^{10} X^{10} \\
&+ \ldots + C_m
\end{aligned}
\tag{39}
$$

and

$$
\begin{aligned}
\xi_{m,n} &= C_0{}^n + \frac{n}{10} C_0^{2n-1} \phi^2 + \frac{48n^2 - 30n}{7!} C_0^{3n-2} \phi^4 \\
&+ \frac{122n^3 - 183n^2 + 70n}{27 \times 7!} C_0^{4n-3} \phi^6 \\
&+ \frac{10064n^4 - 25284n^3 + 21610n^2 - 6300n}{3 \times 11!} C_0^{5n-4} \phi^8 \\
&+ \ldots + \frac{3}{\phi^2} \frac{dC_m}{dX} \Big|_{X=1}
\end{aligned}
\tag{40}
$$

where C_0, is the root of polynomial, namely the concentration of reactant at the core of the spherical catalyst pellet. It should be noted, the root C_0 can be obtained analytically when the ADM approximate solution term $m \leq 4$ for reaction orders $n = 0.5$, 1.0, 2.0,[10,11] otherwise the Newton iteration method will be needed for C_0.

The solution method, the Adomian Decomposition Method (ADM), is mechanized for solving the nonlinear models according to the principle of "Parameter Decomposition".[2,3] A Mathematica code of the ADM,[12] for general order reactions in planar or spherical catalyst pellets, is given in more detail in the Appendix. Thus, the algebraic expressions of the approximate solutions and the computed data of results can all be easily obtained.

2. Concentration Profiles and Effectiveness[10,11]

(i) Concentration Profiles

Figures 3a and 3b show the variation in dimensionless concentration in the catalyst slab as a function of Thiele modulus for second order and half order reactions respectively. In the case of these reaction orders there is no analytical solution to the

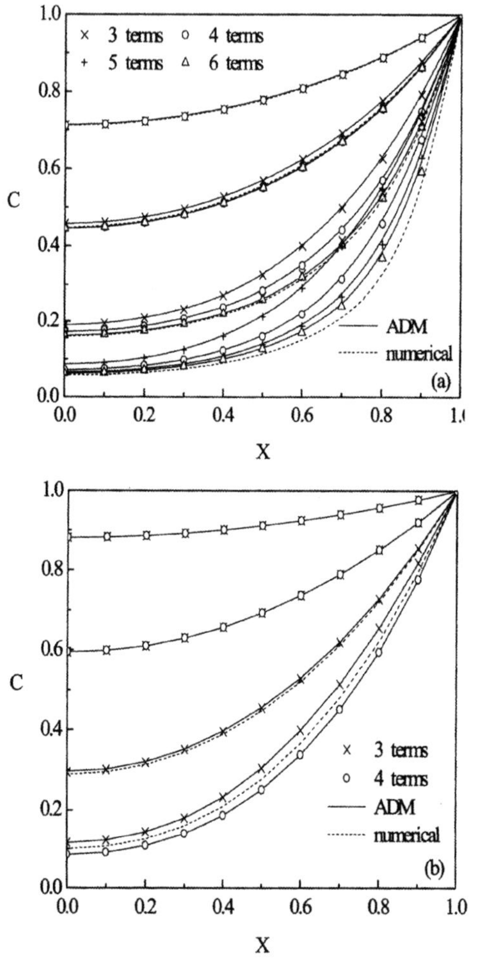

Figure 3. Variation of concentration in porous catalyst slab for (a) a second order reaction $\phi = 1, 2, 5, 10$ and (b) a half order reaction $\phi = 0.5, 1.0, 2.0$.

problem and thus comparisons with a numerical method can only be made. The numerical method is a finite difference procedure based on Newman's BAND method.[7] The BAND method has proven to be very effective for solution of non-linear problems of mass transfer and reaction in electrochemical systems. For the second order reaction there is good agreement between the decomposition method, with three terms, and the numerical solution for a Thiele modulus of 1.0. As the Thiele modulus increases the accuracy (agreement with the numerical solution) decreases with a three-term decomposition solution. In the case of a Thiele moduli above 5, a six-term decomposition is required to achieve reasonable agreement. Similarly for the half order reaction, at a relative higher value of Thiele modulus (2.0), either the three-term or four-term decomposition method give acceptable agreement with the numerical method.

Figures 4a and 4b show the variation in dimensionless concentration in the spherical porous catalyst as a function of Thiele modulus for second order and half order reactions respectively. More terms in the approximate solution are used here to obtain greater accuracy, and to prove the validity of the ADM.

Comparing those concentration profiles in Figures 3 and 4, it can be seen that, although the concentration profiles for a spherical shape ($a_g = 3$) are similar to those for a planar catalyst ($a_g = 1$), at the same value of Thiele modulus ϕ, the concentration gradients at $X = 1$, $dC/dX|_{X=1}$, for a sphere are smaller than those for a plane. It can be concluded that $dC/dX|_{X=1}$ for a cylinder shape ($a_g = 2$) will be in between the gradients of the plane and the spherical catalysts.[13] Also it can be seen that, with the same shape and at the same value of Thiele modulus ϕ, $dC/dX|_{X=1}$ will be greater for a smaller value of reaction order n.

(ii) Effectiveness

To assess the generality of the decomposition solution method we have applied it to several arbitrary reaction orders, for example, $n = 1.73, 0.67, -0.5, -1.0$ etc.[10,11] These reaction orders have been chosen to represent typical cases of kinetics in heterogeneous catalysis and electrocatalysis where adsorption phenomena play a major role. Values of effectiveness of a plane catalyst pellet for the different reaction orders are shown in Figure 5. Clearly all of the data for positive reaction orders show the expected trend of a decrease in effectiveness with increase in Thiele modulus. Effectiveness values determined for reaction

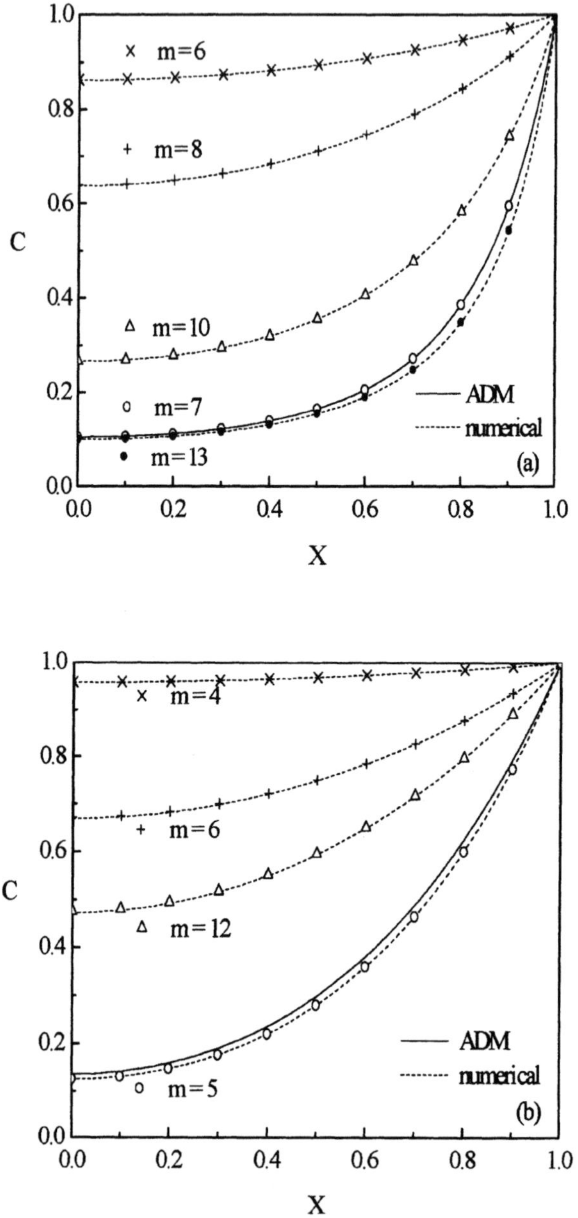

Figure 4. Variation of concentration in spherical porous catalyst for a second order reaction (a) ϕ = 1, 2, 5, 10 and a half order reaction (b) ϕ = 0.5, 1.5, 2.5, 3.0.

orders less than 1.0, i.e., 0.5 and 0.67 (Figure 5a), show agreement between finite difference solution and the four-term decomposition, up to Thiele moduli of 2.0 (Figure 5b). Effectiveness, for apparent reaction order less than 0, i.e., –0.5 and –1.0 (Figure 5c), as expected increase with Thiele modulus, at lower concentrations in the pores the reaction rate is suppressed to a smaller extent.

Figure 6 presents the effectiveness factors of a spherical catalyst pellet for reaction orders of 1.0, 2.0 (Figure 6a) and 0.5 (up curve in Figure 6b). A reasonable agreement between the four-term decomposition solutions and the finite difference method is achieved over most of the range of Thiele modulus.

Comparing the effectiveness factors in Figure 5 and Figure 6, it can be seen, at the same value of Thiele modulus ϕ, the effectiveness factors for a sphere ($a_g = 3$) is greater than that for a plane ($a_g = 1$), (down curve in Figure 6b). It can be concluded that effectiveness factors for a cylinder shape ($a_g = 2$) will be in

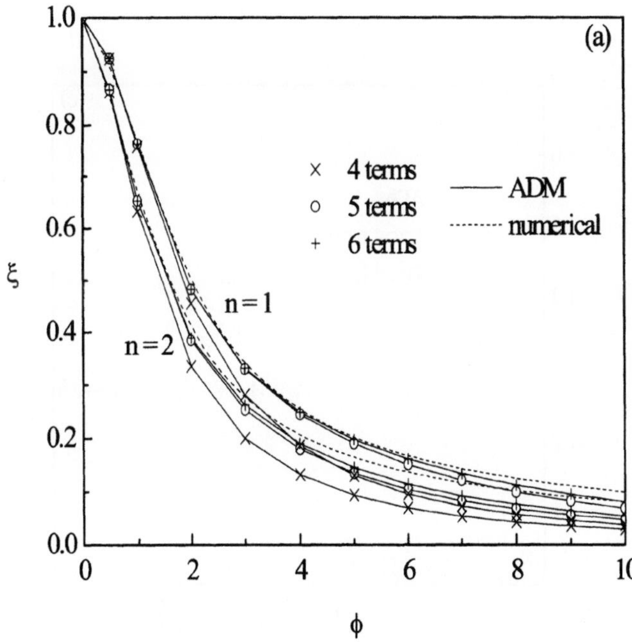

Figure 5. Effectiveness of catalyst slab for catalytic reactions of different orders. (a) $n = 1, 2$ (b) $n = 0.5, 0.67$, (c) $n = -0.5, -1.0$

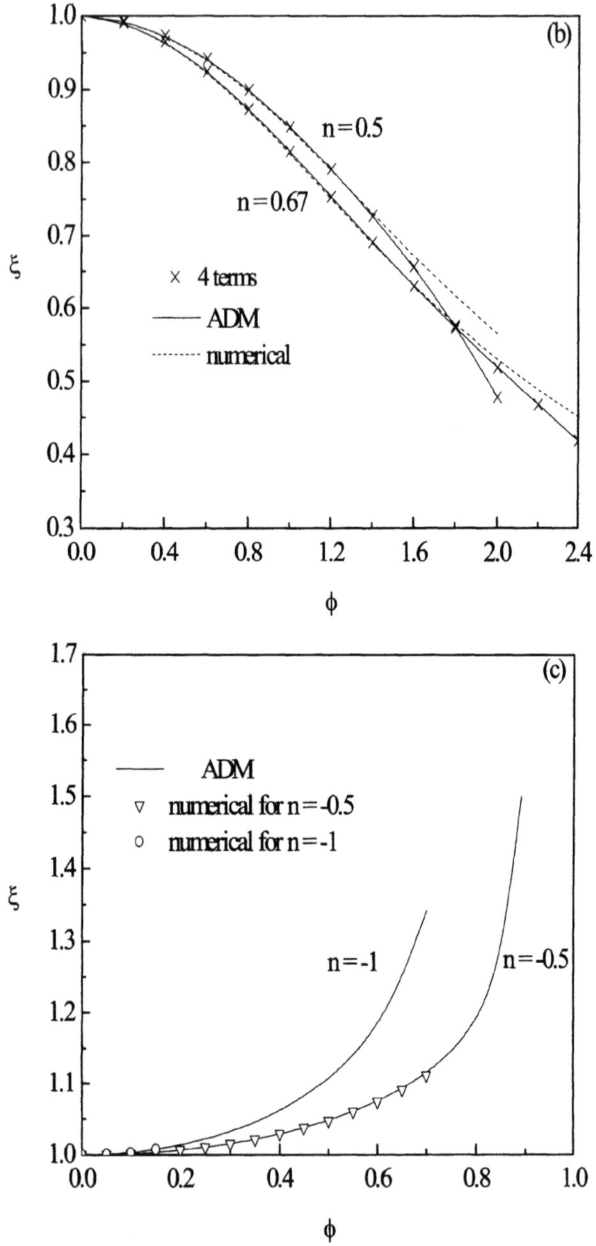

Figure 5. Continuation.

between those of the plane and the spherical catalysts.[13] Also it can be seen, with the same shape and at the same value of Thiele modulus ϕ, effectiveness factors will be greater for a smaller value of reaction order n.

In conclusion, the ADM is a more convenient method to obtain approximate solutions for the models with arbitrary reaction orders, n, so that the relationship between effectiveness ξ and Thiele modulus ϕ can be quantitatively understood in more details for the nonlinear cases.

IV. APPLICATION TO THE INFLUENCE OF MASS TRANSPORT IN ELECTROCATALYSTS

For electrocatalysts we generally define the reaction rate in terms of an electrode polarization equation. The polarization equation describes the transfer of charge from the matrix to the solution. For an electrochemical reaction $O + ne^- \leftrightarrow R$, this can be written in the Butler-Volmer form[1,7]

$$i = i_0\left[\frac{c_O}{c_O^0}\exp\left(-\frac{\alpha_c nF\eta}{R_gT}\right) - \frac{c_R}{c_R^0}\exp\left(\frac{\alpha_a nF\eta}{R_gT}\right)\right] \tag{41}$$

where, c_O^0 and c_R^0 represent the concentrations in the bulk, c_O and c_R represent the concentrations in matrix, which are related to inner diffusion; α_c, and α_a are the transfer coefficients, and n is the number electrons transferred in electrode reaction.

The overpotential can be written as

$$\eta = \phi^m - \phi^l - U' \tag{42}$$

where, ϕ^m and ϕ^l are the potentials in the matrix and solution respectively, U' is the open-circuit value of $\phi^m - \phi^l$ when the concentrations of the reactant and product are c_O^0 and c_R^0 in the bulk, and i_0 is a constant representing the corresponding exchange current density at the concentrations c_O^0, c_R^0, which might conveniently be taken to be the initial concentrations external to the electrode.[7] The reaction is often taken to be first order with respect to the reactant and product, at a given electrode potential.

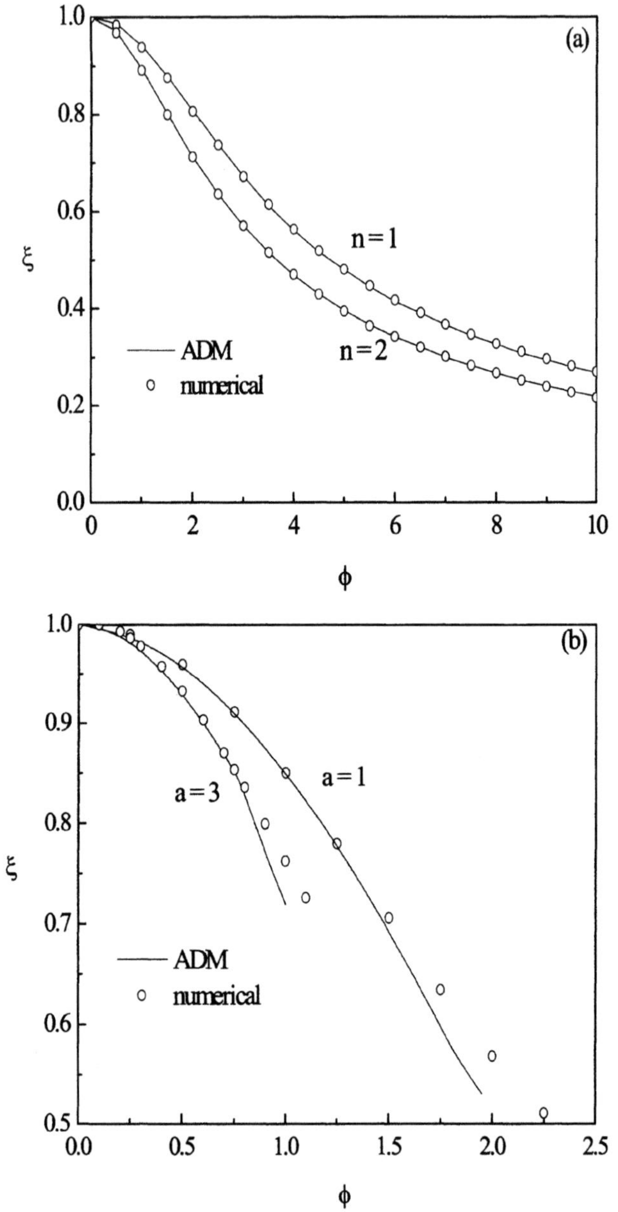

Figure 6. Effectiveness of catalyst sphere for catalytic reactions of different orders: $n = 1, 2$ for catalyst sphere, (b) $n = 0.5$, for sphere ($a = 3$) and slab ($a = 1$).

The problem of diffusion in electrocatalysts is analogous to that of diffusion in heterogeneous catalysis. The assumptions adopted in the model are as follows

1. the model of three dimensional electrodes can be simplified into one dimensional,
2. isothermal and steady state conditions apply,
3. ionic migration of the reacting species is neglected,
4. the electrode is characterized by a uniform specific area, a (m^{-1}), and
5. the potential distribution is uniform in the electrocatalysts.

A solution-phase reactant will be depleted during the operation of a three-dimensional electrode (electrocatalysts), and diffusion of this species from a reservoir at the face of the electrode represents a loss by concentration difference.

This process involves a redox reaction obeying the Eq. (41). The rate of reaction per unit volume for species O can be written in the form[1]

$$R_t = \frac{ai_0}{nF}[\frac{c_O}{c_O^0}\exp(-\frac{\alpha_c nF\eta}{R_g T}) - \frac{c_R}{c_R^0}\exp(\frac{\alpha_a nF\eta}{R_g T})] \tag{43}$$

The diffusion in the porous electrode can be described by Fick's law. We have for a redox reaction, the divergences for species O and R as

$$\nabla \cdot \vec{N_O} = -D_{eO}\nabla \cdot \nabla c_O = \frac{s_i a i_0}{nF}[\frac{c_O}{c_O^0}\exp(-\alpha_c \Phi)] \tag{44}$$

$$\nabla \cdot \vec{N_R} = -D_{eR}\nabla \cdot \nabla c_R = -\frac{s_i a i_0}{nF}[\frac{c_R}{c_R^0}\exp(\alpha_a \Phi)] \tag{45}$$

where s_i is the stoichiometric coefficient of species i in electrode reaction and the dimensionless overpotential is

$$\Phi = \beta\eta \tag{46}$$

where $\beta = nF/R_g T$.

The analysis considers a three-dimensional planar electrode of thickness L, bounded on one side by a current feeder ($x = 0$)and on the other side by a free solution ($x = L$). For the electrochemical reaction O + $ne^- \leftrightarrow$ R, $s_i = 1$, the model of the concentration distribution in one dimension can be written as:

$$\frac{d^2 c_O}{dx^2} = \frac{-a i_0}{D_{eO}} [\frac{c_O}{c_O^0} \exp(\alpha_c \Phi)] \tag{47}$$

$$\frac{d^2 c_R}{dx^2} = \frac{a i_0}{D_{eR}} [\frac{c_R}{c_R^0} \exp(\alpha_a \Phi)] \tag{48}$$

For an anodic current, Eq. (48) should be used, and the dimensionless model will be

$$\frac{d^2 C}{dX^2} = sC \exp(\alpha_a \Phi) \tag{49}$$

where the dimensionless distance, X and the dimensionless concentration, C, in the pore are

$$X = \frac{x}{L}, \qquad C = \frac{c_R}{c_R^0} \tag{50}$$

The dimensionless diffusion modulus is defined as

$$s = \frac{s_i a i_0 L^2}{nFD_e c_R^0} \tag{51}$$

where the group $s_i a i_0 L/nF$ is a characteristic property for electrode reaction, while $L/D_e c_R^0$ is a characteristic property for diffusion.

Suppose the overpotential is uniform in the three dimensional electrode

$$\Phi = \text{constant} \tag{52}$$

we set the dimensionless transfer coefficient as

$$\alpha = \alpha_a \Phi \tag{53}$$

Thus Eq. (48) becomes

$$\frac{d^2C}{dX^2} = sC\exp(\alpha) \tag{54}$$

If $c_R{}^0$ at $X = 1$ is given, we solve the model by using the boundary conditions

$$X = 0, \qquad \frac{dC}{dX} = 0 \tag{55}$$

$$X = 1, \qquad C = 1 \tag{56}$$

With the value of α constant, Eq. (54) is a linear problem and analytical solutions can be obtained. The situation is similar to the role of diffusion in a porous catalyst pellet if Eq. (54) is written as

$$\frac{d^2C}{dX^2} = \omega^2 C \tag{57}$$

where

$$\omega^2 = s\exp(\alpha) \tag{58}$$

which represents the Thiele modulus, for an anode current reaction with diffusion in a plane porous electrode.

Then the analytical solution of the dimensionless concentration[13] is

$$C = \frac{\cosh(\omega X)}{\cosh(\omega)} \tag{59}$$

and the effectiveness is

$$\xi = \frac{1}{\omega^2}\frac{dC}{dX}\bigg|_{X=1} = \frac{1}{\omega}\tanh(\omega) \tag{60}$$

We define the dimensionless total current density as

$$I = \frac{i_T}{ai_0L} \tag{61}$$

where i_T is the total current density for an electrocatalyst slab,

$$i_T = \frac{nFD_e c_R^0}{L}(\frac{dC}{dX})_{X=1} = \frac{nFD_e c_R^0}{L}\omega \tanh(\omega) = ai_0 LI \tag{62}$$

$$I = \frac{1}{s}(\frac{dC}{dX})_{X=1} = \frac{1}{s}\omega \tanh(\omega) \tag{63}$$

The local dimensionless total current distribution will be

$$I_{loc} = \frac{1}{s}(\frac{dC}{dX}) = \frac{1}{s}\omega \tanh(\omega X) \tag{64}$$

1. Internal Diffusion and Film Mass Transport

To model a porous electrocatalyst we may consider a second type of mass transport (in addition to diffusion) locally within the electrode, i.e., a mass transport resistance between the electrode surface and the solution. This situation may arise, for example, when the electrode surface is covered by a thin layer of polymer electrolyte or as in a fuel cell electrode in which the electrocatalyst is also covered by a thin water layer.

The mass transfer conditions between the fluid phase and fixed phase are assumed to be uniform, as defined by a single mass transport coefficient. Then Eq. (43) should be rewritten as

$$R_t = \frac{ai_0}{nF}[\frac{c_O^s}{c_O^0}\exp(-\frac{\alpha_c nF\eta}{R_g T}) - \frac{c_R^s}{c_R^0}\exp(\frac{\alpha_a nF\eta}{R_g T})] \tag{65}$$

where $c_O{}^S$ and $c_R{}^S$ represent the concentrations at the internal surface, which depend on local mass transfer.

For an anodic current, Eq. (57) becomes

$$\frac{d^2C}{dX^2} = sC_s \exp(\alpha) \tag{66}$$

where, C_s, is a dimensionless concentration on the inner surface of pore

$$C_s = \frac{c_R^s}{c_R^0} \tag{67}$$

Introducing the mass transfer parameter, for the Tafel type reaction, assuming migration is negligible, the local flux will be

$$N_{loc} = k_l(c_R - c_R^s) = \frac{i_a}{nF} = i_0 \frac{c_R^s}{c_R^0} \exp(\alpha_a \Phi) \tag{68}$$

where k_l is the average mass transfer coefficient in the electrode and i_a is the anodic current density. We can obtain the local anode current[1]

$$i_a = i_0 C_s \exp(\alpha_a \Phi) = \frac{C}{\dfrac{1}{i_0 \exp(\alpha_a \Phi)} + \dfrac{1}{i_l}} \tag{69}$$

where, the limiting current density $i_l = nFk_l c_R^0$. The dimensionless surface concentration will be

$$C_s = \frac{C}{\dfrac{i_0}{nFk_l c_R^0} \exp(\alpha_a \Phi) + 1} = \frac{C}{\dfrac{i_0}{i_l} \exp(\alpha_a \Phi) + 1} = C\left(\frac{1}{\dfrac{\exp(\alpha)}{\gamma} + 1}\right) \tag{70}$$

where the dimensionless limit current density is given by

$$\gamma = \frac{i_l}{i_0} \tag{71}$$

Substituting Eq. (70) into Eq. (66), the model equation can be rewritten as

$$\frac{d^2 C}{dX^2} = s \frac{C}{\dfrac{1}{\exp(\alpha)} + \dfrac{1}{\gamma}} \tag{72}$$

With the constant values of both α and γ, let

$$i_0' = i_0\left(\frac{1}{1+\frac{\exp(\alpha)}{\gamma}}\right) \tag{73}$$

which represents the appreciable exchange current density reduced by the internal mass transfer, then from Eq. (72)

$$s' = s\left(\frac{1}{1+\frac{\exp(\alpha)}{\gamma}}\right) \tag{74}$$

Equation (66) becomes

$$\frac{d^2C}{dX^2} = s'C\exp(\alpha) \tag{75}$$

With the value of α constant, Eq. (75) is a linear problem for which an analytical solution can be obtained

The situation is also similar to the role of diffusion in a porous catalyst pellet if Eq. (75) is written as

$$\frac{d^2C}{dX^2} = \omega'^2 C \tag{76}$$

where

$$\omega'^2 = s'\exp(\alpha) = \frac{s}{\frac{1}{\exp(\alpha)} + \frac{1}{\gamma}} \tag{77}$$

which represents the Thiele modulus, for an anode current reaction in a plane porous electrode under the influence of diffusion and local mass transfer.

The analytical solution of dimensionless concentration and effectiveness are

$$C = \frac{\cosh(\omega' X)}{\cosh(\omega')} \tag{78}$$

$$\xi = \frac{1}{\omega'^2} \frac{dC}{dX}\bigg|_{X=1} = \frac{1}{\omega'} \tanh(\varpi') \tag{79}$$

For this case of both diffusion and local mass transfer in the porous electrode, we define the dimensionless total current density as

$$I = \frac{i_T}{a i_0' L} \tag{80}$$

$$i_T = \frac{nFD_e c_R^0}{L} \left(\frac{dC}{dX}\right)_{X=1} = \frac{nFD_e c_R^0}{L} \omega' \tanh(\omega') = a i_0' L I \tag{81}$$

where

$$I = \frac{1}{s'} \left(\frac{dC}{dX}\right)_{X=1} = \frac{1}{s'} \omega' \tanh(\omega') \tag{82}$$

The local total current distribution will be

$$I_{loc}' = \frac{1}{s'} \left(\frac{dC}{dX}\right) = \frac{1}{s'} \omega' \tanh(\omega' X) \tag{83}$$

In addition, for an electrochemical reaction, $p\mathrm{O} + ne^- \leftrightarrow q\mathrm{R}$, this can be written in the general Butler-Volmer form[1,7]

$$i = i_0\left[\left(\frac{c_O}{c_O^0}\right)^p \exp\left(-\frac{\alpha_c nF\eta}{R_g T}\right) - \left(\frac{c_R}{c_R^0}\right)^q \exp\left(\frac{\alpha_a nF\eta}{R_g T}\right)\right] \tag{84}$$

Similar to the nonlinear form of Eq. (18), Eqs. (57) and (76) will be changed into

$$\frac{d^2C}{dX^2} = \omega^2 C^q \tag{86}$$

and

$$\frac{d^2C}{dX^2} = \omega'^2 C^q \tag{87}$$

Unlike the Thiele modulus ϕ related to a reaction order n for porous catalyst pallets, ω and ω' will maintain keep their form in this nonlinear case, due to the dimensionless concentrations used in the general B-V equations (Eqs. 41 and 84). The approximate general solutions of Eq. (18) can be used for Eqs. (86) and (87) as long as the reaction order n and Thiele modulus ϕ are replaced by q and ω or ω' respectively. The expressions for the concentration Λ and the effectiveness factor ξ for a qth-order reaction by using m term approximating can be obtained from Eqs. (25) and (26).

2. Agglomerate Model of Electrocatalysis

In the modeling of porous electrodes as, for example, used in fuel cells the concept of an agglomerate model is frequently used, e.g. in polymer electrolyte fuel cell (PEFC) and the direct methanol fuel cell (DMFC) models.[14,15] In this approach the structure of the electrode is described by agglomerates of supported catalyst, approximated as a sphere, with a homogenous content of ionomer (solid ionic conducting polymer) and catalyst on a carbon carrier (Figure 7). The pore structure is assumed continuous as is the agglomerate structure. Mass transport of reactants and product are governed by Fick's Law of radial diffusion for a sphere.

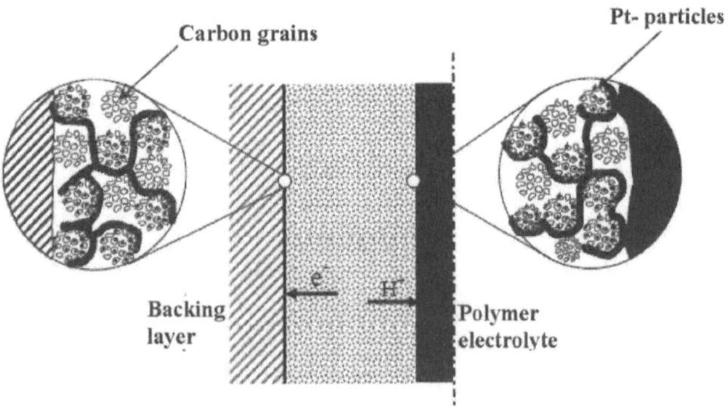

Figure 7. Agglomerate model of electrocatalyst.

Thus, according to Eq. (84) for an electrochemical reaction, $p\text{O} + n\,e^- \leftrightarrow q\text{R}$, the divergences for species O and R within an agglomerate is given by

$$\nabla \cdot \vec{N_O} = -D_{eO}\nabla \cdot \nabla c_O = \frac{pai_0}{nF}[(\frac{c_O}{c_O^0})^p \exp(-\frac{\alpha_c nF\eta}{R_g T})] \qquad (87)$$

$$\nabla \cdot \vec{N_R} = -D_{eR}\nabla \cdot \nabla c_R = -\frac{qai_0}{nF}[(\frac{c_R}{c_R^0})^q \exp(\frac{\alpha_a nF\eta}{R_g T})] \qquad (88)$$

For a qth-order anodic reaction then we have

$$\frac{d^2C}{dX^2} + \frac{2}{X}\frac{dC}{dX} - \omega^2 C^q = 0 \qquad (89)$$

$$X = 0, \qquad \frac{dC}{dX} = 0 \qquad (90)$$

$$X = 1, \qquad C = 1 \qquad (91)$$

where $X = x/r_0$, r_0 is the radius of the agglomerate sphere; $C = c_R/c_R^0$, c_R^0 is the concentration on the outer surface of agglomerate sphere;

$$\omega^2 = s\exp(\alpha), \qquad (92)$$

where $s = s_i a i_0 r_0^2/nFD_e c_R^0$.

For a linear case, in which reaction order $q = 1$, Eq. (89) gives the analytical solution

$$C = \frac{\sinh(\omega X)}{X\sinh(\omega)} \qquad (93)$$

$$\xi = \frac{3}{\omega^2}\frac{dC}{dX}\bigg|_{X=1} = \frac{3}{\omega}(\frac{1}{\tanh(\omega)} - \frac{1}{\omega}) \qquad (94)$$

where

$$\left.\frac{dC}{dX}\right|_{X=1} = \omega(\frac{1}{\tanh(\omega)} - \frac{1}{\omega}) \tag{95}$$

Here, we define the dimensionless total current density as

$$I = \frac{i_t}{ai_0 r_0} \tag{96}$$

where t_t is the total current density for an agglomerate sphere,

$$i_t = \frac{nFD_e c_R^0}{r_0}(\frac{dC}{dX})_{X=1} \tag{97}$$

$$I = \frac{1}{s}(\frac{dC}{dX})_{X=1} = \frac{1}{s}\omega(\frac{1}{\tanh(\omega)} + \frac{1}{\omega}) \tag{98}$$

The local total current distribution should be

$$I_{loc} = \frac{1}{s}(\frac{dC}{dX}) = \frac{1}{s}\omega(\frac{1}{X\tanh(\omega X)} + \frac{C}{\omega X}) \tag{99}$$

For an arbitrary reaction order, $q \neq 0, 1$, the approximate forms of general solutions of Eq. (31) can be used for the solutions to Eq. (89) as long as the reaction order n and Thiele modulus ϕ are replaced by q and ω respectively. The expressions of concentration Λ and effectiveness factor ξ for a qth-order reaction, by using m term approximating can be obtained as Eqs. (39) and (40) above.

The total current density of one agglomerate sphere can also be evaluated by

$$\frac{s_i i_t}{nF} = \frac{D_e c_R^0}{r_0}(\frac{dC}{dX})_{X=1} = \frac{D_e c_R^0}{3r_0}\omega^2\xi \tag{100}$$

Then the Weisz modulus which only includes observables is[13]

$$\phi_W = \omega^2\xi = \frac{3s_i i_t r_0}{nFD_e c_R^0} \tag{101}$$

and

$$i_t = \frac{r_0}{3} a i_0 \exp(\frac{\alpha_a nF}{RT}\eta)\xi \quad (102)$$

These relations are useful for analysis.

V. APPLICATION TO MODELS FOR THREE-DIMENSIONAL ELECTRODES

1. The General Form of Model of Three-Dimension Electrodes

Three-dimensional electrodes may include packed bed electrodes as well as porous electrodes. The most important character of 3-D electrode is that there is a potential or current distribution related to the electrode reaction, coupled with concentration distributions within the whole electrode. The mathematical model for a fixed 3-D electrode will be a set of coupled differential equations.

The charge balance and the material balance for species i in a volume element are the divergences of the fluxes of current and material respectively as

$$\nabla \cdot \vec{i} = ai(\eta, c_i) \quad (103)$$

$$\nabla \cdot \vec{N_i} = -\frac{s_i}{nF} ai(\eta, c_i) \quad (104)$$

where a is the special area of three dimensional electrode, η is the local surface overpotential (see Eq. 42).

The coupled equations, Eq. (103) and Eq. (104), will be the general model for all three-dimensional electrodes. The relationship between the charge and material balances can be derived as

$$\nabla \cdot \vec{N_i} = -\frac{s_i}{nF} \nabla \cdot \vec{i} \quad (105)$$

which can be used for decoupling later.

2. Porous Electrode Reactor

The analysis is concerned with a one-dimensional model of electrodes in which reaction rates are distributed unevenly due to diffusion as well as a variation in electrode potential.[4,5] The treatment of the problem of a simultaneous variation in electrolyte concentration and potential distribution in the electrode is treated in an analogous manner to that of non-isothermal chemical reactions in porous catalysts.[16] The results show that several dimensionless groups or numbers control the electrode behavior. Figure 8 shows a back fed porous anode used in the model.

Assumptions adopted in this analysis are as follows:

1. The porous electrode is one dimensional and both solid and electrolyte phase are continuous media with uniform effective conductivities.
2. Isothermal and steady state conditions apply.
3. Ionic migration of the reacting species is neglected.
4. The electrode is characterized by a uniform specific area, *a*.

According to the coupled equation, Eqs. (103) and (104), with the polarization Eq. (41), a charge balance coupled with a material balance for a redox pair in a volume element will be

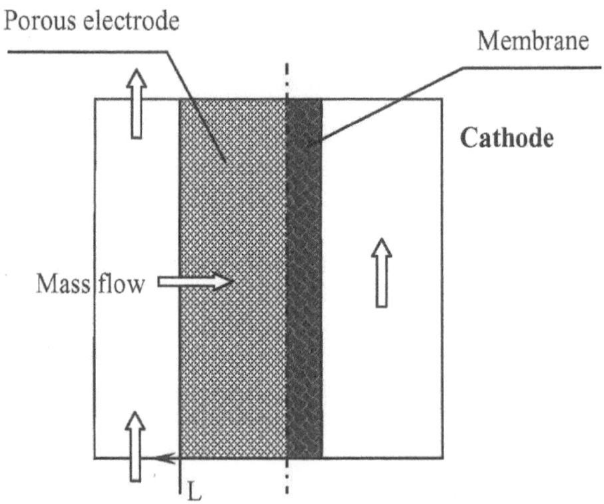

Figure 8. Schematic diagram of the porous anode reactor.

$$\nabla \cdot \vec{i} = ai_0[\frac{c_O}{c_O^0}\exp(-\frac{\alpha_c nF\eta}{R_g T}) - \frac{c_R}{c_R^0}\exp(\frac{\alpha_a nF\eta}{R_g T})] \quad (106)$$

$$\nabla \cdot \vec{N_O} = \frac{-s_i ai_0}{nF}[\frac{c_O}{c_O^0}\exp(-\frac{\alpha_c nF\eta}{R_g T})] \quad (107)$$

$$\nabla \cdot \vec{N_R} = \frac{-s_i ai_0}{nF}[-\frac{c_R}{c_R^0}\exp(\frac{\alpha_a nF\eta}{R_g T})] \quad (108)$$

where η is the local surface overpotential, and c_R and c_O are the concentrations on the local inner surface of the pores.

Assuming the potential in conductive solid phase and the potential of open-circuit, ϕ^m and U' in Eq. (42), are constant. According to Ohm law, the vector of local current density in the solution phase can be written as

$$\vec{i} = -\kappa \nabla \phi^l \quad (109)$$

where $\nabla \phi^l$ is the gradient of potential of electrolyte and κ is the effective conductivity of the electrolyte.

The divergence of $\vec{i}$ should be

$$\nabla \cdot \vec{i} = -\kappa \nabla^2 \phi^l \quad (110)$$

and when diffusion is dominant in solution in a porous electrode, the flux density can be written in a form of Fick law

$$\vec{N_i} = -D_{ei} \nabla c_i \quad (111)$$

where ∇c_i is the gradient of concentration of species i and D_{ei} is an effective diffusion coefficient for specie i.

The divergence of $\vec{N_i}$ should be

$$\nabla \cdot \vec{N_i} = -D_{ei} \nabla^2 c_i \quad (112)$$

Let $s_i = 1$ for $O + n\,e^- \leftrightarrow R$ and considering an anodic current only, the charge balance Eq. (106) and material balance Eq. (108) become

$$-\kappa\nabla^2\phi^l = -ai_0\left[\frac{c_R}{c_R^0}\exp\left(\frac{\alpha_a nF}{R_g T}\eta\right)\right] \tag{113}$$

$$-D_e\nabla^2 c_R = \frac{-ai_0}{nF}\left[\frac{c_R}{c_R^0}\exp\left(\frac{\alpha_a nF}{R_g T}\eta\right)\right] \tag{114}$$

where $\eta = \phi^m - \phi^l - U'$ with assuming of constant ϕ^m and U'.

For a simplified one-dimensional model, namely, the transfer process in the porous electrode is dominated by diffusion with the effective coefficient D_e, Eqs. (113) and (114) become

$$\frac{d^2\eta}{dx^2} = \frac{ai_0}{\kappa}\left[\frac{c_R}{c_R^0}\exp\left(\frac{\alpha_a nF}{R_g T}\eta\right)\right] \tag{115}$$

$$\frac{d^2 c_R}{dx^2} = \frac{ai_0}{nFD_e}\left[\frac{c_R}{c_R^0}\exp\left(\frac{\alpha_a nF}{R_g T}\eta\right)\right] \tag{116}$$

To generalize the model, Eqs. (115) and (116), the dimensionless model will be

$$\frac{d^2\Phi}{dX^2} = v^2 C\exp(\alpha_a\Phi) \tag{117}$$

$$\frac{d^2 C}{dX^2} = sC\exp(\alpha_a\Phi) \tag{118}$$

where

$$\Phi = \beta\eta, \beta = \frac{nF}{R_g T} \tag{119}$$

$$C = \frac{c_R}{c_R^0}$$

$$X = \frac{x}{L}$$

$$v^2 = \frac{ai_0 L^2 \beta}{\kappa}$$

$$s = \frac{s_i a i_0 L^2}{nFD_e c_R^0}$$

The Φ is a dimensionless overpotential variable; and v^2 is a dimensionless conduction parameter, in which the group $ai_0L\beta$ is a characteristic property for electrode reaction, while L/κ is a characteristic property for conduction resistance.

To solve the model, boundary conditions are needed at $X = 1$, on the membrane side. If η^0 and $c_R{}^0$ at $X = 1$ are given, we solve the model by using the boundary conditions

$$X = 1, \quad \Phi = \Phi^0, \quad C = 1 \tag{120}$$

For an easier decomposition solution and for faster convergence to the roots, we scale the boundary condition values of dimensional potential to 1.0. Let

$$\Psi = \Phi / \Phi^0, \quad \mu = v^2 / \Phi^0 \quad \text{and} \quad \alpha = \alpha_a \Phi^0 \tag{121}$$

where Ψ is the relative dimensionless overpotential variable, μ is the dimensionless conduction modulus, and α is the dimensionless transfer coefficient.

Hence a set of ordinary different equations with boundary conditions, which describes the potential and concentration distributions, can be obtained as

$$\frac{d^2\Psi}{dX^2} = \mu C \exp(\alpha\Psi) \tag{122}$$

$$\frac{d^2C}{dX^2} = sC \exp(\alpha\Psi) \tag{123}$$

$$X = 0, \qquad \frac{d\Psi}{dX} = 0, \qquad \frac{dC}{dX} = 0 \tag{124}$$

$$X = 1, \qquad \Psi = 1, \qquad C = 1 \tag{125}$$

In this case the overpotential and concentration changes across the electrode structure are related by the three dimensionless parameters, μ, s and α.

To assist in the solution of these differential equations, we can use Eq. (105) to decouple the model Eqs. (122) and (123), to obtain a relationship between Ψ and C,

$$\frac{d^2\Psi}{dX^2} = \frac{\mu}{s}\frac{d^2C}{dX^2} \tag{126}$$

which on integration results in

$$(\Psi - 1) = \frac{\mu}{s}(C - 1) \tag{127}$$

Hence the variation in potential and concentration across the electrode structure are related by the three parameters μ, s and α.

Combining Eqs. (122), (123) and (127) gives two independent different equations

$$\frac{d^2\Psi}{dX^2} = \mu[1 + \frac{s}{\mu}(\Psi - 1)]\exp(\alpha\Psi) \tag{128}$$

$$X = 0, \qquad \frac{d\Psi}{dX} = 0 \tag{129}$$

$$X = 1, \qquad \Psi = 1 \tag{130}$$

or

$$\frac{d^2C}{dX^2} = sC\exp\{\alpha[1 + \frac{\mu}{s}(C - 1)\} \tag{131}$$

$$X = 0, \qquad \frac{dC}{dX} = 0 \tag{132}$$

$$X = 1, \qquad C = 1 \tag{133}$$

According to the definition, the effectiveness factor E of the porous electrode is given by

$$E = \frac{i_T}{i_a(c_R^0, \eta^0)} \tag{134}$$

where i_T is the total current density based on the cross-sectional area of the porous electrode, and $i_a(c_R^0, \eta^0) = i_0 \exp(\frac{\alpha_a nF}{RT}\eta^0) = i_0\exp(\alpha_a\Phi^0)$ is the calculated intrinsic current density without the diffusion influence, so we have

$$E = \frac{(\frac{d\Psi}{dX})_{X=1}}{\mu\exp(\alpha_a\Phi^0)} = \frac{I}{\exp(\alpha)} \tag{135}$$

or

$$E = \frac{(\frac{dC}{dX})_{X=1}}{s\exp(\alpha_a\Phi^0)} = \frac{I}{\exp(\alpha)} \tag{136}$$

where a dimensionless current density I above is introduced as

$$I = \frac{i_T}{ai_0L} \tag{137}$$

The total current density is given by

$$i_T = \kappa(\frac{d\eta}{dx})_{x=L} \tag{138}$$

or

$$i_T = nFD_e(\frac{dc}{dx})_{x=L} \tag{139}$$

and the dimensionless forms will be

$$I = \frac{1}{\mu}(\frac{d\Psi}{dX})_{X=1} = \frac{1}{s}(\frac{dC}{dX})_{X=1} \tag{140}$$

We thus have

$$i_T = ai_0LI = \frac{ai_0L}{\mu}(\frac{d\Psi}{dX})_{X=1} = \frac{\kappa\eta^0}{L}(\frac{d\Psi}{dX})_{X=1} \tag{141}$$

or

$$i_T = ai_0LI = \frac{ai_0L}{s}(\frac{dC}{dX})_{X=1} = \frac{nFD_ec_R^0}{L}(\frac{dC}{dX})_{X=1} \tag{142}$$

Also, the model can be used for a polarization Eq. (84) with an arbitrary order, then Eq. (128) and Eq. (131) become

$$\frac{d^2\Psi}{dX^2} = \mu C^p \exp(\alpha\Psi) \tag{143}$$

$$\frac{d^2C}{dX^2} = sC^p \exp(\alpha\Psi) \tag{144}$$

For this three-dimensional electrode model, with or without decoupling of the two coupled equations, the approximate solutions can be obtained by using the Mathematica codes of the ADM given in the Appendix.[17] The algebraic expressions of dimensionless potential and concentration are in a series form with even orders as

$$\Psi(X) = \sum_{n=0}^{\infty}\Psi_n(X) \tag{145}$$

$$= \Psi_0 + b_1(\Psi_0)X^2 + b_2(\Psi_0)X^4 + b_3(\Psi_0)X^6 + b_4(\Psi_0)X^8 + ...$$

or

$$C(X) = \sum_{n=0}^{\infty} C_n(X) \tag{146}$$
$$= C_0 + b_1(C_0)X^2 + b_2(C_0)X^4 + b_3(C_0)X^6 + b_4(C_0)X^8 + ...$$

where b_1, b_2, ..., b_n ... are the coefficients of the solution series given by the program codes.

Figure 9 shows the approximate solutions of dimensionless potential and concentration with different terms for a second order reaction in a porous slab electrode, and shows the comparisons between the approximate and numerical solutions. The potential and concentration profiles are obtained by using the coupled equation model with diffusion.

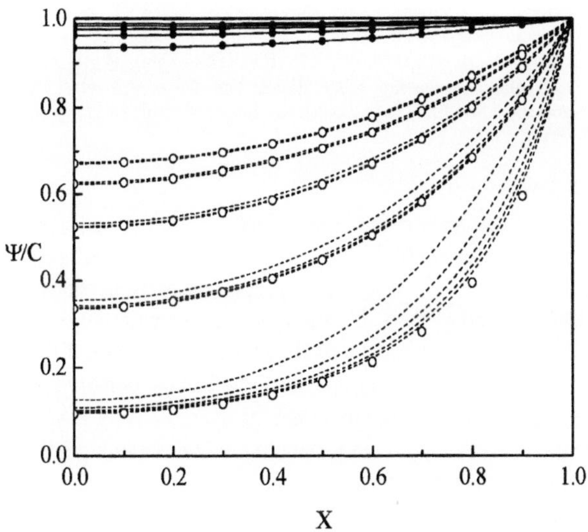

Figure 9. The second order profiles of relative dimensionless potential Ψ and dimensionless concentration C with dimensionless distance X. $\alpha = 0.5$, $\nu^2 = 0.1$, $s = 1.0$. $\Phi^0 = 0.5, 1.0, 2.0, 4.0, 8.0$ (from down to up for Ψ, from up to down for C.) Solid line: relative dimensionless potentials; dashed line: dimensionless concentration; 3,4,5,6,7 term (from down to up but all overlapped for Ψ, from up to down for each set of C.); ●: Numerical result for the relative dimensionless potentials; O: numerical result for the dimensionless concentration.

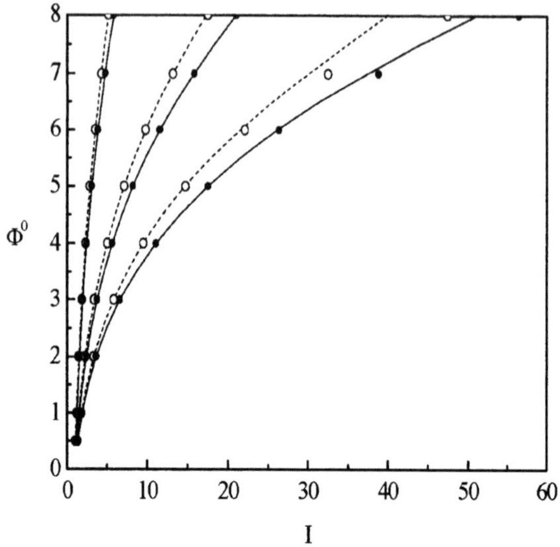

Figure 10. Variations of dimensionless potential Φ^0 with dimensionless current density I for different values of transfer coefficient α_a: α_a = 0.25, 0.5, 0.75 (from left to right); ν^2 = 0.1; s = 0.1; solid line: the first order; dashed line: the second order; ●: numerical result for the first order; O: numerical result for the second order.

Figure 10 shows the approximate solutions with different values of transfer coefficient α_a for first and second order reactions in a porous slab electrode, and gives comparisons between the approximate and numerical solutions. The potential-current curves are obtained by using the coupled equation model with diffusion. If Φ^0 continues to increase in this case, the current density I with the largest α_a will be too large to converge even by using BAND. The value of parameter α_a is very sensitive when obtaining solution to the nonlinear electrode model.

3. Packed-Bed Electrode Reactor[4,5]

Three dimensional packed bed electrodes are generally considered for reactions which operate with low current densities in order to increase localized mass transfer rates and/or increase overall current per unit cell volume. The maximum current density at any position in the electrode structure is limited by the prevailing conditions of mass transfer. The limiting current thus can also have

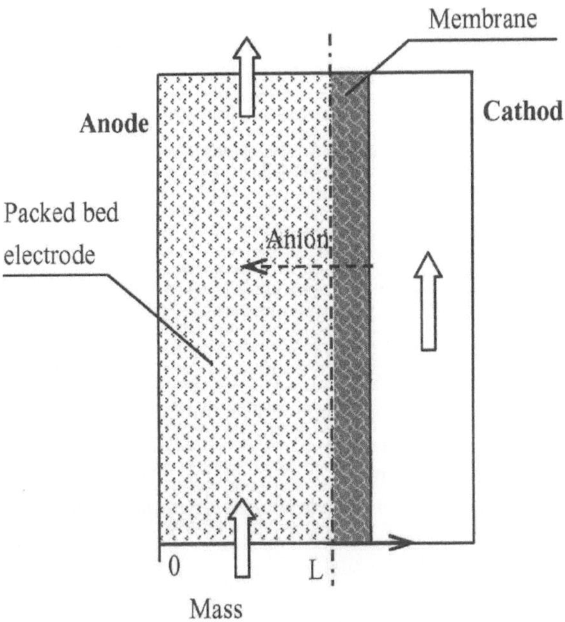

Figure 11. Schematic diagram of an anodic packed-bed anode reactor.

an effect on the reactor performance and particularly the potential distribution.

The three-dimensional or packed bed electrode has a thickness L, bounded on one side by a current feeder ($x = 0$) and on the other side by a membrane or free solution, $x = L$ (Figure 11). Both electrolyte and electrode phases are assumed to be continuous media with uniform effective conductivities and in which the electrical potential obeys Ohm's law.

Assumptions adopted in this analysis are as follows:

1. The packed bed electrode is one dimensional and both solid and electrolyte phase are continuous media with uniform effective conductivities.
2. Isothermal and steady state conditions apply.
3. Ionic migration of the reacting species is neglected.
4. The electrode is characterized by a uniform specific area, a.
5. The mass transfer conditions between the fluid phase and fixed phase are also assumed to be uniform and defined by a single mass transport coefficient, k_l.

The derivation of the model is similar to that derived for the porous electrode model. We have very similar model equations to Eqs. (115) and (116) for an anodic current, given by

$$\frac{d^2\eta}{dx^2} = \frac{ai_0}{\kappa}\left[\frac{c_R^s}{c_R^0}\exp\left(\frac{\alpha_a nF\eta}{RT}\right)\right] \tag{147}$$

$$\frac{d^2 c_R}{dx^2} = \frac{ai_0}{D_l}\left[\frac{c_R^s}{c_R^0}\exp\left(\frac{\alpha_a nF\eta}{RT}\right)\right] \tag{148}$$

Note that in the Eqs. (147) and (148), the surface concentration c_R^s is different from the solution concentration c_R in the second derivative on the left and in Eqs. (115) and (116) due to convection. Furthermore D_l is the lateral dispersion coefficient instead of a lateral diffusion coefficient D_e in the porous electrode case.

The dimensionless model will be

$$\frac{d^2\Phi}{dX^2} = v^2 C_s \exp(\alpha_a \Phi) \tag{149}$$

where

$$C_s = \frac{c_R^s}{c_R^0} \tag{150}$$

and

$$\frac{d^2 C}{dX^2} = sC_s \exp(\alpha_a \Phi) \tag{151}$$

where the dispersion modulus is

$$s = \frac{s_i a i_0 L^2}{nFD_l c_R^0} \tag{152}$$

Introducing the mass transfer parameter, for the Tafel type reaction, the local flux will be

$$N_{loc} = k_l(c_R - c_R^s) = \frac{i_a}{nF} = \frac{i_0}{nF}\frac{c_R^s}{c_R^0}\exp(\alpha_a\Phi) \tag{153}$$

where k_l is the average mass transfer coefficient in the packed bed electrode. Then we can obtain

$$C_s = \frac{C}{\frac{i_0}{nFk_lc_R^0}\exp(\alpha_a\Phi)+1} = \frac{C}{\frac{i_0}{i_l}\exp(\alpha_a\Phi)+1} \tag{154}$$

namely

$$i_a = \frac{C}{\frac{1}{i_0\exp(\alpha_a\Phi)}+\frac{1}{i_l}} \tag{155}$$

where the limiting current density $i_l = nFk_lc_R^0$.

Substituting Eq. (154) into Eqs. (149) and (151), the model equation can be rewritten as

$$\frac{d^2\Phi}{dX^2} = \nu^2\frac{C}{\frac{1}{\exp(\alpha_a\Phi)}+\frac{1}{\gamma}} \tag{156}$$

$$\frac{d^2C}{dX^2} = s\frac{C}{\frac{1}{\exp(\alpha_a\Phi)}+\frac{1}{\gamma}} \tag{157}$$

where $\gamma = i_l/i_0$. For convenience again, let

$$\Psi = \Phi/\Phi^0, \quad \mu = \nu^2/\Phi^0 \quad \text{and} \quad \alpha = \alpha_a\Phi^0 \tag{158}$$

Hence, a set of ordinary different equations with boundary conditions, which describes the potential and concentration distributions, can be obtained as

$$\frac{d^2\Psi}{dX^2} = \mu\frac{C}{\frac{1}{\exp(\alpha\Psi)}+\frac{1}{\gamma}} \tag{159}$$

$$\frac{d^2C}{dX^2} = s\frac{C}{\dfrac{1}{\exp(\alpha\Psi)} + \dfrac{1}{\gamma}} \tag{160}$$

$$X = 0, \qquad \frac{d\Psi}{dX} = 0, \qquad \frac{dC}{dX} = 0 \tag{161}$$

$$X = 1, \qquad \Psi = 1, \qquad C = 1 \tag{162}$$

To assist in the solution of these difference equations we can use the relationship between Ψ and C, i.e., Eq. (127).

Combining Eqs. (159), (160) and (127) gives two independent different equations having the same boundary conditions of Eqs. (128) and (131),

$$\frac{d^2\Psi}{dX^2} = \mu\frac{[1 + \dfrac{s}{\mu}(\Psi - 1)]}{\dfrac{1}{\exp(\alpha\Psi)} + \dfrac{1}{\gamma}} \tag{163}$$

$$X = 0, \qquad \frac{d\Psi}{dX} = 0$$

$$X = 1, \qquad \Psi = 1$$

or

$$\frac{d^2C}{dX^2} = s\frac{C}{\dfrac{1}{\exp\{\alpha[1 + \dfrac{\mu}{s}(C - 1)]\}} + \dfrac{1}{\gamma}} \tag{164}$$

$$X = 0, \qquad \frac{dC}{dX} = 0$$

$$X = 1, \qquad C = 1$$

Hence the overpotential and concentration changes across the electrode structure are related by the four dimensionless parameters, μ, s, α and γ.

Also the efficiency factor E of the three dimensional electrode is given by

$$E = \frac{(\frac{d\Psi}{dX})_{X=1}}{\mu \exp(\alpha_a \Phi^0)} = \frac{I}{\exp(\alpha)} \tag{165}$$

or

$$E = \frac{(\frac{dC}{dX})_{X=1}}{s \exp(\alpha_a \Phi^0)} = \frac{I}{\exp(\alpha)} \tag{166}$$

The Adomian solution of this model are also in the series forms of Eqs. (145) and (146), and can be obtained, with or without decoupling of the two coupled equations, by the program codes in the Appendix.[12,17] The algebraic expressions of dimensionless potential and concentration has been given in References[4,5] and is not shown here for the sake of brevity.

Figure 12(a) shows the typical distributions in local current for a first order reaction with different values of ν^2 and applied dimensionless overpotentials Φ^0 for the coupled anode model, including mass transfer parameter γ. The values of Φ^0, in the range of 0.5 to 16 typically represent overpotentials in the approximate range of 25 to 800 mV. The total current flows from $X = 0$ (anode fed plane) to $X = 1$ (membrane). Current is much higher at the face of the electrode adjacent to the membrane or free electrolyte solution and decreases towards the current collector. An increase in potential increases the local current density and thereby increases the overall variation in current density throughout the electrode.

With a higher value of ν (higher surface area or thinner electrode) there is a reduction in the variation of dimensionless current density. An increase in conductivity, which decreases the value of ν, will increase the dimensionless current density through a more uniform potential distribution and a greater utilization of the available electrode area. At higher values of electrode potential the effect of the parameter ν on the current distribution is less pronounced because more of the electrode is significantly affected by mass transport limitations.

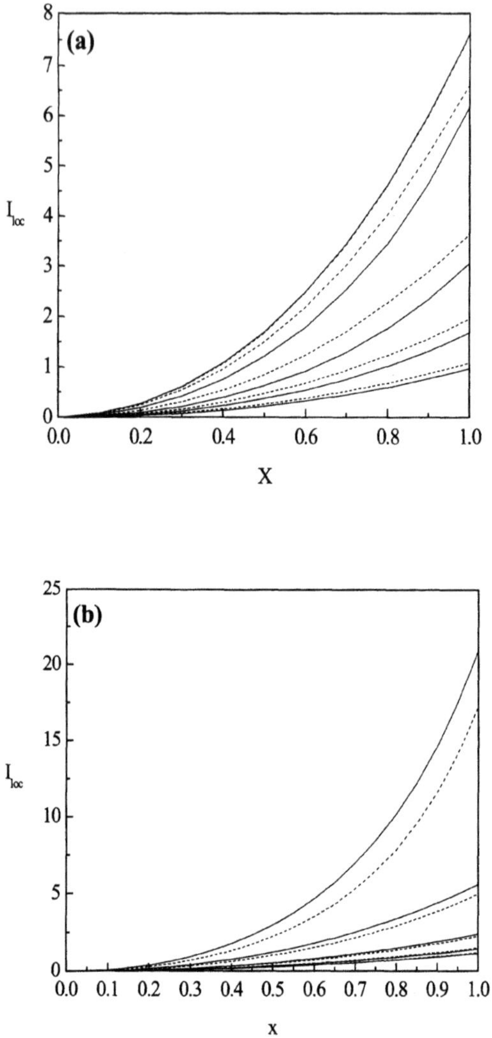

Figure 12. Distribution of local dimensionless current densities I_{loc} with dimensionless distance X: (a) $\alpha = 0.5$, $\gamma = 10$, $s = 0.1$, $\Phi^0 = 0.5, 1.0, 2.0, 4.0, 16.0$ (from top to bottom). Solid line: $\nu^2 = 1.0$; dotted line: $\nu^2 = 0.1$; (b) $\alpha = 0.5$, $\nu^2 = 0.1$, $s = 0.1$, $\Phi^0 = 0.5, 1.0, 2.0, 4.0, 8.0$ (from top to bottom). Solid line: the first order; dotted line: the second order.

The effect of pore diffusion is described through the term, s, which represents a ratio of the kinetic rate to the diffusion rate in the electrode. In general increasing the value of s, i.e., decreasing the diffusion rate relative to the kinetic rate, has the effect of causing a significant reduction in the local current densities and that more of the electrochemical activity of the electrode is focused closer to the membrane. This is a consequence of the reduced concentration of reactant away from the membrane due to, for example, a slower diffusion rate (lower diffusion coefficient).

Figure 12(b) shows the local current distribution of first and second order reactions and applied over potentials Φ^0 for the coupled anode model without the mass transfer parameter γ. The figure also shows the effect of a change in the electrode kinetics, in terms of an increase in the reaction order (with respect to reactant concentration) to 2.0, on the current distribution. Essentially a similar variation in current density distribution is produced, to that of a first order reaction, although the influence of mass transport limitations is more severe in terms of reducing the local current densities.

Figure 13 shows the potential and concentration distributions for different values of dimensionless potential under conditions when internal pore diffusion (s = 0.1) and local mass transport (γ = 10) are a factor. As expected the concentration and relative overpotential decrease further away from the free electrolyte (or membrane) due to the combined effect of diffusion mass transport and the poor penetration of current into the electrode due to ionic conductivity limitations. The major difference in the data is with respect to the variation in reactant concentrations. In the case when an internal mass transport resistance occurs (γ = 10) the fall in concentration, at a fixed value of electrode overpotential, is not as great as the case when no internal mass transport resistance occurs. This is due to the resistance causing a reduction in the consumption of reactant locally, and thereby increasing available reactant concentration; the effect of which is more significant at higher electrode overpotentials.

Figure 14 shows the distributions removing the influence of the internal mass transport resistance on the current distribution, i.e., γ is very high (model Eqs. 117 and 118 apply). In this case much higher local current densities are achieved, although the problem of a non-uniform distribution in current density prevails, with as before much of the electrode not very active.

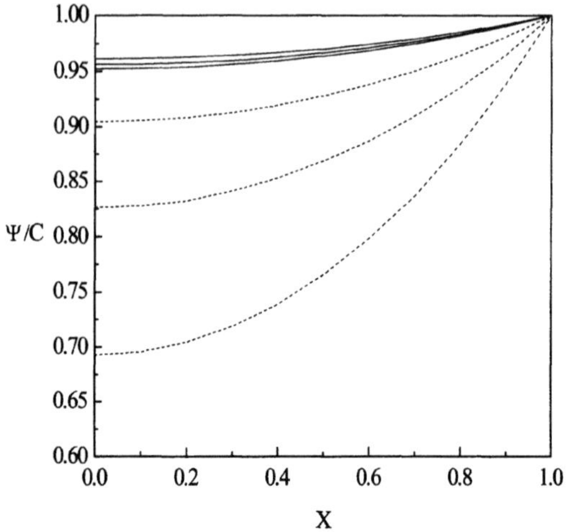

Figure 13. The distributions of relative dimensionless potential Ψ and dimensionless concentration C with values of Φ^0 at membrane $\alpha = 0.5$, $\nu^2 = 0.1$, $\gamma = 10.0$, $s = 0.1$, $\Phi^0 = 2.0, 4.0, 8.0$ (from top to bottom for C, from bottom to top for Ψ). Solid line: relative dimensionless potentials Ψ; dotted line: dimensionless concentration C.

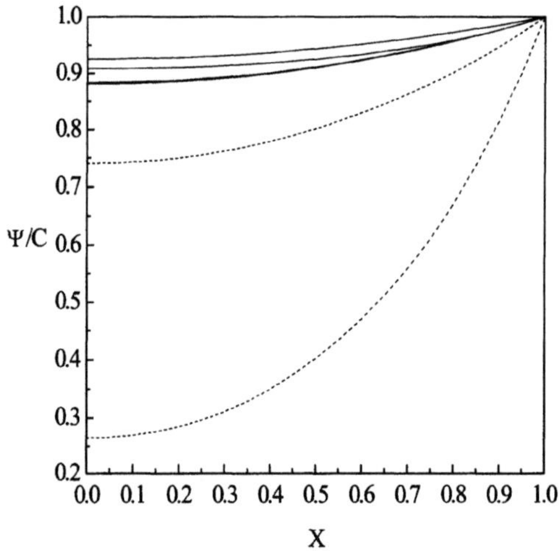

Figure 14. Profiles of relative dimensionless potential Ψ and dimensionless concentration C with dimensionless distance X: $\alpha = 0.5$, $\nu^2 = 0.1$, $s = 0.1$; $\Phi^0 = 2.0, 4.0, 8.0$ (from top to bottom). Solid line: relative dimensionless potentials Ψ; dotted line: dimensionless concentration C.

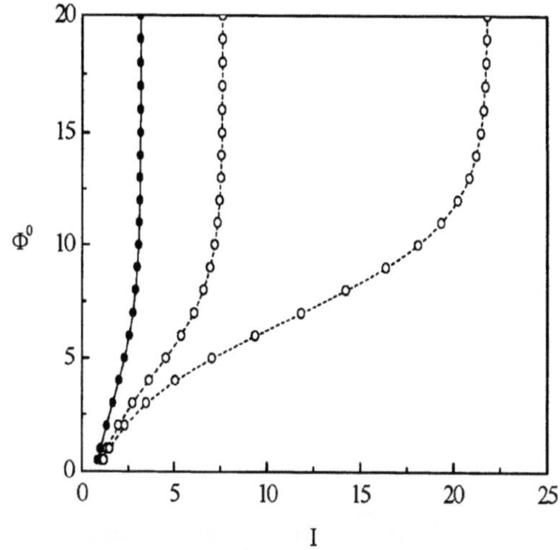

Figure 15. Comparison of decomposition data with numerical data for the dependence of dimensionless potential Φ^0 with the total dimensionless current density I. $\alpha_c = 0.5$; $\gamma = 10.0$, $s = 1.0$; $\gamma = 10.0$, $s = 0.1$; and $\gamma = 50.0$, $s = 0.1$ (from left to right). Solid line: $\nu^2 = 1.0$, •: numerical result; dotted line: $\nu^2 = 0.1$, o: numerical result.

Figure 15 gives typical values of electrode polarization for different values of γ, s and ν^2. As the value of the overpotential at the membrane increases the electrode takes on limiting current density characteristics. As expected as the value of the parameter γ increases (local mass transport coefficient is higher) the electrode approaches a higher mass transport limiting current density as the overpotential rises. As the relative rate of internal diffusion decreases, (s increases) the values of the overall limiting current densities are lower. The overall dimensionless limiting currents are also higher at lower values of the parameter ν. Also shown on Figure 15 is a comparison of the model solution using the decomposition method and a finite difference method, where good agreement is achieved.

Figure 16 shows a comparison of the electrode polarization behavior with that obtained with an assumption of a constant concentration, and shows the influence of the transfer coefficient

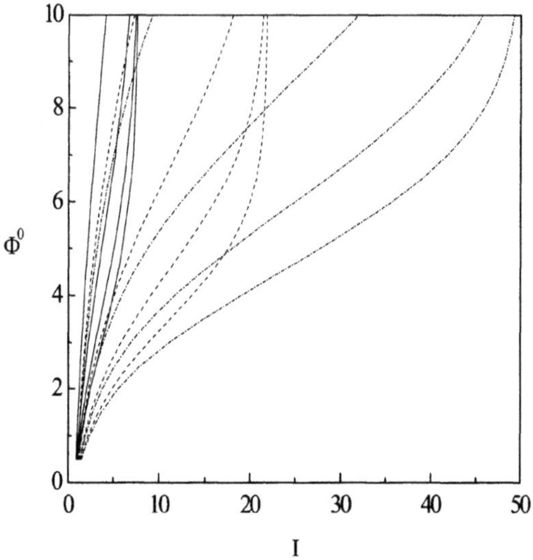

Figure 16. Variations of dimensionless potential Φ^0 with dimensionless current density I for different values of transfer coefficient α_a: $\alpha_a = 0.25, 0.5, 0.75, 1.0$ (from left to right). Solid line: $v^2 = 1.0$, $\gamma = 10.0$, $s = 0.1$; dotted line: $v^2 = 0.1$, $\gamma = 50.0$, $s = 0.1$; dash-dot line: $v^2 = 0.1$, $\gamma = 50.0$, assumed constant concentration.

on the overall cell polarization behavior. As shown the greater value of transfer coefficient, the lower is the electrode polarization, i.e., potentials are lower at a given current density. The values of limiting current reached are effectively independent of the value of the transfer coefficient.

The electrode polarization curve characteristics exhibited above are typical of those seen in many fuel cell electrodes. Thus a potential application of the ADM approximate solutions is determining the key electrochemical and mass transport parameters.

4. Simplification of Packed-Bed Electrode with a Low Conversion

In a special case i.e., low reactant conversion, in which the bulk electrolyte reactant concentration is assumed constant, the coupled Eqs. (156) and (157) can be simplified into a single differential equation (as $C = 1$). For a cathodic reaction, the charge balance Eq. (106) becomes

$$\nabla \cdot \vec{i} = ai_0\left[\frac{c_O^s}{c_O^0}\exp\left(-\frac{\alpha_c nF\eta}{RT}\right)\right] \tag{167}$$

It's similar to the Eq. (156), the model will be

$$\frac{d^2\Phi}{dX^2} = \frac{-\nu^2}{\dfrac{1}{\exp(-\alpha_c\Phi)} + \dfrac{1}{\gamma}} \tag{168}$$

where Φ, is still the dimensionless overpotential variable. In this case, the potential drop across the electrode structure is related by the three parameters, ν^2, α_c and γ.

To solve the simple model without decoupling, there are two choices for the boundary condition, at $X = 1$, on the membrane side. If the dimensionless current I can be measured and the dimensionless potential Φ^0 is unknown, we solve the model to obtain Φ^0 by using the boundary condition

$$X = 1, \qquad \frac{d\Phi}{dX} = -\nu^2 I \tag{169}$$

If Φ^0 can be measured and I is unknown, we solve the model to obtain $(d\Phi/dX)_{X=1}$ by using the boundary condition

$$X = 1, \qquad \Phi = \Phi^0 \tag{170}$$

where still, $I = i_T / ai_0L$ with i_T the total current density based on the cross-sectional area of the electrode, given by

$$i_T = ai_0LI = \frac{-ai_0L}{\nu^2}\left(\frac{d\Phi}{dX}\right)_{X=1} = \frac{-\kappa}{L\beta}\left(\frac{d\Phi}{dX}\right)_{X=1} \tag{171}$$

The solution and its derivative are in series forms in which the constant root Φ^0, namely the value of the potential at $X = 0$, is undetermined yet.

The $m = + 1$ term approximates for the dimensionless potential are

$$\Theta_{m+1}(X) = \sum_{n=0}^{m} \Phi_n(X) \tag{172}$$

$$(\frac{d\Theta_{m+1}}{dX})_{X=1} = \sum_{n=0}^{m} (\frac{d\Phi_n}{dX})_{X=1} \tag{173}$$

serve as the approximate solution and its derivatives, and should satisfy the boundary conditions. To determine the unknown constant Φ^0, we impose the boundary condition $\Phi = \Phi^0$ or $d\Phi/dX = -v^2 I$ at $X = 1$ on the approximates. This will lead to an algebraic equation for each approximation with different terms.

For an easier decomposition solution and for faster convergence to the roots, the model equation can be rewritten in a form, which scales the boundary condition values to 1.0. This re-arrangement is shown next.

When I is known, this means that the boundary condition with a derivative will be used rather than the boundary condition with a constant; we then have

$$\frac{d^2\Omega}{dX^2} = \rho \frac{1}{(\frac{1}{\exp(\alpha\Omega)} + \frac{1}{\gamma})} \tag{174}$$

$$X = 0, \qquad \frac{d\Omega}{dX} = 0 \tag{175}$$

$$X = 1, \qquad \frac{d\Omega}{dX} = 1 \tag{176}$$

where Ω, is a relative dimensionless resistance variable

$$\Omega = \frac{\Phi}{-v^2 I} \quad \text{and} \quad \rho = \frac{1}{I}, \quad \alpha' = -\alpha_c v^2 I \tag{177}$$

In this case, the resistant drop across the electrode structure is related by the three dimensionless parameters, ρ, α' and γ.

The effectiveness factor will be

$$E = \frac{I}{\exp(\alpha'\Omega(1))} \tag{178}$$

The $m+1$ term approximates for the dimensionless resistance are

$$Z_{m+1}(X) = \sum_{n=0}^{m} \Omega_n(X) \tag{179}$$

and

$$(\frac{dZ_{m+1}}{dX})_{X=1} = \sum_{n=0}^{m} (\frac{d\Omega_n}{dX})_{X=1} \tag{180}$$

hence

$$E_m = \frac{I}{\exp(\alpha' Z_{m+1}(1))} \tag{181}$$

When Φ^0 is evaluated, we have

$$\frac{d^2\Psi}{dX^2} = \mu \frac{1}{(\frac{1}{\exp(\alpha\Psi)} + \frac{1}{\gamma})} \tag{182}$$

$$X = 0, \qquad \frac{d\Psi}{dX} = 0 \tag{183}$$

$$X = 1, \qquad \Psi = 1 \tag{184}$$

where Ψ, is still the relative dimensionless overpotential variable

$$\Psi = \frac{\Phi}{\Phi^0} \quad \text{and} \quad \mu = \frac{-\nu^2}{\Phi^0}, \qquad \alpha = -\alpha_c \Phi^0 \tag{185}$$

In this case, the potential drop across the electrode structure is related by the three dimensionless parameters μ, α and γ. The dispersion modulus s related to the concentration has disappeared.

The effectiveness factor will be

$$E = \frac{(\frac{d\Psi}{dX})_{X=1}}{\mu \exp(\alpha)} = \frac{I}{\exp(\alpha)} \tag{186}$$

The $m + 1$ term approximates for the dimensionless potential are

$$\Theta_{m+1}(X) = \sum_{n=0}^{m} \Psi_n(X) \tag{187}$$

and

$$(\frac{d\Theta_{m+1}}{dX})_{X=1} = \sum_{n=0}^{m} (\frac{d\Psi_n}{dX})_{X=1} \tag{188}$$

By an inverse operation of ADM, the local dimensionless current I_{loc} can be calculated by

$$I_{loc} = \frac{1}{\mu}(\frac{d\Theta_{m+1}}{dX}) = \frac{1}{\mu}\int_0^X (\sum_{n=0}^{m} A_n) dX \tag{189}$$

The approximate of the total dimensionless current I_m will be

$$I_m = \frac{1}{\mu}(\frac{d\Theta_{m+1}}{dX})_{X=1} = \frac{1}{\mu}\int_0^1 (\sum_{n=0}^{m} A_n) dX \tag{190}$$

where

$$\int_0^1 (\sum_{n=0}^{m} A_n) dX = A_0 + \sum_{n=1}^{m} \frac{A_n}{(2n-1)!} \tag{191}$$

hence

$$E_m = \frac{I_m}{\exp(\alpha)} \tag{192}$$

The dimensionless models keep the same form as with the original although there are differences at $X = 1$ on the membrane side only. The two boundary conditions, used to find the roots, Ω_0 or Ψ_0, are simplified.

VI. EXAMPLES OF PACKED BED ELECTRODES APPLICATIONS

1. Electrochemical Reduction of Nitrobenzene in a Packed-Bed Electrode Reactor[18]

The electrochemical reduction of nitrobenzene to produce p-aminophenol has attracted industrial interest for several decades. However, some limitations may be met in this process regarding overall reaction rate, selectivity and current efficiency using a two-dimensional electrode reactor. These restrictions are due to the organic electrode reaction rate being slow and to the low solubility of nitrobenzene in an aqueous solution. In this example, a packed bed electrode reactor (PBER), which has a large surface area and good mass transfer properties, was used in order to achieve a high selectivity and a reasonable reaction rate for the production of p-aminophenol. The reaction mechanism in an acid solution can be simplified as

at the cathode:

$$\underset{(A_s)}{C_6H_5NO_2} \underset{k_1}{\overset{+4e^-}{\rightarrow}} \underset{(B_s)}{C_6H_5NHOH} \underset{k_2}{\overset{+2e^-}{\rightarrow}} \underset{(D_s)}{C_6H_5NH_2} \tag{193}$$

in the solution:

$$\underset{(B_s)}{C_6H_5NHOH} \underset{k_1}{\overset{+H^+}{\rightarrow}} \underset{(B)}{C_6H_5N^+H_2OH} \underset{k_C}{\overset{-H^+}{\rightarrow}} \underset{(C)}{p-HOC_6H_4NH_2} \tag{194}$$

In this scheme, nitrobenzene (A_s) is first cathodicaly reduced to the intermediate phenyl-hydroxylamine (B_s) at the electrode surface. The electrode reaction from B_s to aniline (D_s) and the homogeneous reaction from B to p-aminophenol (C) are parallel reactions. The selectivity of the reaction system depends on the competition between these two reactions. During the operation of the PBER, excessive nitrobenzene was added into the electrolyte to maintain it at a saturation concentration.

Mathematical expressions of concentrations, c_A^s and c_B^s at the surface of cathode, can be obtained.[18] The total current density in the electrolyte is described as

$$i = 4Fk_1c_A^S + 2Fk_2c_B^S \tag{195}$$

where k_1 is the rate constant of electrode reaction from A_s to B_s , k_2 is the rate constant of electrode reaction from B_s to D_s.

The ratio of aniline to p-aminophenol in the solution, namely the selectivity, can be written as

$$\frac{c_D}{c_C} = k_2\left(\frac{a}{k_c} + \frac{1}{k_l}\right) \tag{196}$$

where a is the special area of the PBER, k_c is the rate constant of homogeneous reaction from B to C and k_l is the mass transfer coefficient.

The mathematical model for the PBE reactor should in general be a two-dimensional model describing the potential and concentration distributions within the packed bed electrode. However, the model can be simplified in certain cases. Under a recycle flow operation, for example, the conversation per pass through the packed bed is small, so that the PBER can be treated as a differential reactor, the potential distribution only in the lateral direction is considered. In this case which is similar to the case in 5.4, the two-dimensional model can be written in a one-dimensional Poisson equation form as

$$\frac{d^2\phi^l}{dx^2} = -\frac{ai}{\kappa} \tag{197}$$

and the boundary conditions are (see Figure 11):

$$x = 0, \qquad \frac{d\phi^l}{dx} = 0 \qquad \text{at the feeder} \tag{198}$$

$$x = L, \qquad \phi^l = \text{constant} \qquad \text{at the membrane} \tag{199}$$

where ϕ^l is the potential of the electrolyte phase, a is the electrode area per unit reactor volume, i is the total current density and κ is the effective conductivity of the electrolyte.

In the reactor model, the reaction rates were assumed constant in the axial direction at any position x. Hence, the height of the packed bed is only related to the conversation per pass through the cell. On the other hand, the thickness of the packed bed influences the local reaction rate as well as the selectivity, and becomes a key parameter for the PBER design. To achieve a high selectivity and a reasonable reaction rate, the suitable thickness can be obtained by using computer iteration. The predictions of the model are in reasonable agreement with the experiment results.[18]

A key point of this work is to solve the nonlinear model Eq. (197) to obtain the potential distribution in the packed bed. We should pay attention to Eq. (195) in which the current density has contributions from the preferred electrode reaction as well as the side electrode reaction. Accordingly, Eqs. (42) and (195) can be combined into the equation for overpotential distribution as

$$\frac{d^2\eta}{dx^2} = \frac{aF(n_1k_1c_A^s + n_2k_2c_B^s)}{\kappa} \tag{200}$$

$$x = 0, \qquad \frac{d\eta}{dx} = 0 \tag{201}$$

$$x = L, \qquad \eta = \eta^0 \tag{202}$$

where

$$k_1 = k_{01}\exp(-\frac{n_1F\alpha_{c1}\eta}{R_gT}) \tag{203}$$

$$k_2 = k_{02}\exp(-\frac{n_2F\alpha_{c2}\eta}{R_gT}) \tag{204}$$

This model was solved numerically by BAND program.[18] Numerical solution of this model was used to determine lateral distributions of overpotential for calculating the overall selectivity $\bar{S}$:

$$\bar{S} = \frac{(-r_A)-(-r_B)}{-r_c} = \frac{1}{L}\int_0^L S_l(x)dx \tag{205}$$

where S_l is the local selectivity.

Comparisons of the experimental and theoretical performance with different thickness of the packed bed electrode are shown in Figure 17. Selectivity $\bar{S}$ generally decreases with an increase in bed thickness, x_0, and/or with a decrease in flow rate, Q, due to a greater non-uniformity in electrode potential distribution.

As a example of solving the nonlinear model by ADM, we assume that the value of k_l is very large, so

$$c_A^s \approx c_A^0 \tag{206}$$

and for a steady state

$$r_A - r_B - r_C = 0 \tag{207}$$

$$ak_1c_A^0 - ak_2c_A^s - k_cc_B^s = 0 \tag{208}$$

we have

$$c_B^s = \frac{ak_1c_A^0}{ak_2 + k_c} \tag{209}$$

and the selectivity

$$\frac{C_D}{C_C} = \frac{ak_2}{k_C} \tag{210}$$

Introducing C_{Bs} into the model equation gives

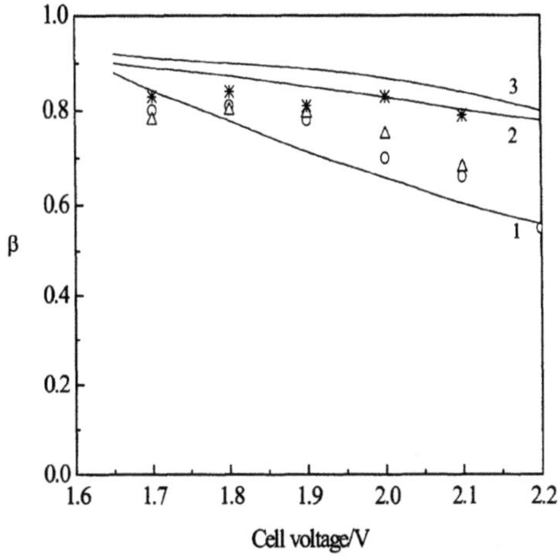

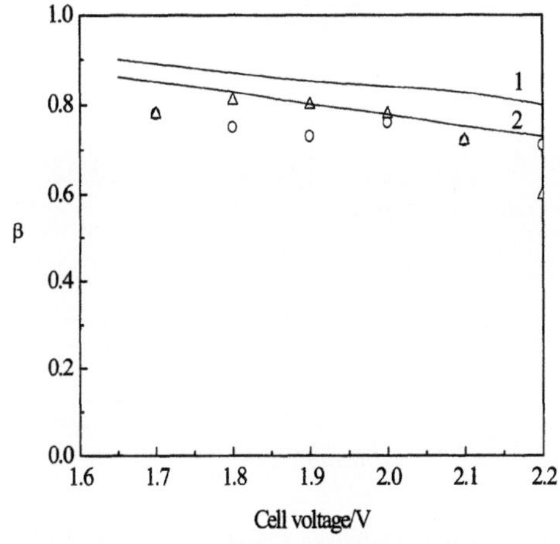

Figure 17. Plot of theoretical and experimental selectivity vs. applied cell voltage. (a) $x_0 = 0.02$ m; $Q = 1.22 \times 10^{-5}$ m^3 s^{-1}, solid line 1 = calculated, o = experimental; $Q = 2.75 \times 10^{-5}$ m^3 s^{-1}, solid line 2 = calculated, Δ = experimental; $Q = 4.30 \times 10^{-5}$ m^3 s^{-1}, solid line 3 = calculated, * = experimental. (b) $Q = 2.75 \times 10^{-5}$ m^3 s^{-1}; $x_0 = 0.02$ m, solid line 1 = calculated, o = experimental; $x_0 = 0.03$ m, solid line 2 = calculated, Δ = experimental.

$$\frac{d^2\eta}{dx^2} = \frac{aFc_A^0\left[n_1k_1 + an_2k_2k_1/(ak_2 + k_c)\right]}{\kappa} \tag{211}$$

$$= \frac{aFc_A^0\left[n_1k_{01}\exp(\beta_1\eta) + an_2k_{01}k_{02}\exp\frac{(\beta_1+\beta_2)\eta}{ak_{02}\exp(\beta_2\eta)+k_c}\right]}{\kappa}$$

$$= \varpi_1\exp(\beta_1\eta) + \frac{\varpi_{12}\exp(\beta_{12}\eta)}{\varpi_2\exp(\beta_2\eta)+k_c}$$

where

$$\varpi_1 = \frac{n_1k_{01}aFc_A^0}{\kappa} \tag{212}$$

$$\varpi_{12} = \frac{an_2k_{01}k_{02}aFc_A^0}{\kappa}$$

$$\varpi_2 = ak_{02}$$

$$\beta_1 = \frac{-n_1F\alpha_{c1}}{R_gT}$$

$$\beta_2 = \frac{-n_2F\alpha_{c2}}{R_gT}$$

$$\beta_{12} = \beta_1 + \beta_2$$

We use the known lateral distant L and overpotential η^0:

$$\frac{d^2\Psi}{dX^2} = \frac{L^2}{\eta^0}\left[\varpi_1\exp(\beta_1\eta_1\Psi) + \frac{\varpi_{12}\exp(\beta_{12}\eta_1\Psi)}{\varpi_2\exp(\beta_2\eta_1\Psi)+k_c}\right] \tag{213}$$

$$= \mu_1\exp(\alpha_1\Psi) + \mu_{12}\exp\frac{(\alpha_1+\alpha_2)\Psi}{\varpi_2\exp(\alpha_2\Psi)+k_c}$$

where

$$\Psi = \frac{\eta}{\eta^0}, \qquad \mu_1 = \frac{L^2 \varpi_1}{\eta^0}, \qquad \mu_{12} = \frac{L^2 \varpi_{12}}{\eta^0} \tag{214}$$

and

$$\alpha_1 = \beta_1 \eta^0, \qquad \alpha_2 = \beta_2 \eta^0, \qquad \alpha_{12} = \alpha_1 + \alpha_2 \tag{215}$$

The dimensionless model will be

$$\frac{d^2\Psi}{dX^2} = \mu_1 \exp(\alpha_1 \Psi) + \frac{\mu_{12} \exp(\alpha_{12}\Psi)}{\varpi_2 \exp(\alpha_2 \Psi) + k_c} \tag{216}$$

$$X = 0, \qquad \frac{d\Psi}{dX} = 0 \tag{217}$$

$$X = 1, \qquad \Psi = 1 \tag{218}$$

The first term on the right side of Eq. (216) is related to the preferred electrode reaction, and the second term is related to the side electrode reaction as well as the solution reaction.

The approximate solution of the dimensionless potential distribution along the lateral direction can be obtained by the program code in the Appendix:

$$\begin{aligned} Z_{m+1} = {} & \Psi_0 + \frac{1}{2}[\mu_1 \exp(\alpha_1 \Psi_0) + \frac{\mu_{12} \exp(\alpha_{12}\Psi_0)}{k_c + \varpi_2 \exp(\alpha_2 \Psi_0)}]X^2 \\ & + \{ [\mu_1 k_c \exp(\alpha_1 \Psi_0) + (\mu_{12} + \mu_1 \varpi_2) \exp(\alpha_{12}\Psi_0)] \\ & \bullet \{\mu_1 \alpha_1 k_c{}^2 \exp(\alpha_1 \Psi_0) \\ & + k_c(\mu_{12}\alpha_{12} + 2\alpha_1 \mu_1 \varpi_2) \exp(\alpha_{12}\Psi_0) \\ & + \left(\mu_{12}\alpha_1\varpi_2 + \mu_1\alpha_1\varpi_2^2\right) \exp[(\alpha_{12} + \alpha_2)\Psi_0] X^4 \}\} \\ & / 24(k_c + \varpi_2 \exp(\alpha_2 \Psi_0))]^3 + \ldots + \Psi_m \end{aligned} \tag{219}$$

which is the usual algebraic expression like

$$\begin{aligned} Z_{m+1}(X) &= \sum_{n=0}^{m} \Psi_n(X) \\ &= \Psi_0 + b_1(\Psi_0)X^2 + b_2(\Psi_0)X^4 + b_3(\Psi_0)X^6 + \ldots + \Psi_m \end{aligned} \tag{220}$$

Integrating of $c_D/c_C = ak_2/k_c$, the total average selectivity will be

$$\bar{S} = \frac{a}{k_c}\int_0^1 k_2(X)dX \tag{221}$$

where

$$k_2 = k_{02}\exp(-\frac{n_2 F\alpha_2\eta}{RT}) = k_{02}\exp(\beta_2\eta) = k_{02}\exp[\frac{\beta_2}{\eta^0}\Psi(X)] \tag{222}$$

It is clear that the model and ADM can be used for the analysis of the PBE reactor in the more complex reaction system.

2. Direct Electrochemical Oxidation of Propylene in a Sparged Packed-Bed Electrode Reactor[19,20]

The oxidation of alkenes has attracted interest over decades as a route for organic electro-synthesis. This has included attempts to carry out direct anodic oxidation to eliminate problems associated with the in situ use of electrochemically generated oxidants. A sparged packed bed electrode reactor (SPBER), with propylene oxidation separated from the generation of hydrogen gas, was used for the three-phase electrochemical process in which there is a high volume fraction of alkene gas in the inter-electrode gap. The reactant, propylene, undergoes direct oxidation to the epoxide, which undergoing further chemical reaction (saponification), in solution, to the glycol. The reactions are

$$C_3H_6 + 2OH^- \rightarrow C_3H_6O + H_2O + 2e^- \quad \text{at the anode} \tag{223}$$

$$C_3H_6O + H_2O \rightarrow CH_3CHOHCH_3OH \quad \text{in the solution} \tag{224}$$

A special feature of this reaction is that the first order kinetics is not related to the propylene concentration fed by convection and gas-liquid mass transfer (a saturated propylene concentration assumed in electrolyte), but related to the OH^- ion concentration, because the OH^- ion pass through the membrane from the cathode side:

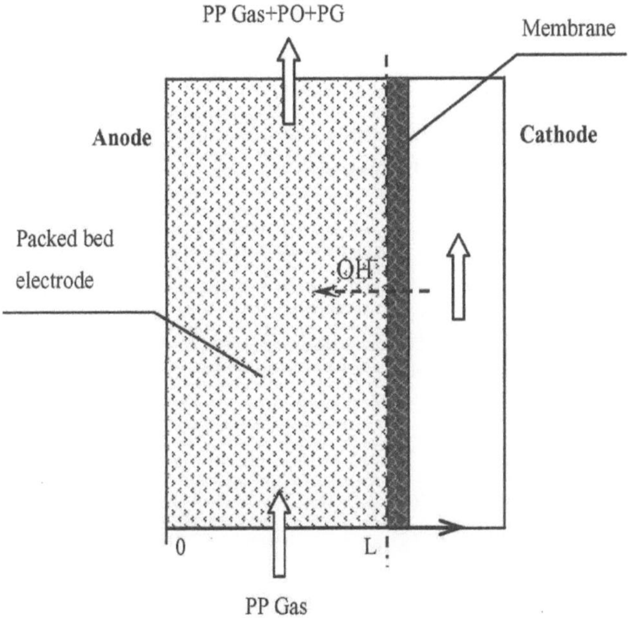

Figure 18. Schematic diagram of the sparged-packed bed anode reactor.

$$r_1 = \frac{a_p}{2F}\left\{91.9c^s_{OH^-} \exp\frac{-11.8-0.194F\eta}{R_gT}\right\} \tag{225}$$

The highest value of concentration of hydroxide was assumed as a constant at the anion exchange membrane, and the one dimension distribution of hydroxide ion in the packed anode was modeled. In this case, the mixed-flow reactor was changed into a flow-by type as shown in Figure 18.

A non-linear mathematical model, which is a set of ordinary differential equations, for the process in the SPBER was developed.[19] The model accounts for the heterogeneous electrochemical reaction and homogeneous reaction in the bulk solution. The lateral distributions of potential, current density and concentration in the reactor were studied. The potential distribution in the lateral dimension, x, of the packed bed was described by a one dimensional Poisson equation as:

$$\frac{d^2\phi^l}{dx^2} = \frac{ai}{\kappa} \tag{226}$$

This equation can be rewritten into the following form describing the distribution of overpotential:

$$\frac{d^2\eta}{dx^2} = \frac{ai}{\kappa} \tag{227}$$

where $\eta = \phi^m - \phi^l - U'$ according to Eq. (42).

The model was formulated on the basis of a set of material balances, for each of the species involved in the aqueous phase:

$$D_l \frac{d^2 c_i}{dx^2} - \Sigma r_i + r_i^a = 0 \tag{228}$$

where D_l is the lateral dispersion coefficient and c_i is the concentration of species i.

The first term in the equation represents the mass transport rate, Σr_i is the total rate of chemical reaction, r_i^a is the rate of gas absorption given by

$$r_i^a = k_{gi} a_l (c_{gi} - c_i / m_i)/(1 - \varepsilon_g) \tag{229}$$

where k_{gi} is the gas–liquid specific mass transport coefficient of species, a_l is the interfacial area, c_{gi} is the concentration in the gas phase, m_i is the phase distribution coefficient and ε_g is the gas void fraction.

Under steady-state conditions the equivalent material balance for the gas phase is:

$$(c_{gi}^0 - c_{gi}) / \tau_g \varepsilon_g - r_i^a = 0 \tag{230}$$

where c_{gi}^0 is the inlet gas phase concentration and τ_g is the holding time for the gas phase.

Solution of this model was used to determine lateral distributions of overpotential, current density and concentrations of products in the SPBER, and to simulate the effect of important operation parameters, such as gas sparing rate and applied

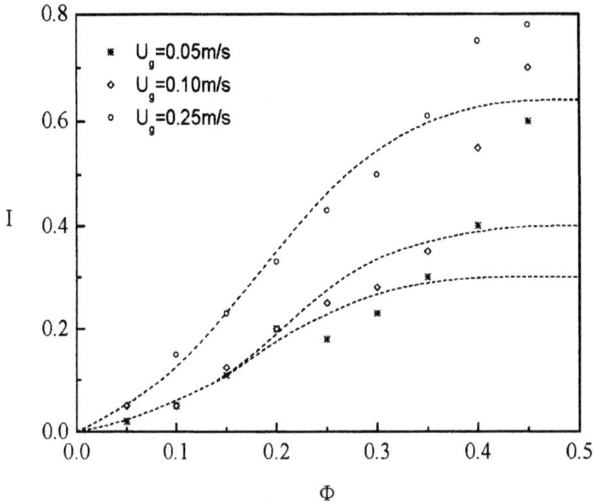

Figure 19. Experimental steady-state cell current electrode potential behavior.

overpotential, on these distributions. Figure 19 compares the numerical results and experimental data.[19]

As another example of solving the nonlinear model by ADM, we assume that at high overpotentials the electrochemical kinetics can be represented by the following high field Butler-Volmer approximation, the current density in Eq. (227) is rewritten as:

$$i = i_0 \frac{c_A^s}{c_A^0} \exp(\frac{\alpha_a F}{R_g T} \eta) \tag{231}$$

where c_A^s is the concentrations of OH^- ions at the surface of the electrode and c_A^0 is the concentration at the membrane. The parameters i_0 and α_a can be obtained from the experimental kinetics Eq. (225) by regression.

Then the potential equation (Eq. 227) can be changed into the form of Eq. (147) as

$$\frac{d^2\eta}{dx^2} = \frac{a i_0}{\kappa} [\frac{c_A^s}{c_A^0} \exp(\frac{\alpha_a F \eta}{R_g T})] \tag{232}$$

Secondly assuming that hydrolysis, reaction of propylene oxide to propylene glycol, is only significant at high pH, and at high concentration, with the condition that this reaction in solution is neglected. The concentration distribution of hydroxide ion in the bed of the SPBE reactor can be given in the form of Eq. (148) by

$$\frac{d^2c_A}{dx^2} = \frac{ai_0}{D_l}\left[\frac{c_A^s}{c_A^0}\exp\left(\frac{\alpha_a F\eta}{R_g T}\right)\right] \tag{233}$$

With the procedure of generalization, we obtained the same form of the coupled Eqs. (159) and (160) that have decoupled and solved by ADM.[20]

The current distribution in the packed bed can be calculated by Eq. (155) as

$$i_a(X) = \frac{C(X)}{\dfrac{1}{i_0\exp[\alpha_a\Phi(X)]} + \dfrac{1}{i_l}} \tag{224}$$

where $C = c_A/c_A^0$, $\Phi = nF\eta/R_gT$, $X = x/L$, the limiting current density is given by $i_l = nFk_lc_A^0$, and a dimensionless form

$$\frac{i_a}{i_0} = \frac{C(X)}{\dfrac{1}{\exp[\alpha_a\Phi(X)]} + \dfrac{1}{\gamma}} \tag{225}$$

where $\gamma = i_l/i{=}_0$.

The total current density will be

$$i_t = \int_0^1 i_a(X)dX \tag{226}$$

Due to the complexity of mass transfer between gas-liquid-solid phases, it is difficult to evaluate the average value of mass transfer coefficient k_l from the literature. A realistic way to evaluate k_l is to use the algebraic expression of solution and by regression to obtain the experimental data rather than by regression with solving the set of non-linear differential equations.

The two models in Sections VI.1 and VI.2 have been solved by a numerical method based on a finite difference routine BAND (j).[7,18,20] To solve a non-linear model, iteration with trail values is required. Furthermore, double iterations are needed in cases, for example, when it is required to optimize the thickness of the PBER, or to regress the key parameters from experimental data. These complex situations make the convergences of the solution difficult.

In general it is desirable to solve the non-linear models by an approximate analysis to give analytical solutions that can be used for checking the numerical procedures and for evaluating the numerical results. A role of ADM is to take a place just in between the roles of analytical and numerical methods.

3. Two-Dimensional Model of Packed-Bed Electrodes

In this Section so far, ADM is used to solve theoretical generalized models in the forms of ordinary differential equation(s). For diffusion-convection problems, the distributions along the axial direction of the packed bed electrode were neglected in certain cases, and mass transfer in the three dimensional electrodes were characterized by an average coefficient k_l.

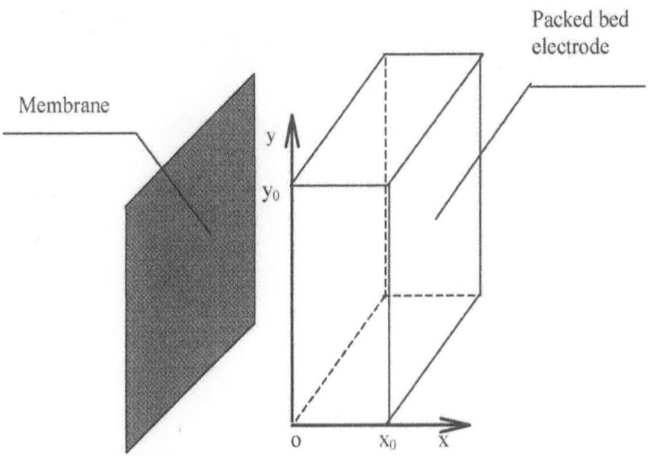

Figure 20. Schematic diagram of the packed-bed electrochemical reactor for 2-D model.

A model to describe potential and concentration distributions in both axial and lateral directions of PBE reactor (Figure 20) was developed.[21] It uses the charge balance and the material balance, i.e., Eqs. (103) and (104):

$$\nabla \cdot \vec{i} = ai(\eta, c_i) \tag{103}$$

$$\nabla \cdot \vec{N_i} = -\frac{s_i}{nF} ai(\eta, c_i) \tag{104}$$

The dimensionless model equations for an electrochemical reduction reaction in a PBE are in a form of a set of partial differential equations:

$$\frac{\partial^2 \Phi}{\partial X^2} + \zeta \frac{\partial^2 \Phi}{\partial Y^2} = \sigma[C_s \exp(-\alpha_c \Phi) - \exp(\alpha_a \Phi)] \tag{227}$$

$$\frac{\partial^2 C}{\partial X^2} + Pe \cdot \zeta \frac{\partial C}{\partial Y} = Sh_m (C - C_s) \tag{228}$$

where

$$(C - C_s) = Da_m [C_s \exp(-\alpha_c \Phi) - \exp(\alpha_a \Phi)] \tag{229}$$

and

$$X = \frac{x}{L}, \quad Y = \frac{y}{Z}, \quad \Phi = \frac{nF\eta}{R_g T}, \quad C = \frac{c}{c^0}, C_s = \frac{c^s}{c^0} \tag{230}$$

The boundary conditions are

$$X = 0, \quad 0 \le Y \le 1, \quad \frac{\partial C}{\partial X} = 0, \quad \Phi = \Phi^0 \tag{231}$$

$$X = 1, \quad 0 \le Y \le 1, \quad \frac{\partial C}{\partial X} = 0, \quad \frac{\partial \Phi}{\partial X} = 0$$

$$Y = 0, \quad 0 \le X \le 1, \quad C = 1, \quad \frac{\partial \Phi}{\partial Y} = 0$$

$$Y = 1, \quad 0 \le X \le 1, \quad \frac{\partial C}{\partial Y} = 0, \quad \frac{\partial \Phi}{\partial Y} = 0$$

There are five dimensionless groups in the model equations:

- Aspect ratio: $\zeta = \dfrac{x_0}{y_0}$
- Dimensionless resistance $\sigma = \dfrac{a i_0 L^2 nF}{\kappa R_g T}$
- Peclet number $Pe = \dfrac{\overline{v_y} L}{D_l}$
- Modified Sherwood number $Sh_m = \dfrac{a k_l L^2}{D_l}$
- Modified Damkohler number $Da_m = \dfrac{i_0}{nFk_l c^0}$

The model was solved by a numerical method based on BAND and orthogonal collocation.[21] This method was suitable for solving this model, but in the case of the highest applied potential exhibited oscillatory, fluctuating behavior. Figure 21 shows the typical potential and concentration distributions under medium potential conditions for 2-D model.

If the electrode kinetic expression is simplified by neglecting the reverse reaction, the model equation for a cathodic reaction only can be rewritten as

$$\frac{\partial^2 \Phi}{\partial X^2} + \zeta \frac{\partial^2 \Phi}{\partial Y^2} = -\sigma C_s \exp(-\alpha_c \Phi) = -\sigma \frac{C}{\dfrac{1}{\exp(-\alpha_c \Phi)} + Da_m} \tag{232}$$

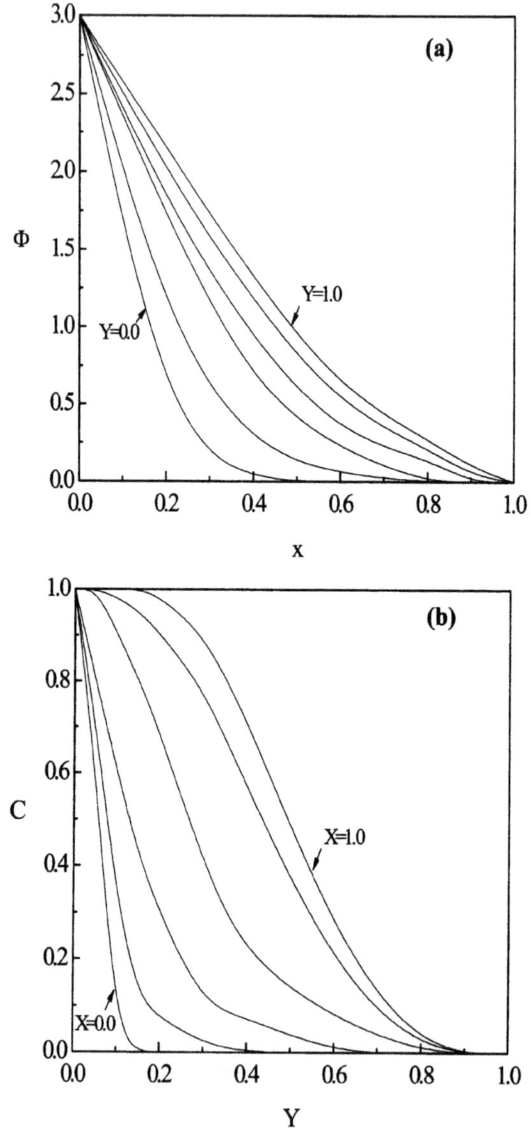

Figure 21. (a) The distribution of dimensionless potential at a medium applied potential, values of Y vary from 0.0 to 1.0 in increments of 0.2. (b) The distribution of dimensionless concentration at a medium applied potential, values of X vary from 0.0 to 1.0 in increments of 0.2.

$$\frac{\partial^2 C}{\partial X^2} + Pe \cdot \zeta \frac{\partial C}{\partial Y} = -Sh_m Da_m C_s \exp(-\alpha_c \Phi)$$
$$= -Sh_m Da_m \frac{C}{\dfrac{1}{\exp(-\alpha_c \Phi)} + Da_m} \tag{233}$$

where

$$(C - C_s) = Da_m C_s \exp(-\alpha_c \Phi) \tag{234}$$

$$C_s = \frac{C}{1 + Da_m \exp(-\alpha_c \Phi)} \tag{235}$$

By using the dimensionless variables the model equations become

$$\frac{\partial^2 \Phi}{\partial X^2} + \zeta \frac{\partial^2 \Phi}{\partial Y^2} = -\nu^2 \frac{C}{\dfrac{1}{\exp(-\alpha_c \Phi)} + \dfrac{1}{\gamma}} \tag{236}$$

$$\frac{\partial^2 C}{\partial X^2} + Pe \cdot \zeta \frac{\partial C}{\partial Y} = -s \frac{C}{\dfrac{1}{\exp(-\alpha_c \Phi)} + \dfrac{1}{\gamma}} \tag{237}$$

Furthermore in a more generalized form

$$\frac{\partial^2 \Psi}{\partial X^2} + \zeta \frac{\partial^2 \Psi}{\partial Y^2} = -\mu \frac{C}{\dfrac{1}{\exp(\alpha \Psi)} + \dfrac{1}{\gamma}} \tag{238}$$

$$\frac{\partial^2 C}{\partial X^2} + Pe \cdot \zeta \frac{\partial C}{\partial Y} = -s \frac{C}{\dfrac{1}{\exp(\alpha \Psi)} + \dfrac{1}{\gamma}} \tag{239}$$

The boundary conditions are similar to Eqs. (161) and (162), viz.

$$X = 0, \quad 0 \le Y \le 1, \quad \frac{\partial C}{\partial X} = 0, \quad \Psi = 1 \tag{240}$$

$$X = 1, \quad 0 \le Y \le 1, \quad \frac{\partial C}{\partial X} = 0, \quad \frac{\partial \Psi}{\partial X} = 0$$

$$Y = 0, \quad 0 \le X \le 1, \quad C = 1, \quad \frac{\partial \Psi}{\partial Y} = 0$$

$$Y = 1, \quad 0 \le X \le 1, \quad \frac{\partial C}{\partial Y} = 0, \quad \frac{\partial \Psi}{\partial Y} = 0$$

In this case, the potential drop across the electrode structure is related by the six dimensionless parameters μ, s, α, γ as well as the aspect ratio ζ and Peclet numberPe that character the convection in the axial direction. It is possible to approximately analyze this generalized theoretical nonlinear model using the set of partial differential equations with boundary conditions by ADM.[22]

VII. CONCLUSIONS

The model solution of porous or particulate three-dimensional electrodes is obtained using the Adomian's "inverse operator method" (IOM) or "Decomposition Method". Data calculated by the decomposition method, are comparable with that obtained by a finite difference method using the BAND program. In general the Adomian method gives faster convergence, than that of the finite difference method, for the model over a wide range of parameters.

An aim of the model is to determine the influence of the various mass transport parameters and show how they influence the polarization behavior of three-dimensional electrodes. In the model we have adopted relatively simple electrode kinetics, i.e., Tafel type, The approach can also be applied to more complicated electrode kinetics which exhibit non-linear dependency of reaction rate (current density) on reactant concentration.

In summary, for three dimensional electrode reactors, where the theoretical dimensionless model developed depends on six dimensionless parameters (μ, s, α, γ as well as the aspect ratio ζ and Peclet number Pe), the nonlinear differential equation(s) with boundary conditions can be analyzed approximately by ADM. The ADM solution mechanism is easier to use and faster to converge than conventional numerical methods.

ACKNOWLEDGEMENT

The authors acknowledge the support of this British-Chinese joint research project provided by the British Royal Society and China National Natural Science foundation.

SYMBOLS

1. ADM's Nomenclature

A_n	Adomian Polynomials
$\widetilde{F}y$	Adomian's general form of non-linear differential equation
f_i	the terms arising from integrating $g(x)$
$g(x)$	inhomogeneous term
L	the highest order derivative of non-linear differential equation
L^{-1}	inverse operator of L
N	nonlinear operator of an analytic nonlinearity
n	number of terms
m	number of finite terms
R	differential operator of order less than L
u	dependent variable
v	dependent variable
u_n	component n of the series of solution $u(x)$
v_n	component n of the series of solution $v(x)$
u_{m+1}	approximate solution with $m + 1$ terms
v_{m+1}	approximate solution with $m + 1$ terms
x	independent variable
λ	parameter for extending of Adomian polynomial

2. Nomenclatures in this Paper

a	special area of three dimensional electrode (m^{-1})
a_g	geometry parameter
b_n	coefficients of the solution series given by ADM
c_i	concentration of species i per unit volume of solution, (mol m^{-3})
c^0	reference or bulk concentration, (mol m^{-3})
c_O	concentration of oxidized species in bulk (mol m^{-3})
c_R	concentration of reduced species in bulk (mol m^{-3})

c^s	surface concentration of the key species (mol m^{-3})
C	dimensionless concentration variable = c/c^0
C_n	components of finite term approximate of dimensionless C
C_s	dimensionless surface concentration = c^s/c^0
D_l	lateral dispersion coefficient, ($m^2\ s^{-1}$)
D_e	effective diffusivity, ($m^2\ s^{-1}$)
E	effectiveness of three dimensional electrodes
E_m	approximate of m terms for E
F	Faraday constant (Coulombs mol^{-1})
i_0	exchange current density (Am^{-2})
i_a	anodic current density (Am^{-2})
i_l	limiting current density (Am^{-2})
i_T	total current density (Am^{-2})
I	dimensionless total current density = i_T/ai_0L
I_{loc}	local dimensionless current density
I_m	approximate of m terms for dimensionless current density I
k	reaction rate constant
k_0	pre-exponential factor
k_l	average coefficient of mass transfer (m s^{-2})
L	length in the lateral direction (m),
n	number of electrons involved in reaction
n	chemical reaction order
N_i	flux of species i (mol $m^{-2}\ s^{-1}$)
N_{loc}	local flux in the three dimensional electrode (mol $m^{-2}\ s^{-1}$)
p	order of cathodic reaction, order of reaction
q	order of anodic reaction
r_0	radius of sphere or cylinder catalyst, radius of the agglomerate sphere
R_g	universal gas constant, 8.3143 ($Jmol^{-1}K^{-1}$)
$R(c_i)$	consuming rate of reactant i per unit volume (mol $m^{-3}s^{-1}$)
R_t	total elelctrochemical reaction rate (mol $m^{-3}s^{-1}$)
s	diffusion (or dispersion) modulus = $s_i a i_0 L^2/nFD_e c^0$
s'	dimensionless parameter = $s\{1/[1+\exp(\alpha)/\gamma]\}$
s_i	stoichiometric coefficient of species i in electrode reaction
T	temperature (K)
U'	the open-circuit value of $\phi^m - \phi^l$ (V)
$\bar{v}_y$	average fluid velocity in the vertical direction (m s^{-1})
x	distance in the lateral direction (m)
X	dimensionless distance = x/L in the lateral direction, r/r_0 in the radial direction

y	distance in the vertical direction (m)
Y	dimensionless distance = y/y_0 in the vertical direction
α_a	anodic charger transfer coefficient (dimensionless)
α_c	catholic charger transfer coefficient (dimensionless)
α	dimensionless transfer coefficient = $\alpha_a\Phi$ or = $-\alpha_a\Phi$
α'	dimensionless parameter $\alpha' = -\alpha_c \nu^2 I$
β	$= nF/R_gT$ (V^{-1})
ϕ	Thieie mediums = $\sqrt{k_0L^2{c^0}^{n-1}/D_e}$
ϕ_W	Weisz modulus = $\varpi^2\xi$
ϕ^m	potential in conductive solid phase (V)
ϕ^l	potential in solution phase (V)
Φ	dimensionless potential variable = $\beta\eta$
Φ^0	dimensionless potential at $X = 1$
γ	dimensionless limiting current density = i_l/i_0
η	overpotential (V)
η^0	applied overpotential (V)
κ	effective conductivity of electrolyte phase ($\Omega^{-1}m^{-1}$)
μ	conduction modulus = $-\nu^2/\Phi^0$
ρ	dimensionless parameter = $1/I$
σ	dimensionless resistance = $ai_0L^2nF/\kappa R_gT = \nu^2$
ν^2	dimensionless parameter = $ai_0L^2\beta/\kappa$
Υ	dimensionless variable = CX for spherical geometry
$\Lambda_{m+1,n}$	approximate dimensionless concentration with $m + 1$ terms for n-th reaction
Θ_{m+1}	approximate of $m + 1$ terms for dimensionless overpotentials
ω	Thiele mediums for electrocatalysts = $\sqrt{s\exp(\alpha)}$
ω'	Thiele mediums for electrocatalysts = $\sqrt{s'\exp(\alpha)}$
ξ	effectiveness of porous catalyst or porous electrode
$\xi_{m,n}$	approximate of m terms for effectiveness of n-th reaction
Ω	relative dimensionless resistance variable = $\Phi/(-\nu^2 I)$
Ψ	relative dimensionless potential variable = Φ/Φ^0
Ψ_n	components of approximate of dimensionless overpotential
ζ	aspect ratio = x_0/y_0

Z_{m+1} approximate of $m + 1$ terms for relative dimensionless resistance Ω

APPENDIX: ADM MATHEMATICA CODES

1. ADM to Solve One ODE[12]

The model equations for a porous catalyst pellet or electrode in a planar form is

$$\frac{d^2C}{dX^2} - \phi^2 C^p = 0 \tag{A.1}$$

where p is the reaction order, and in a spherical form is

$$\frac{d^2C}{dX^2} - \phi^2 C^p X^{1-p} = 0 \tag{A.2}$$

According to ADM, Eqs. (A.1) and (A.2) can be decomposed as:

$$LC + NC = 0 \tag{A.3}$$

where, L is the highest order derivative, $L = d/dX^2$; N is the nonlinear operator.

Applying L^{-1} to Eq. (A.3) gives:

$$L^{-1}LC = L^{-1}NC \tag{A.4}$$

For Eq. (A.1) in a planar form

$$NC = f(C) = C^p = \sum_{n=0}^{\infty} A_n \tag{A.5}$$

and for Eq. (A.2) in a spherical form

$$NC = f(C) = C^p X^{1-p} = \sum_{n=0}^{\infty} A_n \tag{A.6}$$

Here A_n are the Adomian polynomials defined as

$$A_n = \frac{1}{n!}\frac{d^n}{d\lambda^n} f(u(\lambda))\Big|_{\lambda=0} \tag{A.7}$$

where

$$u = \sum_{i=0}^{\infty} u_i \lambda^i \tag{A.8}$$

The approximate analytical solutions are in the form:

$$C = \sum_{n=0}^{\infty} C_n = C_0 + L^{-1}\sum_{n=0}^{\infty} A_n \tag{A.9}$$

The approximate of the nonlinear term with finite m terms will be

$$NC \simeq \sum_{n=0}^{m-1} A_n \tag{A.10}$$

which can be obtained by an finite m-term approximate of Adomian polynomials.

The approximate solution with $m + 1$ terms should be in a series form as

$$C(x) \simeq \Lambda_{m+1} = \sum_{n=0}^{m} C_n = C_0 - L^{-1}\sum_{n=0}^{m-1} A_n \tag{A.11}$$

The constants C_0 can be determined from the boundary conditions.

By applying symbolic computation of Mathematica software, a mathematical mechanization of ADM based on the principle of parameterization, was carried out to reduce the solution for approximate expressions of model equations (A.1) and (A.2). The code of parameterized ADM is listed below.

The process is as follows: the term number m is first input, then two one-dimensional arrays will be created as $a = Array[t, m]$, $u = Array[l, m]$; a circle sentence $Do[c = u[[i]]*\lambda^{i-1} + c,\{i,m\}]$ with introducing the parameter λ is used to produce the expression

$u = \sum_{i=0}^{m} u_i \lambda^i$. In this code, *f*[*c*_] is the function expression of the nonlinear term *N*, *l*[1] = $u_0 = c_0$ is the root of Adomian polynomial. The component of solution, c_n, can be calculated by using a function *polyadomian*[*i*_].

If the boundary conditions are applied to the approximate expression $c = \sum_{n=0}^{m} c_n$, a one variable algebraic equation can be obtained to determine the unknown constant c_0. However, different methods may be used for other kinds of boundary conditions. It should be noted, the root c_0 can be obtained analytically when the ADM approximate solution term $m \leq 5$ for reaction orders $n = 0.5$, 1.0, 2.0,[10,11] otherwise the Newton iteration method will be needed for c_0.

PAMC (Parameterized ADM Mathematica Code) for diffusion–reaction models:

```
Clear [λ, a, u, n, c]
m: = 8
a = Array [t, m]
u = Array[l, m]
c = 0
Do[c = u[[i]]*λ^(i-1) + c,{i, m}]
f[c_] = c^n
l[1] = c0
λ = 0
a[[1]] = f[c]
Clear[λ]
polyadomian[i_]:= [ l[i+1] = ∫∫φ^2 * a[[i]]dxdx; Clear[i];
                    a[[i+1]] = 1/(i)! * D[f[c],{λ,i}]; λ = 0 ]
Do[polyadomian[i], {i = m - 1}]
c = 0
Do[c = u[[i]] + c,{i, m}]
c
```

Output example of the algebraic expression of the approximate solution *C*(*X*) with eight terms for the model of catalyst slab is as:

$$
\begin{aligned}
C = {} & c0 + \frac{1}{2} c0^{n} x^{2} \phi^{2} + \frac{1}{24} c0^{-1+2n} n x^{4} \phi^{4} + \\
& \frac{1}{720} c0^{-2+3n} (-3n + 4n^{2}) x^{6} \phi^{6} + \\
& \frac{c0^{-3+4n} (30n - 63n^{2} + 34n^{3}) x^{8} \phi^{8}}{40320} + \\
& \frac{c0^{-4+5n} (-630n + 1689n^{2} - 1554n^{3} + 496n^{4}) x^{10} \phi^{10}}{3628800} + \\
& \frac{1}{479001600} (c0^{-5+6n} \\
& \quad (22680n - 71910n^{2} + 88089n^{3} - 49914n^{4} + 11056n^{5}) x^{12} \phi^{12}) + \\
& \frac{1}{87178291200} (c0^{-6+7n} (-1247400n + 4498470n^{2} - 6687387n^{3} + \\
& \quad 5170518n^{4} - 2083704n^{5} + 349504n^{6}) x^{14} \phi^{14})
\end{aligned}
$$

2. ADM to Solve the Coupled ODE's[17]

Similarly, any two coupled ordinary differential equations can be written for the operator equations of ADM

$$(L_1 + R_1 + N_1)u = g_1(x) \tag{B.1}$$

$$(L_2 + R_2 + N_2)v = g_2(x)$$

where L_1, L_2 are two highest order derivatives, R_1, R_2 are the derivatives of order less than L, N_1, N_2 are nonlinear terms, and $g_1(x)$, $g_2(x)$ are inhomogeneous terms.

We apply the inverse invertible linear operators L_1^{-1}, L_2^{-1} to Eq. (B.1), then

$$L_1^{-1}L_1u = L_1^{-1}g_1(x) - L_1^{-1}R_1u - L_1^{-1}N_1u \tag{B.2}$$

$$L_2^{-1}L_2v = L_2^{-1}g_2(x) - L_2^{-1}R_2u - L_2^{-1}N_2v$$

ADM defines the solution $u(x)$ and $v(x)$ by the infinite series

$$u = \sum_{n=0}^{\infty} u_n \tag{B.3}$$

$$v = \sum_{n=0}^{\infty} v_n$$

the components should be

$$u_0 = f_{i1} + L_1^{-1} g_1(x) \tag{B.4}$$

$$v_0 = f_{i2} + L_2^{-1} g_2(x)$$

$$u_{n+1} = -L^{-1}(N_1 u_n), n \geq 0$$

$$v_{n+1} = -L_2^{-1}(N_2 v_n), n \geq 0$$

Decomposing the nonlinear terms N_1, N_2 by the Adomian polynomials

$$N_1 u = f_1(u(x), v(x)) = \sum_{n=0}^{\infty} A_n(u_n(x), v_n(x)) \tag{B.5}$$

$$N_2 u = f_2(u(x), v(x)) = \sum_{n=0}^{\infty} B_n(u_n(x), v_n(x))$$

Introducing the parameter λ, the polynomials A_n and B_n can be identified as

$$A_n = \frac{1}{n!} \frac{d^n}{d\lambda^n} f[u(\lambda), v(\lambda)]\Big|_{\lambda=0} \tag{B.6}$$

$$B_n = \frac{1}{n!} \frac{d^n}{d\lambda^n} f[u(\lambda), v(\lambda)]\Big|_{\lambda=0}$$

Consequently, all components of the decomposition are identified and can be calculated.

For the boundary value problems in this paper, the L_1 and L_2 are linear Laplacian operators, the R_1 and R_2 disappear, the inhomogeneous terms $g_1(x)$ and $g_2(x)$ also equal to zero. The function f_{i1}, f_{i2} arising from integrating should be $f_{i1} = a_0 + a_1x$ and $f_{i2} = b_0 + b_1x$.

Also for a symmetry geometry, the catalyst center as a boundary usually has a boundary condition $x = 0$, $du/dx = 0$ and $dv/dx = 0$, this leads to $f_{i1} = a_0$ and $f_{i2} = b_0$ which mean $u_0 = a_0$ and $v_0 = b_0$, therefore it's much easier since only two undetermined constants a_0 and b_0 need to be evaluated.

The approximate solutions with $m + 1$ term will be

$$u(x) \simeq u_{m+1} = \sum_{n=0}^{m} u_n = u_0 - L_1^{-1} \sum_{n=0}^{m-1} A_n \tag{B.7}$$

$$v(x) \simeq v_{m+1} = \sum_{n=0}^{m} v_n = v_0 - L_2^{-1} \sum_{n=0}^{m-1} B_n$$

The process of calculating the Adomian polynomials can be well mechanized by applying symbolic computation of Mathematica, based on the parameterization principle of Eq. (B.6). Then, we apply Eq. (B.7) to generate partial components of the sum $u = \sum_{n=0}^{\infty} u_n$ and $v = \sum_{n=0}^{\infty} v_n$. Thus, if the series converges the $m + 1$ term partial sum $u_{m+1} = \sum_{n=0}^{m} u_n$ and $v_{m+1} = \sum_{n=0}^{m} v_n$ would be the approximate solutions since $\lim_{n\to\infty} u_n = u$ and $\lim_{n\to\infty} v_n = v$ have been identified.

The integrating constants a_0 and b_0 can be evaluated by letting the partial sum u_{m+1} and v_{m+1} to satisfy the boundary-conditions respectively. That is, we will solve two coupled algebraic equations by the general numerical method, for example, regressing by Newton method, to evaluate two integration constants as long as the catalyst with a symmetry geometry.

The model for the three-dimensional or packed bed electrode, which describes the overpotential and concentration distributions, is a set of ordinary different equations with boundary conditions as

$$\frac{d^2\Psi}{dX^2} = \mu \frac{C}{\frac{1}{\exp(\alpha\Psi)} + \frac{1}{\gamma}} \tag{B.8}$$

$$\frac{d^2C}{dX^2} = s \frac{C}{\frac{1}{\exp(\alpha\Psi)} + \frac{1}{\gamma}} \tag{B.9}$$

$$X = 0, \qquad \frac{d\Psi}{dX} = 0, \qquad \frac{dC}{dX} = 0 \tag{B.10}$$

$$X = 1, \qquad \Psi = 1, \qquad C = 1 \tag{A-2-11}$$

The two-coupled ordinary differential equations have been decomposed by parameterized ADM, and the Mathematica code is listed below.

PAMC for coupled ODE's of tree-dimentional electrode model:

```
Clear [λ, a, b, u, v, c, φ]
m: = 4
a = Array [t, m]
b = Array [p, m]
u = Array [l, m]
v = Array [q, m]
c = 0
Do[c = u[[i]]*λ^(i-1) + c,{i, m}]
φ = 0
Do[φ = v[[i]]*λ^(i-1) + φ,{i, m}]
f1[c_,φ_] = c / (1/exp(α*φ) + 1/γ)
f2[c_,φ_] = c / (1/exp(α*φ) + 1/γ)
l[1] = c0
q[1] = φ0
a[[1]] = f1[c_,φ_]
b[[1]] = f2[c_,φ_]
```

$$polyadomian[i\_]:=\begin{bmatrix} \lambda = 0; \\ l[i+1] = \iint s * a[[i]]dxdx; \\ q[i+1] = \iint \mu * b[[i]]dxdx; \\ Clear[i]; \\ a[[i+1]] = \frac{1}{(i)!} * D[f1[c,\phi],\{\lambda,i\}]; \\ b[[i+1]] = \frac{1}{(i)!} * D[f2[c,\phi],\{\lambda,i\}]; \end{bmatrix}$$

$Do[polyadomian[i], \{i = m - 1\}]$

$c = 0$

$Do[c = u[[i]] + c, \{i, m\}]$

$\phi = 0$

$Do[\phi = v[[i]] + \phi, \{i, m\}]$

c

ϕ

REFERENCES

[1]K. Scott, *Electrochemical Reaction Engineering*, Academic Press, London, 1991.

[2]G. Adomian, *Solving Frontier Problems of Physics: The Decomposition Method*, Kluwer, Boston, 1994.

[3]G. Adomian, *Stochastic Systems*, Academic Press, New York, 1983.

[4]Y. P. Sun and K. Scott, An analysis of the influence of mass transfer on porous electrode performance, *J. Chem. Eng.* **1024**(1) (2004) 83–91.

[5]Y. P. Sun and K. Scott, The influence of mass transfer on a porous fuel cell electrode, *Fuel Cell*, **4**(1-2) (2004) 30–38.

[6]B. A. Finlayson, *Nonlinear Analysis in Chemical Engineering*, McGraw-Hill Inc., USA, 1980.

[7]J. S. Newman, *Electrochemical Systems*, 2nd ed., Prentice Hall, New Jersey, 1991.

[8]Y. Cherrault, Convergence of Adomian's method, *Kybernetes*, **18**(2) (1989) 31–38.

[9]A. Wazwaz and S. El-Sayed, A new modification of the Adomain deposition method for linear and nonlinear operators, *Appl. Math and Computing* **122** (2001) 393–405.

[10]Y. P. Sun, S. B. Liu, and K. Scott, Approximate solution for the nonlinear model of diffusion and reaction in porous catalysts by the decomposition method, *J. Chem. Eng.* **102** (1) (2004) 1–10.

[11]S. B. Liu, Y. P. Sun, and K. Scott, Analytic solution of diffusion reaction in spherical porous catalyst, *Chem. Eng. Technol.* **26** (1) (2003) 87–95.

[12]B. L. Hou, and Y. P. Sun, Applying mechanization for soving nonlinear boundary value problem of ordinary differential equation in reaction engineering, *J. Computers and Applied Chemistry* **23**(3) (2006) 255–259.

[13]C. N. Sattersfield and T. K. Sherwood, *The Role of Diffusion in Catalysis*, Addison-Wesley Publishing Co. Inc., Massachusetts, 1963.

[14]M. Eikerling and A. Kornyshev, Modeling the performance of the cathode catalyst layer of polymer electrolyte fuel cells, *J. Electroanal. Chem.* **435** (1998) 98–106.

[15]J. Nordlund and G. Lindbergh, A model for a porous DMFC anode, *J. Electrochem. Soc.* **149** (9) (2002) A1107.
[16]E. Petersen, Non-isochronal chemical reaction in porous catalysts, *Chem. Eng. Sci.* **17** (1962) 987–995.
[17]Y. P. Sun, B. L. Hou and K. Scott, Solving Generalized Models of Porous Catalysts and Electrodes by Automation of the Adomian Decomposition Method, Submitted to Chem. Eng. Technol.
[18]Y. P. Sun, W. L. Xu, and K. Scott, A study of the electrochemical reaction of nitrobenzene to p-aminophenol in a packed bed electrode reaction, *Electrochim. Acta* **38**(13) (1993) 1753–1759.
[19]Y. P. Sun, O. Qiu, W. L. Xu, and K. Scott, A study of the performance of a sparged packed-bed electrode reactor for the direct electrochemical oxidation of propylene, *J. Electroanal. Chem.* **503** (2001) 36–44.
[20]Y. P. Sun, B. T. Wu, O. Qiu, W. L. Xu, and K. Scott, An approximate solution to the model of a sparged packed bed electrode reactor, *J. Appl. Electrochem.* **32** (2002) 1219–1227.
[21]Y. P. Sun, W. L. Xu, and K. Scott, An efficient method for solving the model equations of a two-dimensional packed bed electrode, *J. Appl Electrochem.* **25** (1995) 755–763.
[22]B. T. Wu, Y. P. Sun, A new exploration of the inverse operator solution for a kind of strong nonlinear partial differential equations with initial–boundary value conditions, *Acta Mathematica Scientia* **19**(3) (1999) 347–355.

Volume 41
Index

MIX
Papier aus verantwortungsvollen Quellen
Paper from responsible sources
FSC® C105338

If you have any concerns about our products,
you can contact us on
ProductSafety@springernature.com

In case Publisher is established outside the EU,
the EU authorized representative is:
Springer Nature Customer Service Center GmbH
Europaplatz 3, 69115 Heidelberg, Germany

Printed by Libri Plureos GmbH
in Hamburg, Germany